AF738273

Theory of Topological Structures

Mathematics and Its Applications

Managing Editor:

M. HAZEWINKEL

Centre for Mathematics and Computer Science, Amsterdam, The Netherlands

Editorial Board:

F. CALOGERO, *Universita degli Studi di Roma, Italy*
Yu. I. MANIN, *Steklov Institute of Mathematics, Moscow, U.S.S.R.*
A. H. G. RINNOOY KAN, *Erasmus University, Rotterdam, The Netherlands*
G.-C. ROTA, *M.I.T., Cambridge Mass., U.S.A.*

Gerhard Preuss

Department of Mathematics,
Free University Berlin, F.R.G.

Theory of Topological Structures

An Approach to Categorical Topology

Translated by Andreas Behling

D. Reidel Publishing Company

A MEMBER OF THE KLUWER ACADEMIC PUBLISHERS GROUP

Dordrecht / Boston / Lancaster / Tokyo

Library of Congress Cataloging in Publication Data

Preuss, Gerhard, 1944–
Theory of topological structures.

(Mathematics and its applications.)
Translation from the author's German manuscript.
Bibliography: p.
Includes index.
1. Categories (Mathematics) 2. Algebraic topology. I. Title.
II. Series: Mathematics and its applications (D. Reidel Publishing Company).
QA169.P734 1987 514'.2 87-26390

ISBN-13: 978-94-010-7786-6 e-ISBN-13: 978-94-009-2859-6
DOI: 10.1007/978-94-009-2859-6

Published by D. Reidel Publishing Company,
P.O. Box 17, 3300 AA Dordrecht, Holland.

Sold and distributed in the U.S.A. and Canada
by Kluwer Academic Publishers,
101 Philip Drive, Norwell, MA 02061, U.S.A.

In all other countries, sold and distributed
by Kluwer Academic Publishers Group,
P.O. Box 322, 3300 AH Dordrecht, Holland.

All Rights Reserved
© 1988 by D. Reidel Publishing Company, Dordrecht, Holland

Softcover reprint of the hardcover 1st edition 1988

No part of the material protected by this copyright notice may be reproduced or
utilized in any form or by any means, electronic or mechanical
including photocopying, recording or by any information storage and
retrieval system, without written permission from the copyright owner

SERIES EDITOR'S PREFACE

Approach your problems from the right end and begin with the answers. Then one day, perhaps you will find the final question.

'The Hermit Clad in Crane Feathers' in R. van Gulik's *The Chinese Maze Murders.*

It isn't that they can't see the solution. It is that they can't see the problem.

G.K. Chesterton. *The Scandal of Father Brown* 'The point of a Pin'.

Growing specialization and diversification have brought a host of monographs and textbooks on increasingly specialized topics. However, the "tree" of knowledge of mathematics and related fields does not grow only by putting forth new branches. It also happens, quite often in fact, that branches which were thought to be completely disparate are suddenly seen to be related.

Further, the kind and level of sophistication of mathematics applied in various sciences has changed drastically in recent years: measure theory is used (non-trivially) in regional and theoretical economics; algebraic geometry interacts with physics; the Minkowsky lemma, coding theory and the structure of water meet one another in packing and covering theory; quantum fields, crystal defects and mathematical programming profit from homotopy theory; Lie algebras are relevant to filtering; and prediction and electrical engineering can use Stein spaces. And in addition to this there are such new emerging subdisciplines as "experimental mathematics", "CFD", "completely integrable systems", "chaos, synergetics and large-scale order", which are almost impossible to fit into the existing classification schemes. They draw upon widely different sections of mathematics. This programme, Mathematics and Its Applications, is devoted to new emerging (sub)disciplines and to such (new) interrelations as exempla gratia:

- a central concept which plays an important role in several different mathematical and/or scientific specialized areas;
- new applications of the results and ideas from one area of scientific endeavour into another;
- influences which the results, problems and concepts of one field of enquiry have and have had on the development of another.

The Mathematics and Its Applications programme tries to make available a careful selection of books which fit the philosophy outlined above. With such books, which are stimulating rather than definitive, intriguing rather than encyclopaedic, we hope to contribute something towards better communication among the practitioners in diversified fields.

Topology, a relatively new branch of mathematics, tries to capture such ideas as nearness and limits. It is of course immensely useful in virtually all branches of pure and applied mathematics including algebra and logic which, at first sight, seem far removed from the ideas at the basis of topology. The best known definition embodying neighborhoods, nearness, and limit ideas is probably that of a topological space. This one is far from satisfactory in many settings and thus other notions appeared (sometimes restricted classes of topological spaces) which attempt to describe some class of topological structures at once small enough to have lots of nice properties and large enough so that all kinds of naturally occurring topological structures in (functional) analysis, algebra, probability, ... would fall under it and such that all kinds of natural constructions (product, spaces of maps, limits, ...) would not take one out of it. For the systematic investigation of this sort of balancing problem, category theory is extremely useful and thus categorical topology arose.

The subject now seems to have reached a certain plateau of maturity, terminology has stabilized and it is definitely time for a first systematic (unifying) textbook on the subject of topological structures, written by one of the active experts in the field. Hence this book: which I hope and expect will be of natural interest to those engaged in research in categorical topology, and will also benefit all those who use topological structures in their work (i.e. almost all mathematicians) but are not necessarily directly active in research in this field itself.

The unreasonable effectiveness of mathematics in science ...

Eugene Wigner

Well, if you know of a better 'ole, go to it.

Bruce Bairnsfather

What is now proved was once only imagined.

William Blake

As long as algebra and geometry proceeded along separate paths, their advance was slow and their applications limited.

But when these sciences joined company they drew from each other fresh vitality and thenceforward marched on at a rapid pace towards perfection.

Joseph Louis Lagrange.

Bussum, September 1987

Michiel Hazewinkel

CONTENTS

PREFACE

This book is based on lectures the author has given at the Free University Berlin over many years. The first course of this kind took place in summer 1978 in order to prepare my students for attending the International Conference on Categorical Topology (Berlin, August 27-th to September 2-nd, 1978) organized by H. Herrlich and myself. Since Categorical Topology is a fairly young discipline there was no textbook on this subject up to now. The presented course is written for graduate students and interested mathematicians who know already the basic facts on General Topology. Nevertheless some definitions and theorems on uniform spaces are listed up in Chapter 0. Concerning the last chapter of this book the reader is supposed to be acquainted with Algebraic Topology, especially with Čech cohomology theory.

After some preliminary remarks on Set Theory and Category Theory (Chapter 0) topological categories (Chapter 1) are introduced. Then the theory of reflections and coreflections is developped which is also applicable to non-topological categories (Chapter 2). In the following, concrete structures, especially nearness structures, are studied. The interactions of sub- and supercategories of the category Near of nearness spaces (and uniformly continuous maps) are investigated (Chapter 3). Cartesian closedness is studied in Chapter 4 as well as in the more general setting of Chapter 5. Completions are also studied twice, namely for concrete categories as well as for nearness spaces (Chapter 6). Last not least the beautiful relations between dimension theory and cohomology theory known from classical topology are generalized (Chapter 7). In order to be selfcontained representable functors are treated at the end of the book (appendix).

Concerning the presented material not all research areas of Categorical Topology have been included. Especially, I stopped sometimes when the material flowed over to general category theory. Furthermore, I did not treat all applications to other branches of mathematics (e.g. functional analysis or topological algebra); I restricted myself mainly to the field of algebraic topology. Nevertheless I hope that the methods presented will enable the reader to understand all publications on Categorical Topology.

I am very grateful to my friend and colleague Horst Herrlich for his encouragement to publish my lecture notes on Categorical Topology and for his research work that made possible most parts of the book. Further I would like to thank Dipl.-Math. Andreas Behling for translating the main parts of my German manuscript and the Fachbereich Mathematik of the Free University Berlin for paying him. Additionally I thank Mr. Behling for drawing the figures and for preparing the index. I thank Priv.-Doz. Dr. Dr. T. Marny for discussions on the subject presented. Furthermore, I am grateful to Dr. J. Schröder and Dipl.-Math. Olaf Zurth for several parts of the exercises. I thank too Mr. Carsten Scheuch for proofreading. Last not least I thank Mrs. Christa Siewert for her patience in typing the manuscript as well as Mrs. Margrit Barret for assisting her.

Berlin, June 1987

Gerhard Preuß

LIST OF SYMBOLS

Special categories

Notations of some special sets

Let X,Y be sets, $A \subset X$, $B \subset Y$, $f: X \to Y$ a map, $\mathcal{X}$ a topology on X and R an equivalence relation on X .

$\emptyset$ denotes the empty set

$CA := \{x \in X: x \notin A\}$

Sometimes we write $X \setminus A$ instead of CA

$\mathcal{P}(X) := \{A: A \subset X\}$

$f[A] := \{f(x): x \in A\}$

$f^{-1}[B] := \{x \in X: f(x) \in B\}$

$X/_R := \{[x]: [x]$ is an equivalence class with respect to $R\}$

$|X|$ denotes the cardinality of X

$\mathcal{U}_{\mathcal{X}}(x)$ set of neighbourhoods of x with respect to $(X,\mathcal{X})$

$\overset{\circ}{\mathcal{U}}_{\mathcal{X}}(x)$ set of open neighbourhoods of x with respect to $(X,\mathcal{X})$

$\bar{A}^{\mathcal{X}}$ closure of A with respect to $(X,\mathcal{X})$

Sometimes we write $\mathcal{U}(x)$, $\overset{\circ}{\mathcal{U}}(x)$, $\bar{A}$ instead of $\mathcal{U}_{\mathcal{X}}(x)$, $\overset{\circ}{\mathcal{U}}_{\mathcal{X}}(x)$, $\bar{A}^{\mathcal{X}}$.

A^{o} interior of A with respect to $(X,\mathcal{X})$

$\mathbb{N}$ set of natural numbers

$\mathbb{Z}$ set of integers

$\mathbb{Q}$ set of rational numbers

$\mathbb{R}$ set of real numbers

$(a,b) := \{x \in \mathbb{R}: a < x < b\}$

$(a,b] := \{x \in \mathbb{R}: a < x \leq b\}$

$[a,b) := \{x \in \mathbb{R}: a \leq x < b\}$

$[a,b] := \{x \in \mathbb{R}: a \leq x \leq b\}$

Further symbols

INTRODUCTION

In order to handle problems of a topological nature, various attempts have been made in the past to introduce suitable concepts, e.g. topological spaces, uniform spaces, proximity spaces, limit spaces, uniform convergence spaces etc. Since this situation was unsatisfactory, new methods were needed to unify all these theories. Thus a new discipline - called *Categorical Topology* - was created (about 1971). It deals with the investigation of topological categories and their relationships to each other. Nevertheless, the search for a suitable 'structure' by means of which any topological concept or idea can be expressed went on. In 1974 H. Herrlich invented nearness spaces, a very fruitful concept which enables one to unify topological and uniform aspects. But to understand the real meaning of this approach a categorical interpretation is useful. Thus decisive parts of this theory belong to Categorical Topology.

What are the problems we want to treat in this book?

1) Does there exist a categorical framework for the many kinds of spaces topologists are interested in?
2) What is the categorical background of famous constructions such as the Stone-Čech compactification or the completion of a uniform space?
3) Does there exist some kind of 'structure' which leads to a better approach of topological phenomena than the theory of topological spaces?
4) If the answer to 3) is yes, try to solve the following problems:
   a) Any product of paracompact spaces is paracompact.
   b) Any subspace of a paracompact space is paracompact.
   c) Any subspace of a normal space is normal.
   d) Any product of the unit interval [0,1] with a normal space is normal.
   e) Do the Čech cohomology groups with respect to finite covers fulfill the Eilenberg-Steenrod axioms?

f) Does there exist a cohomological characterization of dimension for topological spaces, uniform spaces and proximity spaces simultaneously?

g) Is it possible to obtain well-known extensions and compactifications of topological spaces by means of the construction of a suitable completion?

5) Find classes of spaces such that the product of two quotient maps is a quotient map and that there is a natural function space structure. Furthermore such a class should not be too 'big' or too 'small' and it should be described by suitable axioms.

6) What is the significance of Dedekind's construction of the real numbers for Categorical Topology?

None of the problems 4) a) - g) can be solved within the framework of topological spaces. But the theory of nearness spaces presented in this book solves the problems. Even 5) finds a satisfactory solution within the realm of 'nearness'. Problem 1) is solved by the theory of topological categories (resp. initially structured categories), whereas 2) leads to the theory of reflections. Additionally, 5) is directly connected with the theory of cartesian closed topological (resp. initially structured) categories. Last but not least, the further development of Dedekind's construction of the real numbers is the MacNeille completion of a concrete category (problem 6)), whereas the completion of a nearness space (problem 4) g)) generalizes Cantor's construction of the real numbers.

CHAPTER 0

PRELIMINARIES

0.1 Sets, classes and conglomerates

In Cantor's naive set theory every collection of objects specified by some property was called a set. As well-known this approach leads to contradictions, e.g. to the Russell antinomy of the set R of all sets not members of themselves (we obtain

$$[R \in R] \Leftrightarrow [R \notin R]$$

provided R is a set). In order to block this contradiction we introduce two types of collections: classes and sets. Then a class is a collection of objects specified by some property, whereas a set is a class which is a member of some class. Thus R is no set but a (proper) class and also the concept of the class of all sets makes sense. The axiomatic set theory tries to avoid further antinomies. The axiomatic approach of Gödel, Bernays and von Neumann is suitable to handle classes and sets. For further details the interested reader is referred to Dugundji [25] although it is not necessary for understanding this book to know all the details.
Since, occasionally, we will need to consider collections of classes, we introduce the wider concept of conglomerates. Thus, a conglomerate is a collection having classes as members. Especially, we require that conglomerates are closed under the usual set-theoretic constructions (e.g. formation of pairs, unions, products etc.) and that every class is a conglomerate. Therefore conglomerates may be handled like sets. We are allowed to construct functions between them, equivalence relations on them and so on. Some more hints on the axiomatic treatment of the subject can be found in the appendix of the book "Category Theory" by Herrlich and Strecker [44].

0.2 Some categorical concepts

For each mathematical discipline we define at first objects and then admissible maps for describing the objects. This procedure is formalized by the concept 'category'.

0.2.1 Definition. A category C consists of
(1) a class $|C|$ of *objects* (which are denoted by $A,B,C,\ldots$),
(2) a class of pairwise disjoint sets $[A,B]_C$ for each pair (A,B) of objects (the members of $[A,B]_C$ are called *morphisms* from A to B), and
(3) a *composition* of morphisms, i.e. for each triple (A,B,C) of objects there is a map

$$[A,B]_C \times [B,C]_C \to [A,C]_C$$

$$(f,g) \mapsto g\circ f$$

(where $\times$ denotes the cartesian product) such that the following axioms are satisfied:

Cat_1) (*Associativity*). If $f \in [A,B]_C$, $g \in [B,C]_C$ and $h \in [C,D]_C$, then $h\circ(g\circ f) = (h\circ g)\circ f$.

Cat_2) (*Existence of identities*). For each $A \in |C|$ there is an *identity* (*morphism*) $1_A \in [A,A]_C$ such that for all $B,C \in |C|$, all $f \in [A,B]_C$ and all $g \in [C,A]_C$, $f\circ 1_A = f$ and $1_A\circ g = g$.

0.2.2 Remarks. (1) We write $f\colon A \to B$ or $A \xrightarrow{f} B$ instead of $f \in [A,B]_C$. A (resp. B) is called the domain of f (resp. the codomain of f).

(2) a) The identity 1_A is uniquely determined by A.

b) If $A,A' \in |C|$ with $A \neq A'$, then $1_A \neq 1_{A'}$, because $[A,A]_C \cap [A',A']_C = \emptyset$.

(3) The class of all morphisms of C is denoted by

$$\mathrm{Mor}\ C := \bigcup_{(A,B)\in|C|\times|C|} [A,B]_C ;$$

its elements are called $\mathcal{C}$-morphisms.

(4) The requirement of the disjointness of the morphisms sets is not restrictive because it can always be obtained provided that $[A,B]_{\mathcal{C}}$ is replaced by $[A,B]'_{\mathcal{C}} = \{(A,B,\alpha): \alpha \in [A,B]_{\mathcal{C}}\}$.

0.2.3 Examples of categories. (1) The category Set of sets and maps: $|\underline{\mathrm{Set}}|$ is the class of all sets; $[A,B]_{\underline{\mathrm{Set}}}$ is the set of all maps from A to B for all $A,B \in |\underline{\mathrm{Set}}|$. The composition of morphisms is the usual composition of maps.

Remark. In order to obtain the disjointness of the morphisms sets it is useful to define a map $f: A \to B$ as a triple (A,B,F) where $F \subset A \times B$ has the following property: For each $x \in A$ there exists a unique $y \in B$ such that $(x,y) \in F$.

(2) The category $\underline{\mathrm{Mod}}_R$ of R-modules and R-linear maps (where R denotes a commutative ring with unit): $|\underline{\mathrm{Mod}}_R|$ is the class of all R-modules and Mor $\underline{\mathrm{Mod}}_R$ is the class of all R-linear maps (between any two R-modules). The composition of morphisms is the usual composition of maps.

(3) The category Top of topological spaces (and continuous maps).

(4) The category Ord of ordered sets (and order preserving maps) [An *ordered set* is a pair $(X,\le)$ where X is a set and $\le \subset X \times X$ is a reflexive, antisymmetric and transitive relation].

(5) If $(S,\le)$ is an ordered set, then a category $\mathcal{C}$ is defined as follows:

$$|\mathcal{C}| = S \; ; \; [x,y]_{\mathcal{C}} = \begin{cases} \{(x,y)\} & \text{if } x \le y \\ \emptyset & \text{otherwise} \end{cases}$$

(6) Let $\mathcal{C}$ be a category. Then the dual category $\mathcal{C}^*$ is defined as follows:

(1) $|\mathcal{C}^*| := |\mathcal{C}|$.

(2) $[A,B]_{\mathcal{C}^*} := [B,A]_{\mathcal{C}}$ for all $(A,B) \in |\mathcal{C}^*| \times |\mathcal{C}^*|$.

(3) The composition $\alpha \circ \beta$ in $\mathcal{C}^*$ is defined as the composition $\beta \circ \alpha$ in $\mathcal{C}$.

Hint. If a C-morphism f is considered as a C^*-morphism we write f^* instead of f.

Remark. 1) $(C^*)^* = C$.

2) For each statement in a category C there is a dual statement, namely the corresponding statement in C^* phrased as a statement in C (by reversing all arrows by means of which morphisms are symbolized).

0.2.4 Definition. Let C be a category and $f \in [A,B]_C$ with $(A,B) \in |C| \times |C|$. Then f is called an isomorphism provided that there is some $g \in [B,A]_C$ with $g \circ f = 1_A$ and $f \circ g = 1_B$. If $f \in [A,B]_C$ is an isomorphism then A and B are called isomorphic (denoted by $A \cong B$).

0.2.5 Remarks. ① In 0.2.4 g is uniquely determined by f (if $g' \in [B,A]_C$ with $g' \circ f = 1_A$ and $f \circ g' = 1_B$ then $g = g \circ 1_B = g \circ (f \circ g') = (g \circ f) \circ g' = 1_A \circ g' = g'$) and is denoted by f^{-1}.

② Obviously, an isomorphism in Set is a bijective map (and vice versa), while an isomorphism in Top is a homeomorphism (and vice versa).

③ For every category C, the identity $1_X: X \to X$ is an isomorphism for each $X \in |C|$. If $f: X \to Y$ is an isomorphism in C, then $f^{-1}: Y \to X$ is also an isomorphism. Additionally, the composition of two isomorphisms in C is again an isomorphism. Thus $\cong \subset |C| \times |C|$ is an equivalence relation on $|C|$; the corresponding equivalence classes are called isomorphism classes. A property P for the objects of C is called a C-invariant provided that the following is satisfied: If an object X of C has the property P then all objects of the isomorphism class of X have the property P. (Top-invariants are usually called topological invariants.)

0.2.6 Definitions. Let C be a category. A C-morphism $f: A \to B$ is called

1) a monomorphism provided	1') an epimorphism provided that f^*

that for all pairs (α,β) of $\mathcal{C}$-morphisms with codomain A such that $f\circ\alpha = f\circ\beta$ it follows that $\alpha = \beta$.

2) an <u>extremal monomorphism</u> provided that the following are satisfied:

(1) f is a monomorphism.

(2) If $f = h\circ g$, where g is an epimorphism, then g must be an isomorphism.

is a monomorphism in the dual category $\mathcal{C}^*$ (i.e. for all pairs (α,β) of $\mathcal{C}$-morphisms with domain B such that $\alpha\circ f = \beta\circ f$, it follows that $\alpha = \beta$).

2') an <u>extremal epimorphism</u> provided that f^* is an extremal monomorphism in the dual category $\mathcal{C}^*$ (i.e. the following are satisfied:

(1') f is an epimorphism.

(2') If $f = g\circ h$, where g is a monomorphism, then g must be an isomorphism.).

<u>0.2.7 Proposition</u>. Let $\mathcal{C}$ be a category and $f\colon A \to B$ a $\mathcal{C}$-morphism. Then the following are equivalent:

(1) f is an isomorphism.

(2) f is an epimorphism and an extremal monomorphism.

(3) f is a monomorphism and an extremal epimorphism.

<u>Proof</u>. (1) $\Rightarrow$ (2). a) Let α,β be $\mathcal{C}$-morphisms such that $\alpha\circ f = \beta\circ f$. Then $\alpha\circ f\circ f^{-1} = \beta\circ f\circ f^{-1}$, i.e. $\alpha\circ 1_B = \alpha = \beta\circ 1_B = \beta$. Thus f is an epimorphism.

b)(α) Since f is an isomorphism, f is also a monomorphism (analogously to a)).

(β) Let $f = h\circ g$ where g is an epimorphism

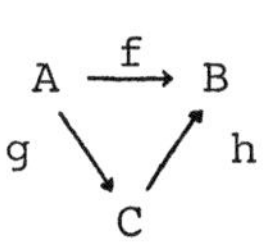

Then $1_A = f^{-1}\circ f = f^{-1}\circ(h\circ g) = (f^{-1}\circ h)\circ g$. Furthermore, $(g\circ(f^{-1}\circ h))\circ g = g\circ((f^{-1}\circ h)\circ g) = g\circ 1_A = g = 1_C\circ g$ which implies $g\circ(f^{-1}\circ h) = 1_C$ since g is an epimorphism.

Therefore, g is an isomorphism.

(2) ⇒ (1) . Since f is an epimorphism and an extremal monomorphism, it follows immediately from $f = 1_B \circ f$ that f is an isomorphism.
Thus the proposition is proved because (2) and (3) are dual statements and (1) is selfdual [more exactly:
f isomorphism ⇔ f^* isomorphism ⇔ f^* epimorphism and extremal monomorphism ⇔ f monomorphism and extremal epimorphism].

0.2.8 Remark. A morphism in a category C is called a bimorphism provided that it is an epimorphism and a monomorphism. Categories in which each bimorphism is already an isomorphism are called balanced; e.g. the categories Set and Mod_R are balanced (the bimorphisms are just those morphisms which are bijective as maps), whereas the category Top is not balanced (e.g. $1_X: (\{0,1\},\mathcal{D}) \to (\{0,1\},\mathcal{I})$, where $\mathcal{D}$ resp. $\mathcal{I}$ denotes the discrete resp. indiscrete topology on the two-point set $\{0,1\}$, is bijective and continuous but the inverse map is not continuous, i.e. the considered map is not an isomorphism). Obviously, in balanced categories there is no distinction between 'epimorphism' and 'extremal epimorphism' (resp. between 'monomorphism' and 'extremal monomorphism').

0.2.9 Definitions. Let C be a category, I a set and $(A_i)_{i\in I}$ a family of objects in C (shortly: a family of C-objects).

A pair $(P,(p_i)_{i\in I})$ with $P \in |C|$ and $p_i \in [P,A_i]_C$ for each $i \in I$ is called a product of the family $(A_i)_{i\in I}$ provided that for each pair $(Q,(q_i)_{i\in I})$ with $Q \in |C|$ and $q_i \in [Q,A_i]_C$ for each $i \in I$ there exists a unique C-morphism q such that the diagram

A pair $(C,(j_i)_{i\in I})$ with $C \in |C|$ and $j_i \in [A_i,C]_C$ for each $i \in I$ is called a coproduct of the family $(A_i)_{i\in I}$ provided that $(C,(j_i^*)_{i\in I})$ is a product of $(A_i)_{i\in I}$ in the dual category C^* (i.e. for each pair $(D,(k_i)_{i\in I})$ with $D \in |C|$ and $k_i \in [A_i,D]_C$ for each $i \in I$ there exists a unique C-morphism k such that the diagram

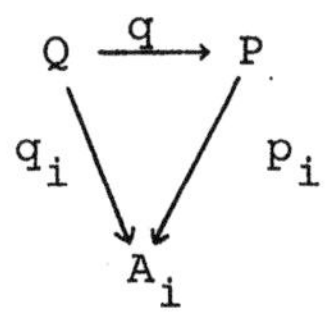

is commutative (i.e. $p_i \circ q = q_i$) for every $i \in I$. We write $\prod_{i \in I} A_i$ instead of P (cf. the following proposition). p_i is called the i-th <u>projection</u>. Sometimes $\prod_{i \in I} A_i$ is already called the product of the family $(A_i)_{i \in I}$.

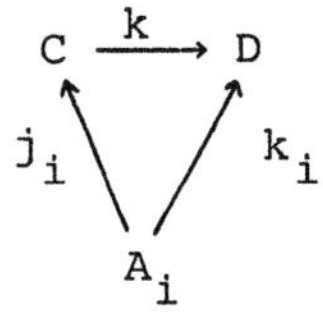

is commutative for every $i \in I$). We write $\coprod_{i \in I} A_i$ instead of C (cf. the following proposition). j_i is called the i-th <u>injection</u>. Sometimes $\coprod_{i \in I} A_i$ is already called the coproduct of the family $(A_i)_{i \in I}$.

<u>0.2.10 Proposition</u>. Let C be a category, I a set and $(A_i)_{i \in I}$ a family of C-objects.

a) If each of $(P,(p_i)_{i \in I})$ and $(P',(p_i')_{i \in I})$ is a product of $(A_i)_{i \in I}$ in C, then there is a unique isomorphism $k \in [P,P']_C$ such that $p_i' \circ k = p_i$ for each $i \in I$.

b) If each of $(C,(j_i)_{i \in I})$ and $(C',(j_i')_{i \in I})$ is a coproduct of $(A_i)_{i \in I}$ in C, then there is a unique isomorphism $j \in [C',C]_C$ such that $j \circ j_i' = j_i$ for each $i \in I$.

<u>Proof</u>. Since $(P',(p_i')_{i \in I})$ is a product of $(A_i)_{i \in I}$, there is a unique $k \in [P,P']_C$ such that $p_i' \circ k = p_i$ for each $i \in I$. Furthermore, since $(P,(p_i)_{i \in I})$ is a product of $(A_i)_{i \in I}$, there is a unique $h \in [P',P]_C$ such that $p_i \circ h = p_i'$ for each $i \in I$. That is, the diagram

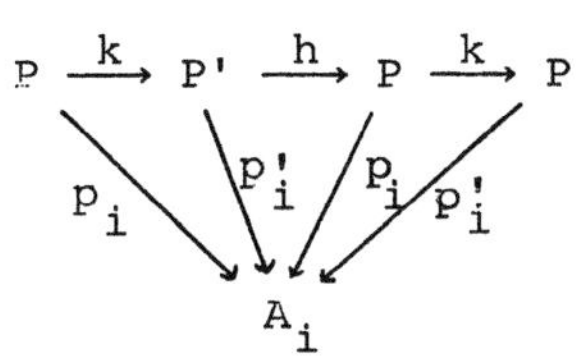

is commutative for all $i \in I$. There is a unique morphism from P to P such that the triangle formed by the two left triangles in the above diagram commutes (note that $(P,(p_i)_{i\in I})$ is a product!). Thus, $h \circ k = 1_P$. Similarly we may conclude that $k \circ h = 1_{P'}$. Therefore k is an isomorphism.

0.2.11 Examples. The products in Set are the cartesian products whereas the coproducts in Set are the disjoint unions. In Top the products are the usual (topological) products and the coproducts are the usual (topological) sums. Usually, in $\underline{Mod}_R$ the products are called direct products and the coproducts are called direct sums.

0.3 Uniform structures

For the convenience of the reader some basic properties of uniform spaces are listed up in this section.

0.3.1 Definition. 1) Let X be a set and $\mathcal{W}$ a filter on $X \times X$ such that the following are satisfied:

U_1) $W \in \mathcal{W}$ implies $\Delta = \{(x,x): x \in X\} \subset W$.

U_2) $W \in \mathcal{W}$ implies $W^{-1} = \{(x,y): (y,x) \in W\} \in \mathcal{W}$.

U_3) For each $W \in \mathcal{W}$ there is some $W^* \in \mathcal{W}$ with
$W^{*2} = \{(x,y):$ there is some $z \in X$ with $(x,z) \in W$ and $(z,y) \in W\} \subset W$

Then $\mathcal{W}$ is called a uniformity on the set X and the pair $(X,\mathcal{W})$ is called a uniform space. The elements of $\mathcal{W}$ are called entourages.

2) For each $V \in \mathcal{W}$, a subset A of X is called V-small provided that $A \times A \subset V$.

3) $\mathcal{B} \subset \mathcal{P}(X \times X)$ is called a base for the uniformity $\mathcal{W}$ on X provided that $\{W \subset X \times X: W \supset B$ for some $B \in \mathcal{B}\} = \mathcal{W}$.

0.3.2 Proposition. Let X be a set. A non-empty collection $\mathcal{B} \subset \mathcal{P}(X \times X)$ is a base for a uniformity $\mathcal{W}$ on X if and only if the following are satisfied:

BU_1) $W \in \mathcal{B}$ implies $\Delta \subset W$.

BU_2) $W \in \mathcal{B}$ implies that there is some $W' \in \mathcal{B}$ with $W' \subset W^{-1}$.

BU_3) $W \in \mathcal{B}$ implies that there is some $W^* \in \mathcal{B}$ with $W^{*2} \subset W$.

BU_4) If $W_1, W_2 \in \mathcal{B}$, then there is some $W_3 \in \mathcal{B}$ such that $W_3 \subset W_1 \cap W_2$.

0.3.3 Remark. If (X,d) is a metric space (resp. a pseudometric space) and $V_\varepsilon = \{(x,y): d(x,y) < \varepsilon\}$ for each $\varepsilon > 0$, then $\mathcal{B} = \{V_\varepsilon: \varepsilon > 0\}$ is a base for a uniformity $\mathcal{W}_d$ on X .

0.3.4 Proposition. Let $(X,\mathcal{W})$ be a uniform space. For each $x \in X$ and each $V \in \mathcal{W}$ let $V(x) = \{y: (x,y) \in V\}$. Then $\mathcal{X}_\mathcal{W} = \{O \subset X:$ for each $x \in X$ there is some $V \in \mathcal{W}$ with $V(x) \subset O\}$ is a topology on X .

0.3.5 Proposition. Let $(X,\mathcal{W})$ be a uniform space. Then the following are valid:

(1) $\mathcal{B} = \{V \in \mathcal{W}: V = V^{-1}\}$ is a base for $\mathcal{W}$, i.e. the symmetric entourages form a base for $\mathcal{W}$.

(2) For each natural number $n \geq 1$ and each base $\mathcal{B}$ for $\mathcal{W}$, $\mathcal{B}_n = \{V^n: V \in \mathcal{B}\}$ is a base for $\mathcal{W}$ where $V^1 = V$ and $V^n = V^{n-1} \circ V$ for $n > 1$ (note that $\circ$ denotes the usual composition of relations).

0.3.6 Definition. Let $(X,\mathcal{W})$, $(X',\mathcal{W}')$ be uniform spaces. A map $f: X \to X'$ is called uniformly continuous provided that one of the following two equivalent conditions is satisfied:

(1) For each $W' \in \mathcal{W}'$ there is some $W \in \mathcal{W}$ such that $(f(x),f(y)) \in W'$ provided that $(x,y) \in W$.

(2) $(f \times f)^{-1}[W'] \in \mathcal{W}$ for each $W' \in \mathcal{W}'$, where $f \times f: X \times X \to X' \times X'$ is defined by $(f \times f)(x,y) = (f(x),f(y))$ for every $(x,y) \in X \times X$.

0.3.7 Remarks. ① If $f: (X,\mathcal{W}) \to (X',\mathcal{W}')$ is uniformly continuous, then $f: (X,\mathcal{X}_\mathcal{W}) \to (X',\mathcal{X}_{\mathcal{W}'})$ is continuous.

(2) The uniform spaces together with the uniformly continuous maps form a category denoted by Unif (composition is the usual composition of maps).

0.3.8 Definition. Let $\mathcal{U}_1, \mathcal{U}_2$ be uniformities on a set X. Then $\mathcal{U}_1$ is called finer than $\mathcal{U}_2$ (and $\mathcal{U}_2$ coarser than $\mathcal{U}_1$) iff $\mathcal{U}_2 \subset \mathcal{U}_1$.

0.3.9 Theorem. Let X be a set, $((Y_i,\mathcal{U}_i))_{i\in I}$ a family of uniform spaces and $f_i\colon X \to Y_i$ maps for each $i \in I$. Put $g_i = f_i \times f_i$ for each $i \in I$. Then all finite intersections of elements of $\{g_i^{-1}[V_i]\colon V_i \in \mathcal{U}_i, i \in I\}$ form a base $\mathcal{B}$ for a uniformity $\mathcal{U}$ on X, more exactly for the coarsest uniformity on X making each f_i uniformly continuous.

0.3.10 Remark. The uniformity $\mathcal{U}$ constructed in 0.3.9 is called the initial uniformity on X with respect to $(f_i\colon X \to (Y_i,\mathcal{U}_i))_{i\in I}$. It has the following property: If $(X',\mathcal{U}')$ is a uniform space, then a map $f\colon (X',\mathcal{U}') \to (X,\mathcal{U})$ is uniformly continuous if and only if all $f_i \circ f$ are uniformly continuous.

0.3.11 Examples. (1) Let $(X,\mathcal{U})$ be a uniform space, $A \subset X$ and $i\colon A \to X$ the inclusion map. Let $\mathcal{U}_A$ be the coarsest uniformity on A making i uniformly continuous. Then $(A,\mathcal{U}_A)$ is called a (uniform) subspace of $(X,\mathcal{U})$. Especially, we obtain $\mathcal{U}_A = \{(A \times A) \cap W\colon W \in \mathcal{U}\}$.

(2) Let $((X_i,\mathcal{U}_i))_{i\in I}$ be a family of uniform spaces, where I is a set. Let $X = \prod_{i\in I} X_i$ be the cartesian product of $(X_i)_{i\in I}$. If $\mathcal{U}$ denotes the coarsest uniformity on X making all projections $p_i\colon\prod X_i \to X_i$ uniformly continuous, then $(X,\mathcal{U})$ is called the (uniform) product space of the family $((X_i,\mathcal{U}_i))_{i\in I}$.

(3) Each family $(\mathcal{U}_i)_{i\in I}$ of uniformities on a set X has a supremum in the set of all uniformities on X ordered by inclusion, i.e. there is a coarsest uniformity $\mathcal{U}$

on X which is finer than each $\mathcal{W}_i$ (namely the coarsest uniformity on X making all identities $1_X^i: X \to (X,\mathcal{W}_i)$ uniformly continuous).

<u>0.3.12 Theorem</u>. For each uniformity $\mathcal{W}$ on a set X there is a family $(d_V)_{V\in\mathcal{W}}$ of pseudometrics on X such that for the corresponding uniformities $\mathcal{D}_V$ (cf. 0.3.3) the following is valid:

$$\mathcal{W} = \sup \{\mathcal{D}_V : V \in \mathcal{W}\} .$$

<u>0.3.13 Remark</u>. For each $V \in \mathcal{W}$, the pseudometric d_V in the above theorem is constructed as follows: Let V_1 be a symmetric entourage such that $V_1 \subset V$ (cf. 0.3.5.(1)). By 0.3.5.(2) there is a sequence $(V_n)_{n\in\mathbb{N}\setminus\{0\}}$ of symmetric entourages such that

$$V_{n+1}^3 \subset V_n \qquad (n = 1,2,\ldots) .$$

Let $h_V: X \times X \to [0,1]$ be defined by

$$h_V(x,y) = \begin{cases} 0 & \text{if } (x,y) \in \bigcap\limits_{n\in\mathbb{N}\setminus\{0\}} V_n \\ 1 & \text{if } (x,y) \in X \setminus V_1 \\ 2^{-k} & \text{if } (x,y) \in (\bigcap\limits_{n=1}^{k} V_n) \cap (X \setminus V_{k+1}) \end{cases}$$

Put $M_{xy} = \{ \sum\limits_{i=1}^{p} h_V(z_{i-1},z_i) : (z_0,\ldots,z_p)$ is a finite sequence of elements of X with $z_0 = x$ and $z_p = y$, $p \in \mathbb{N}\setminus\{0\} \}$ for each $(x,y) \in X \times X$. Then, for each $V \in \mathcal{W}$, a pseudometric d_V is defined by

$$d_V(x,y) = \inf M_{xy} .$$

Furthermore, for each $(x,y) \in X \times X$, the following is valid:

$$\frac{1}{2} h_V(x,y) \leq d_V(x,y) \leq h_V(x,y) .$$

0.3.14 Theorem. A topological space $(X,\mathcal{X})$ is uniformizable (i.e. there is a uniformity $\mathcal{W}$ on X such that $\mathcal{X}_{\mathcal{W}} = \mathcal{X}$) if and only if it is completely regular.

0.3.15 Definition. A uniform space $(X,\mathcal{W})$ is called separated iff $\bigcap_{W\in\mathcal{W}} W = \Delta$.

0.3.16 Theorem. A uniform space $(X,\mathcal{W})$ is separated if and only if one of the following equivalent conditions is satisfied:

(1) $(X,\mathcal{X}_{\mathcal{W}})$ is a T_0-space .

(2) $(X,\mathcal{X}_{\mathcal{W}})$ is a T_1-space .

(3) $(X,\mathcal{X}_{\mathcal{W}})$ is a Hausdorff space .

(4) $(X,\mathcal{X}_{\mathcal{W}})$ is a regular Hausdorff space .

(5) $(X,\mathcal{X}_{\mathcal{W}})$ is a Tychonoff space .

0.3.17 Definition. A uniform space $(X,\mathcal{W})$ is called pseudometrizable (resp. metrizable) provided that there is a pseudometric (resp. metric) d on X such that $\mathcal{W}_d = \mathcal{W}$ (cf. 0.3.3).

0.3.18 Theorem. A uniform space $(X,\mathcal{W})$ is pseudometrizable if and only if $\mathcal{W}$ has a countable base.

0.3.19 Theorem. A uniform space $(X,\mathcal{W})$ is metrizable if and only if it is separated and $\mathcal{W}$ has a countable base.

0.3.20 Definition. 1) A filter $\mathcal{F}$ on a uniform space $(X,\mathcal{W})$ is called a Cauchy filter provided that for each $W \in \mathcal{W}$ there is some $F \in \mathcal{F}$ such that $F \times F \subset W$.

2) A uniform space $(X,\mathcal{W})$ is called complete provided that each Cauchy filter on $(X,\mathcal{W})$ is convergent (in $(X,\mathcal{X}_{\mathcal{W}})$) .

0.3.21 Theorem. Let $(X,\mathcal{W})$ be a separated uniform space. Then there is a dense embedding $r_X: (X,\mathcal{W}) \to (\hat{X},\hat{\mathcal{W}})$ of $(X,\mathcal{W})$ into a complete separated uniform space $(\hat{X},\hat{\mathcal{W}})$ such that for each complete separated uniform space $(Y,\mathcal{R})$ and each uniformly continuous map $f\colon (X,\mathcal{W}) \to (Y,\mathcal{R})$ there is a unique uniformly continuous map $\hat{f}\colon (\hat{X},\hat{\mathcal{W}}) \to (Y,\mathcal{R})$ such that the diagram

$$\begin{array}{ccc} (X,\mathcal{W}) & \xrightarrow{f} & (Y,\mathcal{R}) \\ & {\scriptstyle r_X}\searrow \quad \nearrow{\scriptstyle \hat{f}} & \\ & (\hat{X},\hat{\mathcal{W}}) & \end{array}$$

commutes.

<u>0.3.22 Remark</u>. The uniform space $(\hat{X},\hat{\mathcal{W}})$ in the above theorem is called the *complete hull* of $(X,\mathcal{W})$ and is uniquely determined by the property of containing $(X,\mathcal{W})$ as a dense subspace and of being a complete separated uniform space (up to an isomorphism).

<u>0.3.23 Definition</u>. A uniform space $(X,\mathcal{W})$ is called <u>totally bounded</u> provided that one of the following equivalent conditions is satisfied:

(1) For each $V \in \mathcal{W}$ there is a finite cover of X by V-small sets.

(2) For each $W \in \mathcal{W}$ there is a finite subset E of X such that $W[E] = X$, where $W[E] = \bigcup_{x \in E} W(x)$.

(3) Each ultrafilter on X is a Cauchy filter.

<u>0.3.24 Theorem</u>. A uniform space $(X,\mathcal{W})$ is compact (i.e. $(X,\mathcal{X}_{\mathcal{W}})$ is compact) if and only if it is complete and totally bounded.

CHAPTER I

TOPOLOGICAL CATEGORIES

In topology one is not only interested in the category of topological spaces (and continuous maps), but also in the category of uniform spaces (and uniformly continuous maps), the category of proximity spaces (and δ-maps), the category of limit spaces (and continuous maps) and others. The striking similarities of constructions in the categories mentioned above lead to the question whether it is possible to postulate axioms for a concrete category which may be considered as topological ones. Thus, the problem consists in looking for one or more properties which are independent of the special structure of the considered objects in a concrete category (i.e. properties essentially characterized by morphisms) and which are not satisfied by "algebraic" categories. This claim is fulfilled by the initial structures in the sense of N. Bourbaki provided their unrestricted existence is required. In the category of groups (and homomorphisms) for instance there do not exist arbitrary initial structures, e.g. not every subset of a group is a subgroup. Further conditions may be added for getting the concept "topological category" but they are of a more "technical" nature. In order to obtain final structures simultaneously it is useful (in contrast to N. Bourbaki) to require the existence of initial structures for families of maps which are indexed by a class (instead of a set). After the definition of a topological category and numerous examples (up to measure theory and algebraic topology) in this chapter the categorical properties of topological categories are studied. Finally connectedness and disconnectedness properties are investigated in the realm of topological categories (relations between point-separation axioms and connectedness known from classical topology are generalized!).

1.1 Definitions and Examples

1.1.1. By a concrete category we mean a category C whose *objects* are structured sets, i.e. pairs (X,ξ) where X is a set and ξ is a C-structure on X, whose *morphisms* $f: (X,\xi) \to (Y,\eta)$ are suitable maps between X and Y and whose composition law is the usual composition of maps - in other words: a category C together with a faithful (forgetful) functor[0)] $F: C \to \underline{Set}$ ($\underline{Set}$: category of sets [and maps]).

1.1.2 Definition. A concrete category is called topological iff it satisfies the following conditions:

Cat top_1) *Existence of initial structures.*
For any set X, any family $((X_i,\xi_i))_{i\in I}$ of C-objects indexed by a class I and any family $(f_i: X \to X_i)_{i\in I}$ of maps indexed by I there exists a unique C-structure ξ on X which is initial with respect to $(X,f_i,(X_i,\xi_i), I)$, i.e. such that for any C-object (Y,η) a map $g: (Y,\eta) \to (X,\xi)$ is a C-morphism iff for every $i \in I$ the composite map $f_i \circ g: (Y,\eta) \to (X_i,\xi_i)$ is a C-morphism.

Cat top_2) *Fibre-smallness.*
For any set X, the C-fibre of X, i.e. the class of all C-structures on X, is a set.

Cat top_3) *Terminal separator property.*
For any set X with cardinality one there exists precisely one C-structure on X.

1.1.3 Remarks. (1) Let ξ be the initial structure on X with respect to $(X,f_i,(X_i,\xi_i), I)$. Then $f_i: (X,\xi) \to (X_i,\xi_i)$ is a C-morphism for each $i \in I$ (Hint. Let $(Y,\eta) = (X,\xi)$ and $g = 1_X$ in Cat top_1).).

---

0) cf. 2.1.1 and footnote 28) .

(2) In a topological category C holds:
Let X be a set and ξ, η C-structures on X such that $1_X\colon (X,\xi) \to (X,\eta)$ and $1_X\colon (X,\eta) \to (X,\xi)$ are C-morphisms. Then $\xi = \eta$ (this follows immediately from the uniqueness of initial structures required in Cat top_1)).

1.1.4 Definition. Let C be a topological category, let X be a set, and let ξ, η be C-structures on X. The C-structure ξ is called *finer* than η (and η *coarser* than ξ) [denoted by $\xi \leq \eta$] iff $1_X\colon (X,\xi) \to (X,\eta)$ is a C-morphism.

1.1.5 Proposition. The initial structure ξ on a set X with respect to $(X, f_i, (X_i, \xi_i), I)$ in a topological category C is the coarsest C-structure on X for which each of the maps f_i is a C-morphism.

Proof. By 1.1.3 (1) ξ is a C-structure on X for which all $f_i\colon (X,\xi) \to (X_i,\xi_i)$ are C-morphisms. Let η be a C-structure on X for which all $f_i\colon (X,\eta) \to (X_i,\xi_i)$ are C-morphisms. Since all $f_i \circ 1_X\colon (X,\eta) \to (X_i,\xi_i)$ are C-morphisms, $1_X\colon (X,\eta) \to (X,\xi)$ is a C-morphism, i.e. $\eta \leq \xi$.

1.1.6 Examples (of topological categories).
(1) The category Top of topological spaces (and continuous maps).
(2) The category Unif of uniform spaces (and uniformly continuous maps).
(3) The category Prox of proximity spaces (and δ-maps).
(4) The caterory Lim of limit spaces (and continuous maps)
[Def. Let X be a set and let $F(X)$ be the set of all filters on X.
A subset q of $F(X) \times X$ is called a limit structure on X provided the following are satisfied:
Lim_1) $(\dot{x}, x) \in q$ for each $x \in X$, where $\dot{x} := \{A \subset X\colon x \in A\}$
Lim_2) $(G, x) \in q$ whenever $(F, x) \in q$ and $F \subset G$.
Lim_3) $(F, x) \in q$ and $(G, x) \in q$ imply $(F \cap G, x) \in q$.
(X, q) is called a *limit space* provided q is a limit structure on X.

If (X_1,q_1), (X_2,q_2) are limit spaces, then a map $f: X_1 \to X_2$ is said to be *continuous* iff $(f(\mathcal{F}),f(x)) \in q_2$ for each pair $(\mathcal{F},x) \in q_1$.] . Let X be a set, $(Y_i,q_i)_{i\in I}$ a family of limit spaces and $(f_i: X \to Y_i)_{i\in I}$ a family of maps; then

$$q = \{(\mathcal{F},x) \in \mathcal{F}(X) \times X: (f_i(\mathcal{F}),f_i(x)) \in q_i \text{ for each } i \in I\}$$

is a limit structure, which is initial with respect to $(X,f_i,(Y_i,q_i), I)$.

(5) The category Near of nearness spaces (and uniformly continuous maps)

[Def. A *nearness space* is a pair (X,μ) , where X is a set and μ is a non-empty set of non-empty covers of X satisfying the following axioms:

N_1) $\mathcal{A} < \mathcal{B}$[1] and $\mathcal{A} \in \mu$ imply $\mathcal{B} \in \mu$.

N_2) $\mathcal{A} \in \mu$ and $\mathcal{B} \in \mu$ imply $\mathcal{A} \wedge \mathcal{B} = \{A \cap B: A \in \mathcal{A}$ and $B \in \mathcal{B}\} \in \mu$.

N_3) $\mathcal{A} \in \mu$ imply $\{\mathrm{int}_\mu A: A \in \mathcal{A}\} \in \mu$, where $\mathrm{int}_\mu A = \{x \in X: \{A,X\setminus\{x\}\} \in \mu\}$.

The members of μ are called *uniform covers* of X . If (X,μ) and (Y,η) are nearness spaces, then a map $f: X \to Y$ is called *uniformly continuous* iff $f^{-1}\mathcal{A} = \{f^{-1}[A]: A \in \mathcal{A}\} \in \mu$ for each $\mathcal{A} \in \eta$.] .

Let (X,μ) be a nearness space. A subset μ' of μ is called a *base* for μ iff $\mu = \{\mathcal{A}: \mathcal{A}$ is a cover of X and there exists some $\mathcal{A}' \in \mu'$ which refines $\mathcal{A}\}$.

A *subbase* for μ is any subset μ' of μ such that all finite intersections of elements of μ' form a base for μ (where the intersection $\mathcal{A} \wedge \mathcal{B}$ of two covers $\mathcal{A}$ and $\mathcal{B}$ of X is also a cover of X). Let X be set, $((Y_i,\eta_i))_{i\in I}$ a family of nearness spaces and $(f_i: X \to Y_i)_{i\in I}$ a family of maps, Then $\{f_i^{-1}\mathcal{A}_i: \mathcal{A}_i \in \eta_i$ and $i \in I\}$ is a subbase for a nearness structure on X , which is initial with respect to $(X,f_i,(Y_i,\eta_i), I)$.

---

1) $\mathcal{A} < \mathcal{B} \Leftrightarrow \forall A \in \mathcal{A}\ \exists B \in \mathcal{B}\ \ A \subset B \Leftrightarrow \mathcal{A}$ refines $\mathcal{B}$

(6) The category Bitop of bitopological spaces (and pairwise continuous maps)

[Def. Let X be a set and $\mathcal{P},\mathcal{Q}$ topologies on X. Then $(X,\mathcal{P},\mathcal{Q})$ is called a *bitopological space*. If $(X,\mathcal{P},\mathcal{Q})$, $(Y,\mathcal{R},\mathcal{S})$ are bitopological spaces, then a map $f: X \to Y$ is called *pairwise continuous* iff $f: (X,\mathcal{P}) \to (Y,\mathcal{R})$ and $f: (X,\mathcal{Q}) \to (Y,\mathcal{S})$ are continuous.] .

Let X be a set, $((X_i,\mathcal{P}_i,\mathcal{Q}_i))_{i\in I}$ a family of bitopological spaces and $(f_i: X \to X_i)_{i\in I}$ a family of maps. Then $(\mathcal{P},\mathcal{Q})$ is the initial bitopological structure with respect to $(X,f_i,(X_i,\mathcal{P}_i,\mathcal{Q}_i), I)$, where $\mathcal{P}$ is the initial topology with respect to $(X,f_i,(X_i,\mathcal{P}_i), I)$ and $\mathcal{Q}$ is the initial topology with respect to $(X,f_i,(X_i,\mathcal{Q}_i), I)$.

(7) The category Meas of measurable spaces (and measurable maps)

[Def. Let X be a set. A subset $\mathcal{A}$ of $\mathcal{P}(X)$ is called a σ-algebra on X iff the following are satisfied:

1) $X \in \mathcal{A}$
2) $A \in \mathcal{A}$ implies $CA \in \mathcal{A}$
3) $\bigcup_{n\in\mathbb{N}} A_n \in \mathcal{A}$ for every sequence $(A_n)_{n\in\mathbb{N}}$ in $\mathcal{A}$.

The pair $(X,\mathcal{A})$ is called a *measurable space* provided $\mathcal{A}$ is a σ-algebra on X. If $(X,\mathcal{A})$, $(X',\mathcal{A}')$ are measurable spaces, then a map $f: X \to X'$ is called *measurable* iff $f^{-1}[A'] \in \mathcal{A}$ for each $A' \in \mathcal{A}'$.] .

Let X be a set, $((X_i,\mathcal{A}_i))_{i\in I}$ a family of measurable spaces and $(f_i: X \to X_i)_{i\in I}$ a family of maps. Then the intersection of all σ-algebras on X containing $\bigcup_{i\in I} f_i^{-1} \mathcal{A}_i$ is the initial σ-algebra on X with respect to $(X,f_i,(X_i,\mathcal{A}_i), I)$.

(8) The category Born of bornological spaces (and bounded maps)

[Def. A *bornological space* is a pair $(X,\mathcal{B})$, where X is a set and $\mathcal{B}$ is a subset of $\mathcal{P}(X)$ satisfying

1) $A,B \in \mathcal{B}$ imply $A \cup B \in \mathcal{B}$
2) $B \in \mathcal{B}$ and $A \subset B$ imply $A \in \mathcal{B}$
3) Each finite subset B of X belongs to $\mathcal{B}$.

$\mathcal{B}$ is called a *bornology* on X. The elements of $\mathcal{B}$ are called *bounded sets*.

If $(X,\mathcal{B})$ and $(X',\mathcal{B}')$ are bornological spaces, then a map $f: X \to X'$ is called *bounded* iff $f[B] \in \mathcal{B}'$ for each $B \in \mathcal{B}$.] .

Let X be a set, $((X_i,\mathcal{B}_i))_{i\in I}$ a family of bornological spaces and $(f_i: X \to X_i)_{i\in I}$ a family of maps. Then $\mathcal{B} = \{B \subset X: f_i[B] \in \mathcal{B}_i$ for each $i \in I\}$ is a bornology on X which is initial with respect to $(X,f_i,(X_i,\mathcal{B}_i), I)$.

(9) The category Rere of reflexive relations (and relation preserving maps)

[The objects of Rere are pairs (X,ρ) , where X is a set and ρ is a reflexive relation on X .

$f: (X,\rho) \to (X',\rho')$ is a morphism in Rere provided $(f(x),f(y)) \in \rho'$ for each $(x,y) \in \rho$.] .

Let X be a set, $((X_i,\rho_i))_{i\in I}$ a family of Rere-objects and $(f_i: X \to X_i)_{i\in I}$ a family of maps. Then $\rho = \{(x,y) \in X\times X: (f_i(x),f_i(y)) \in \rho_i$ for each $i \in I\}$ is a Rere-structure which is initial with respect to $(X,f_i,(X_i,\rho_i), I)$.

(10) The category Simp of simplicial complexes (and simplicial maps)

[Def. A *simplicial complex* is a pair $(K,\mathcal{K})$, where K is a set and $\mathcal{K}$ is a subset of $\mathcal{P}(K)$ satisfying

Simp$_1$) $\{k\} \in \mathcal{K}$ for each $k \in K$

Simp$_2$) $E \in \mathcal{K}$ implies that E is non-empty and finite

Simp$_3$) $E \in \mathcal{K}$ and $F \subset E$ and $F \neq \emptyset$ imply $F \in \mathcal{K}$.

The elements of K are called *vertices*, the elements of $\mathcal{K}$ are called *simplexes*.

Let $(K,\mathcal{K})$ and $(K',\mathcal{K}')$ be simplicial complexes. Then a map $f: K \to K'$ is called *simplicial* iff $f[E] \in \mathcal{K}'$ for each $E \in \mathcal{K}$.] .

Let K be a set, $((K_i,\mathcal{K}_i))_{i\in I}$ a family of simplicial complexes and $(f_i: K \to K_i)_{i\in I}$ a family of maps. Then $\mathcal{K} = \{E \subset K: E$ is non-empty and $f_i[E] \in \mathcal{K}_i$ for each $i \in I\}$ is a simplicial structure on K which is initial with respect to $(K,f_i,(K_i,\mathcal{K}_i), I)$.

(11) The category Reg (resp. C Reg) of regular[2)] (resp. completely regular[2)]) topological spaces [and continuous maps].

(12) The category L Con (resp. LP Con) of locally connected (resp. locally pathwise connected) spaces [and continuous maps].
(Hint. Determine the final structures and compare with the following Theorem 1.2.1.1.)

1.2 Special categorical properties of topological categories

1.2.1 Completeness and cocompleteness

1.2.1.1 Theorem. For a concrete category C the following are equivalent:

(1) C satisfies Cat top_1) in 1.1.2.

(2) For any set X, any family $((X_i,\xi_i))_{i\in I}$ of C-objects indexed by some class I and any family $(f_i\colon X_i \to X)_{i\in I}$ indexed by I there exists a unique C-structure ξ on X which is final with respect to $((X_i,\xi_i),\ f_i,X,I)$, i.e. for any C-object (Y,η) a map $g\colon (X,\xi) \to (Y,\eta)$ is a C-morphism iff for every $i \in I$ the composite map $g\circ f_i\colon (X_i,\xi_i) \to (Y,\eta)$ is a C-morphism.

Proof. a) (1) ⇒ (2).

α) Let (R_j,ρ_j) be a C-object and let $h_j\colon X \to R_j$ be a map such that $h_j\circ f_i\colon (X_i,\xi_i) \to (R_j,\rho_j)$ is a C-morphism for each $i \in I$.
Let $((R_j,\rho_j),h_j)_{j\in J}$ be the family of all pairs determined in the way above with the index class J. Let ξ be the initial structure with respect to $(X,h_j,(R_j,\rho_j),J)$. Then ξ is the final structure with respect to $((X_i,\xi_i),f_i,X,I)$:
1. Let $g\colon (X,\xi) \to (Y,\eta)$ be a C-morphism and $i \in I$. Since

2) The Hausdorff axiom is not included.

$h_j \circ f_i\colon (X_i,\xi_i) \to (R_j,\rho_j)$ is a C-morphism for each $j \in J$, $f_i\colon (X_i,\xi_i) \to (X,\xi)$ is (because ξ is initial) a C-morphism. Then $g \circ f_i\colon (X_i \xi_i) \to (Y,\eta)$ is a C-morphism as a composite of two C-morphisms.

2. Suppose all $g \circ f_i\colon (X_i,\xi_i) \to (Y,\eta)$ are C-morphisms. Then there is a $j \in J$ such that $g = h_j$ and $(Y,\eta) = (R_j,\rho_j)$. Hence, since ξ is initial, g is a C-morphism.

β) Let ξ' be another C-structure on X , which is final with respect to $((X_i,\xi_i),f_i,X,I)$. Then $1_X\colon (X,\xi) \to (X,\xi')$ and $1_X\colon (X,\xi') \to (X,\xi)$ are C-morphism. Thus $\xi = \xi'$ (cf. 1.1.3 ②).

b) (2) ⇒ (1): analogously to a).

<u>1.2.1.2</u>. By the previous theorem in a topological category there exist arbitrary initial and final structures. In particular, similar to 1.1.5, the following holds:

<u>Proposition</u>. In a topological category C the final structure ξ on a set X with respect to $((X_i,\xi_i),f_i,X,I)$ is the finest C-structure for which every map $f_i\colon X_i \to X$ is a C-morphism.

<u>Proof</u>. Let $(Y,\eta) = (X,\xi)$ and $g = 1_X$ in 1.2.1.1(2). Then each map $f_i\colon (X_i,\xi_i) \to (X,\xi)$ is a C-morphism. Let η be a C-structure on X for which every $f_i\colon (X_i,\xi_i) \to (X,\eta)$ is a C-morphism. Since all $1_X \circ f_i\colon (X_i,\xi_i) \to (X,\eta)$ are C-morphisms, the map $1_X\colon (X,\xi) \to (X,\eta)$ is a C-morphism, i.e. $\xi \leq \eta$.

<u>1.2.1.3 Theorem</u>. Let C be a topological category. Let I be a set and $((X_i,\xi_i))_{i\in I}$ a family of C-objects.

a) If $X = \prod_{i\in I} X_i$ is the cartesian product of the family $(X_i)_{i\in I}$, $p_i\colon X \to X_i$ are the projection maps $(i \in I)$ and ξ is the initial C-structure on X with respect to $(X,p_i,(X_i,\xi_i),I)$, then

b) If $X = \bigcup_{i\in I} X_i \times \{i\}$, $j_i\colon X_i \to X$ are the (canonical) injection maps for each $i \in I$ (i.e. $j_i(y) = (y,i)$ for each $y \in X_i$ and for each $i \in I$) and ξ is the final C-structure on X with respect to $((X_i,\xi_i),j_i,X,I)$, then

$((X,\xi), (p_i)_{i\in I})$ is the product of the family $((X_i,\xi_i))_{i\in I}$ in the category C .

$((X,\xi), (j_i)_{i\in I})$ is the coproduct of the family $((X_i,\xi_i))_{i\in I}$ in the category C .

Proof.

a) Suppose that (Y,η) is a C-object and $p_i^!: (Y,\eta) \to (X_i,\xi_i)$ is a C-morphism for each $i \in I$. Then a map $p: (Y,\eta) \to \left(\prod_{i\in I} X_i,\xi\right)$ is defined by $p_i(p(y)) = p_i^!(y)$ for each $y \in Y$ and each $i \in I$. Since $p_i \circ p = p_i^!$ is a C-morphism for each $i \in I$ and ξ is initial, p is a C-morphism. Obviously any map $p': Y \to \prod_{i\in I} X_i$ satisfying $p_i \circ p' = p_i^!$ for each $i \in I$ coincides with p .

b) Let (Y,η) be a C-object and suppose that $j_i^!: (X_i,\xi_i) \to (Y,\eta)$ is a C-morphism for each $i \in I$. Then for each $x \in X$ there exists a unique $x_i \in X_i$ such that $(x_i,i) = x$ and thus a map $j: (X,\xi) \to (Y,\eta)$ is defined by $j(x) = j((x_i,i)) = j(j_i(x_i)) = j_i^!(x_i) \in Y$. Since $j \circ j_i = j_i^!$ is a C-morphism for each $i \in I$ and ξ is final, j is a C-morphism. Obviously any map $j': X \to Y$ satisfying $j' \circ j_i = j_i^!$ for each $i \in I$ coincides with j .

1.2.1.4 Definition. A category C has products (resp. coproducts) provided that for every set I , each family $(A_i)_{i\in I}$ of C-objects has a product (resp. coproduct) in C .

1.2.1.5 Remark. By the preceding definition 1.2.1.3 means that every topological category has products and coproducts.

1.2.1.6 Definition. Let C be a category and $f,g: A \to B$ C-morphisms.

1) A C-morphism

$k: K \to A$ is called an equalizer of f

$c: B \to C$ is called a coequalizer of f

and g provided that:

(1) $f \circ k = g \circ k$.

(2) For any $D \in |\mathcal{C}|$ and for any $h \in [D,A]_{\mathcal{C}}$ such that $f \circ h = g \circ h$, there exists a unique $h' \in [D,K]_{\mathcal{C}}$ such that the diagram

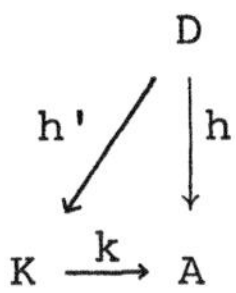

commutes (i.e. $k \circ h' = h$). Instead of k we often write $E(f,g)$ (see the following proposition).

and g provided that:

(1') $c \circ f = c \circ g$.

(2') For any $D \in |\mathcal{C}|$ and for any $h \in [B,D]_{\mathcal{C}}$ such that $h \circ f = h \circ g$, there exists a unique $h' \in [C,D]_{\mathcal{C}}$ such that the diagram

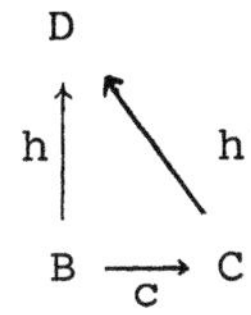

commutes (i.e. $h' \circ c = h$) i.e. c^* is an equalizer of f^* and g^* in the dual category $\mathcal{C}^*$. Instead of c we often write $CE(f,g)$ (see the following proposition).

2) A $\mathcal{C}$-object

$K \in |\mathcal{C}|$ is called an <u>equalizer</u> of $A \in |\mathcal{C}|$ provided that $[K,A]_{\mathcal{C}}$ contains an equalizer.

$C \in |\mathcal{C}|$ is called a <u>coequalizer</u> of $B \in |\mathcal{C}|$ provided that $[B,C]_{\mathcal{C}}$ contains a coequalizer (i.e. C is an equalizer of B in the dual category $\mathcal{C}^*$).

<u>1.2.1.7 Proposition</u>

a) If $k: K \to A$ and $k': K' \to A$ are equalizers of two $\mathcal{C}$-morphisms $f,g: A \to B$ in a category $\mathcal{C}$, then there exists a unique $\mathcal{C}$-isomorphism $i: K \to K'$ such that $k = k' \circ i$.

b) If $c: B \to C$ and $c': B \to C'$ are coequalizers of two $\mathcal{C}$-morphisms $f,g: A \to B$ in a category $\mathcal{C}$, then there exists a unique $\mathcal{C}$-isomorphism $i: C' \to C$ such that $c = i \circ c'$.

Proof. Because of the duality it suffices to prove a): Since k' is an equalizer of f and g, there is a unique $\mathcal{C}$-morphism $i: K \to K'$ such that $k' \circ i = k$. Reversing the roles of k and k', there is a unique $\mathcal{C}$-morphism $i': K' \to K$ such that $k \circ i' = k'$. Hence $k \circ (i' \circ i) = (k \circ i') \circ i = k' \circ i = k = k \circ 1_K$. By 1.2.1.6 (2) there is a unique $h: K \to K$ such that $k \circ h = k$. Thus $h = i' \circ i = 1_K$. Similarly $i \circ i' = 1_{K'}$. Consequently, i is an isomorphism.

1.2.1.8 Theorem. Let $\mathcal{C}$ be a topological category, (X,ξ), (Y,μ) $\mathcal{C}$-objects and $f,g: (X,\xi) \to (Y,\mu)$ $\mathcal{C}$-morphisms.

a) If $K = \{x \in X: f(x) = g(x)\}$ is endowed with the initial $\mathcal{C}$-structure ξ_K with respect to the inclusion map $i: K \to X$, then $i: (K,\xi_K) \to (X,\xi)$ is the equalizer of f and g.

b) Let R be the finest equivalence relation on Y, for which $f(x)$ and $g(x)$ are equivalent for each $x \in X$ (i.e. R is the intersection of all equivalence relations with this property). If $C = Y/_R$ is endowed with the final $\mathcal{C}$-structure μ_R with respect to the natural map $\omega: Y \to C$, then $\omega: (Y,\mu) \to (C,\mu_R)$ is the coequalizer of f and g.

Proof.

a) (1) $f \circ i = g \circ i$ is obvious because f and g coincide on K.

(2) Given a $\mathcal{C}$-morphism $h: (R,\rho) \to (X,\xi)$ such that $f \circ h = g \circ h$. Then $h(y) \in K$ for each $y \in R$ because $f(h(y)) = g(h(y))$ for

b) (1) $\omega \circ f = \omega \circ g$ is obvious because $f(x)$ and $g(x)$ are equivalent for each $x \in X$, i.e. $\omega(f(x)) = \omega(g(x))$ for each $x \in X$.

(2) Given a $\mathcal{C}$-morphism $h: (Y,\mu) \to (R^*,\rho^*)$ such that $h \circ f = g \circ f$. An equivalence relation π_h defined by $(y,y') \in \pi_h$ iff $h(y) = h(y')$

each $y \in R$. Hence a map $h'\colon R \to K$ is defined by $h'(y) = h(y)$ for each $y \in R$. Since $h = i \circ h'$ is a C-morphism and ξ_K is initial, $h'\colon (R,\rho) \to (K,\xi_K)$ is a C-morphism.
Obviously, any map $h''\colon R \to K$ such that $h = i \circ h''$ coincides with h' .

is related to h . Then $f(x)$ and $g(x)$ are equivalent with respect to π_h for each $x \in X$. Thus $R \subset \pi_h$. Hence there exist two maps 1_Y^* and s such that the diagram

$$\begin{array}{ccccc} Y & \xrightarrow{1_Y} & Y & \xrightarrow{h} & R^* \\ \big\downarrow{\scriptstyle\omega} & & \big\downarrow{\scriptstyle\omega^*} & \nearrow{\scriptstyle s} & \\ Y/_R & \xrightarrow{1_Y^*} & Y/_{\pi_h} & & \end{array}$$

commutes (ω^*: natural map). Then $h = h' \circ \omega$, where $h' = s \circ 1_{Y^*}$.
Since μ_R is final with respect to ω , $h'\colon (Y/_R, \mu_R) \to (R^*, \rho^*)$ is a C-morphism. Obviously any map $h''\colon Y/_R \to R^*$ such that $h = h'' \circ \omega$ coincides with h' .

<u>1.2.1.9 Definition</u>. 1) A <u>category</u> C <u>has equalizers (resp. coequalizers)</u> provided that every pair (f,g) of C-morphisms with common domain and common codomain has an equalizer (resp. coequalizer).

2) a) A category C is said to be <u>complete</u> provided that C has products and equalizers.

b) A category C is called <u>cocomplete</u> provided that C has coproducts and coequalizers, i.e. the dual category C^* is complete.

<u>1.2.1.10 Theorem</u>. Every topological category C is complete and cocomplete.

Proof. See 1.2.1.3 and 1.2.1.8.

1.2.2 Special objects and special morphisms

1.2.2.1 Definition. Let C be a topological category and X a set. The initial C-structure ξ_i (resp. final C-structure ξ_d) on X with respect to the empty index class I is called indiscrete (resp. discrete).

1.2.2.2 Remarks. (1) If ξ_i is the indiscrete C-structure on X, then $f: (Y,\eta) \to (X,\xi_i)$ is a C-morphism for every C-object (Y,η) and every map $f: Y \to X$.

(2) If ξ_d is the discrete C-structure, then $f: (X,\xi_d) \to (Y,\eta)$ is a C-morphism for every C-object (Y,η) and every map $f: X \to Y$.

(3) Let X be a one-element set. Then by Cat top_3 there is precisely one C-structure on X. Hence $\xi_i = \xi_d$.

1.2.2.3 Proposition. Let X,Y be sets and let $f: X \to Y$ be a constant map. If (X,ξ) and (Y,η) are objects of a topological category C, then $f: (X,\xi) \to (Y,\eta)$ is a C-morphism.

Proof. Since $f: X \to Y$ is constant, $f[X] = \{y_o\}$ is a one-element set. Suppose that μ is the initial C-structure on $\{y_o\}$ with respect to the inclusion map $i: \{y_o\} \to Y$. Then by Cat top_3) μ is indiscrete. Hence $f: (X,\xi) \to (f[X],\mu)$ defined by $f'(x) = f(x)$ for each $x \in X$ is a C-morphism. Thus $f = i \circ f': (X,\xi) \to (Y,\eta)$ is a C-morphism since the composite of two C-morphisms is a C-morphism.

1.2.2.4 Theorem. In a topological category C a C-morphism $f: (X,\xi) \to (Y,\eta)$ is

a) a monomorphism if and only if $f: X \to Y$ is injective.

b) an epimorphism if and only if $f: X \to Y$ is surjective.

Proof.

a) α) Let $x, y \in X$ such that $f(x) = f(y)$. $\bar{x}: (X,\xi) \to (X,\xi)$ defined by $\bar{x}(z) = x$ for each $z \in X$ and $\bar{y}: (X,\xi) \to (X,\xi)$ defined by $\bar{y}(z) = y$ for each $z \in X$ are C-morphisms (cf. 1.2.2.3) such that $f \circ \bar{x} = f \circ \bar{y}$. Since f is a monomorphism it follows $\bar{x} = \bar{y}$, i.e. $x = y$.

β) Let $\gamma, \delta: (X',\xi') \to (X,\xi)$ be C-morphisms such that $f \circ \gamma = f \circ \delta$. Then $f(\gamma(x')) = f(\delta(x'))$ for each $x' \in X'$. Since f is injective, $\gamma(x') = \delta(x')$ for each $x' \in X$. Thus $\gamma = \delta$.

b) α) (indirect). Suppose that f is not surjective. Then there is a $y' \in Y$ such that $y' \notin f[X]$. $\gamma, \delta: (Y,\eta) \to (\{0,1\},\xi_i)$ defined by $\gamma(y) = 0$ for each $y \in Y$ and

$$\delta(y) = \begin{cases} 0 & \text{for each } y \in f[X] \\ 1 & \text{otherwise} \end{cases}$$

are C-morphisms (cf. 1.2.2.2.(1)) such that $\gamma \circ f = \delta \circ f$ and $\gamma \neq \delta$. Thus f is not an epimorphism.

β) If f is surjective, then for every $y \in Y$ there is an $x \in X$ such that $f(x) = y$. Hence $\alpha \circ f = \beta \circ f$ implies $\alpha(y) = \alpha(f(x)) = \beta(f(x)) = \beta(y)$ for each $y \in Y$, i.e. $\alpha = \beta$. Thus f is an epimorphism.

1.2.2.5 Theorem. In a topological category a C-morphism $f: (X,\xi) \to (Y,\eta)$ is an

a) extremal monomorphism if and only if f is an *embedding*, i.e. $f': (X,\xi) \to (f[X],\eta_{f[X]})$ defined by $f'(x) = f(x)$ for each $x \in X$ is an isomorphism, where $\eta_{f[X]}$ is the initial C-structure on $f[X]$ with

b) extremal epimorphism if and only if f is a *quotient map*, i.e. $f: X \to Y$ is surjective and η is the final C-structure on Y with respect to f.

respect to the inclusion map $i: f[X] \to Y$.

Proof.

a) α) Suppose $f: (X,\xi) \to (Y,\eta)$ is an extremal monomorphism. The diagram

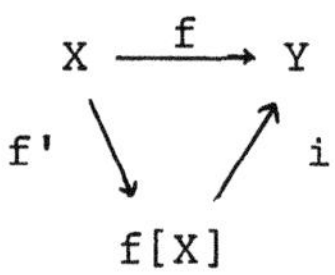

commutes and f' is a surjective C-morphism, i.e. by 1.2.2.4 an epimorphism. Hence (by the definition of an extremal monomorphism) f' is an isomorphism.

β) Let f' be an isomorphism. Then $f = i \circ f'$ is a mono-

b) α) Suppose $f: (X,\xi) \to (Y,\eta)$ is an extremal epimorphism. Then $f: X \to Y$ is surjective. Let the diagram

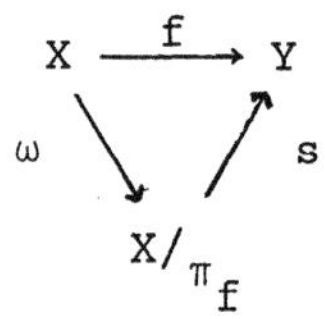

$\left(x \pi_f y \Leftrightarrow f(x) = f(y)\right.$; $X/_{\pi_f}$ is endowed with the final C-structure with respect to the natural map $\left.\omega: X \to X/_{\pi_f}\right)$ commute. So s is an injective (bijective) C-morphism (see the definition of the final C-structure). Hence s is a monomorphism (see 1.2.2.4). Since f is an extremal epimorphism, s is an isomorphism. It follows immediately that η is the final C-structure with respect to f (because $X/_{\pi_f}$ is endowed with a final structure).

β) Let $f: (X,\xi) \to (Y,\eta)$ be a quotient map. Moreover, let $f = g \circ h$, where g is a monomorphism (as a composite of two monomorphisms). If $f = h \circ g$, where g is an epimorphism, then the diagram

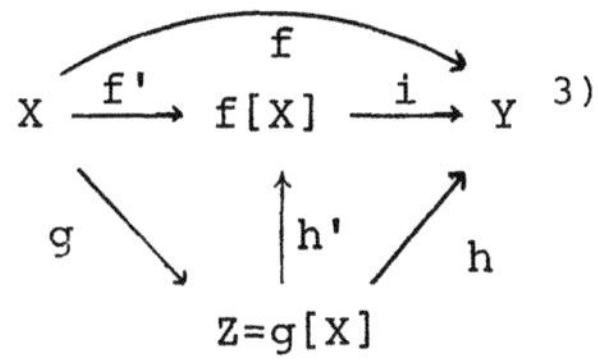

commutes. Especially $f' = h' \circ g$. Since f' is an extremal monomorphism (even an isomorphism), g is an isomorphism. Thus, f is an extremal monomorphism.

morphism, i.e. the diagram

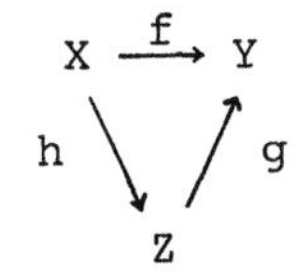

commutes. Since f is an epimorphism, g is an epimorphism (i.e. surjective and a C-morphism). Thus, g is bijective and g is a C-morphism. Since $g^{-1} \circ f = g^{-1} \circ g \circ h = 1_Z \circ h = h$ is a C-morphism and η is the final structure with respect to f, g^{-1} is a C-morphism. Hence, g is an isomorphism. Consequently, f is an extremal epimorphism.

<u>1.2.2.6 Remark</u>. It is a well-known fact that extremal monomorphisms (extremal epimorphisms) are used to define subobjects (quotient objects) in a category C. So $X \in |C|$ is a <u>subobject</u> (resp. <u>quotient object</u>) of $Y \in |C|$ if $[X,Y]_C$ ($[Y,X]_C$) contains an extremal monomorphism (resp. extremal epimorphism).

<u>1.2.2.7 Proposition</u>. If (X,ξ) is an object in a topological category C and $f\colon X \to Y$ is a bijective map, then there exists a unique C-structure η on Y such that $f\colon (X,\xi) \to (Y,\eta)$ is an isomorphism.

<u>Proof</u>. Let η be the final C-structure with respect to f. Then $f\colon (X,\xi) \to (Y,\eta)$ and $f^{-1}\colon (Y,\eta) \to (X,\xi)$ are C-morphisms such that $f^{-1} \circ f = 1_X$ and $f \circ f^{-1} = 1_Y$. Consequently, f is

---

[3] h' is defined by $i \circ h' = h$. The definition makes sense since $h[Z] = f[X]$. h' is a C-morphism because $f[X]$ is endowed with the initial structure with respect to i.

an isomorphism. In order to prove the uniqueness of η let η' be a C-structure on Y such that $f: (X,\xi) \to (Y,\eta')$ is an isomorphism. Obviously, η' is the final C-structure on Y with respect to f. Since final structures are unique (see 1.2.1.1 (2)), we obtain $\eta = \eta'$.

1.2.2.8 Definition. A category C is called

a) well-powered provided that for each $X \in |C|$ there is a set $\{m_i: X_i \to X\}$ of monomorphisms which is *representative* in the following sense:
For each monomorphism $m: Y \to X$ there is an m_i and an isomorphism $h_m: Y \to X_i$ satisfying $m = m_i \circ h_m$.

b) co-well-powered provided that C^* is well-powered, i.e. for every $X \in |C|$ there is a set $\{e_i: X \to X_i\}$ of epimorphisms which is *representative* in the following sense:
For each epimorphism $e: X \to Y$ there is an e_i and an isomorphism $h_e: X_i \to Y$ satisfying $e = h_e \circ e_i$.

1.2.2.9 Theorem. Every topological category C is well-powered and co-well-powered.

Proof. A) Given a cardinal number k. Then there is a set Q of C-objects such that every C-object (Y,η) satisfying $|Y| \le k$ is isomorphic to an object of Q:
Let Z be a set such that $|Z| = k$ and let (Y,η) be a C-object such that $|Y| \le k$. Then Y is equipotent with a subset X of Z, i.e. there is a bijective map $f: Y \to X$. By 1.2.2.7 there exists a unique C-structure ξ on X such that $f: (Y,\eta) \to (X,\xi)$ is an isomorphism. Now let $Q = \{(X,\xi) \in |C|: X \subset Z\}$ and let M_X be the set of all C-structures on X (Note Cat top$_3$)). Then $Q \subset P(Z) \times \bigcup_{X \in P(Z)} M_X$ is a set.

B) α) Let $(X,\xi) \in |C|$ and let $f: (Y,\eta) \to (X,\xi)$ be a C-monomorphism. Then $|Y| \le |X| = k$ ($f: Y \to X$ is injective). Hence, using A), there is a representative set of monomorphisms with codomain (X,ξ). Thus C is well-powered.

β) Let $f: (X,\xi) \to (Y,\eta)$ be an epimorphism in C, i.e. $f: X \to Y$, is surjective. Then $|Y| \leq |X| = k$ (For each $y \in Y$ choose a unique $x \in f^{-1}(y)$. Then an injective map $h: Y \to X$ is defined by $h(y) = x$ for each $y \in Y$.). Hence, using A), there is a representative set of epimorphisms with domain (X,ξ). Thus C is co-well-powered.

1.2.3 Factorization properties

1.2.3.1 Definition. A category C is called

a) (epi, extremal mono)-factorizable provided that for every C-morphism f there are a C-epimorphism e and an extremal monomorphism m in C such that $f = m \circ e$.

b) an (epi, extremal mono)-category provided that
   (1) C is (epi, extremal mono)-factorizable.
   (2) For any C-morphism f and for any two (epi, extremal mono)-factorizations $f = m \circ e = m' \circ e'$, there is an isomorphism j such that the diagram

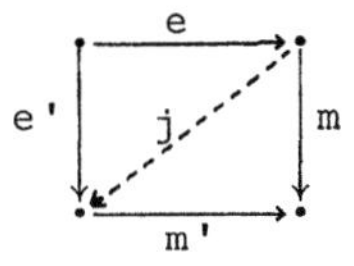

commutes, i.e. every

a') (extremal epi, mono)-factorizable provided that C^* is (epi, extremal mono)-factorizable, i.e. for every C-morphism f there are an extremal epimorphism e in C and a C-monomorphism m such that $f = m \circ e$.

b') an (extremal epi, mono)-category provided that C^* is an (epi, extremal mono)-category, i.e.
   (1') C is (extremal epi, mono)-factorizable.
   (2') For any C-morphism f and for any two (extremal epi, mono)-factorizations $f = m \circ e = m' \circ e'$, there is an isomorphism j such that the diagram

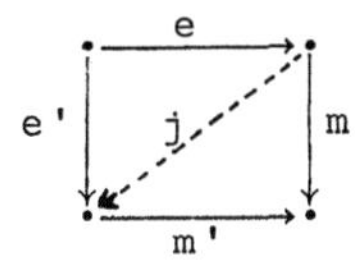

commutes, i.e. every C-mor-

C-morphism is uniquely (epi, extremal mono)-factorizable.

(3) If f,g are extremal monomorphisms and their composite $f\circ g$ is defined, then $f\circ g$ is an extremal monomorphism.

phism is uniquely (extremal epi, mono)-factorizable.

(3') If f,g are extremal epimorphisms and their composite $f\circ g$ is defined, then $f\circ g$ is an extremal epimorphism.

<u>1.2.3.2 Remark</u>. Let E (resp. M) be a class of epimorphisms (resp. monomorphisms) in C which is closed under composition with isomorphisms (C arbitrary category). Then analogous to 1.2.3.1 a) (resp. a')) a category C is called <u>(E,M)-factorizable</u> provided that for every C-morphism f there exist an $e \in E$ and an $m \in M$ such that $f = m\circ e$. C is called <u>uniquely (E,M)-factorizable</u> provided that C is (E,M)-factorizable and for every C-morphism the (E,M)-factorization is unique up to isomorphism (cf. 1.2.3.1 b)(2) [resp. b')(2')]). Finally C is called an <u>(E,M) category</u> provided that C is uniquely (E,M)-factorizable and the classes E and M are closed under composition.

<u>1.2.3.3 Theorem</u>. Every topological category C is an (epi, extremal mono)-category and an (extremal epi, mono)-category.

<u>Proof</u>. 1) If $f\colon (X,\xi) \to (Y,\eta)$ is a C-morphism, then $f = i\circ f'$ is the desired (epi, extremal mono)-factorization of f , where $f'\colon X \to f[X]$ is defined by $f'(x) = f(x)$ for each $x \in X$ and $i\colon f[X] \to Y$ is the inclusion map ($f[X]$ is endowed with the initial C-structure with respect to i !). If $f = m\circ e$ is another (epi, extremal mono)-factorization of f , then by 1.2.2.5 a) $m\colon e[X] \to Y$ may be considered to be an inclusion map, thus $f[X] = e[X]$, i.e. $m = i$. Since i is a monomorphism, $i\circ f' = i\circ e$ implies $f' = e$.
Consequently, the factorization is unique. Since the composition of two embeddings is an embedding, C is an (epi, extremal mono)-category.

2) Let $f: (X,\xi) \to (Y,\eta)$ be a C-morphism. Then an (extremal epi, mono)-factorization of f is given by the commutative diagram

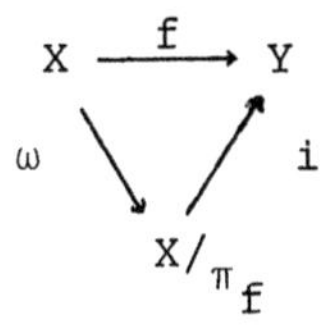

$(x \pi_f x' \Leftrightarrow f(x) = f(x'))$

($X/_{\pi_f}$ is endowed with the final C-structure with respect to the natural map ω !). If $f = m \circ e$ is another factorization of f, then e can be identified with the natural map $X \to X/_{\pi_e}$. Since m is injective, $X/_{\pi_e} = X/_{\pi_f}$, because $x\ \pi_e\ y$, i.e. $e(x) = e(y)$, is equivalent to $m(e(x)) = f(x) = m(e(y)) = f(y)$, i.e. $x\ \pi_f\ y$. Hence $e = \omega$. Since e is an epimorphism, $m \circ e = i \circ e$ implies $m = i$. Consequently, the factorization is unique. Since the composition of two quotient maps is a quotient map, C is an (extremal epi, mono)-category.

<u>1.2.3.4 Remark</u>. It can be shown that every well-powered, complete category C is an (epi, extremal mono)-category and an (extremal epi, mono)-category. Since every topological category C satisfies these conditions (cf. 1.2.2.9 and 1.2.1.10), there is another purely categorical proof of 1.2.3.3 (cf. [31; 7.2.12] resp. [44; 34.1 and 34.5]).

<u>1.2.3.5</u>. Let E and M be as in 1.2.3.2. Then especially the following proposition is true for (epi, extremal mono)-categories and also for (extremal epi, mono)-categories.

<u>Proposition</u>. If C is an (E,M)-factorizable category, then the following are equivalent:

(1) C is an (E,M) - category.

(2) For every commutative diagram in C

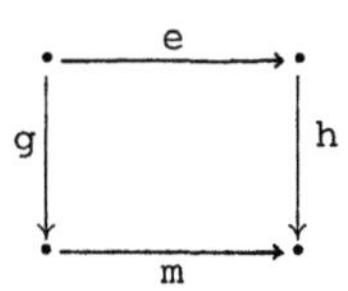

with $e \in \mathcal{E}$ and $m \in \mathcal{M}$, there exists a $\mathcal{C}$-morphism k that makes the diagram

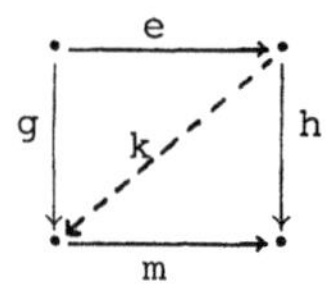

commute

(($\mathcal{E}$,$\mathcal{M}$)-diagonalization property) .

Proof. (1) ⇒ (2): Let $g = m' \circ e'$ and $h = m'' \circ e''$ be ($\mathcal{E}$,$\mathcal{M}$)-factorizations of g and h respectively and let $f = h \circ e = m \circ g$. Then $f = m'' \circ (e'' \circ e)$ and $f = (m \circ m') \circ e'$ are ($\mathcal{E}$,$\mathcal{M}$)-factorizations of f. Thus there exists an isomorphism j such that the diagram

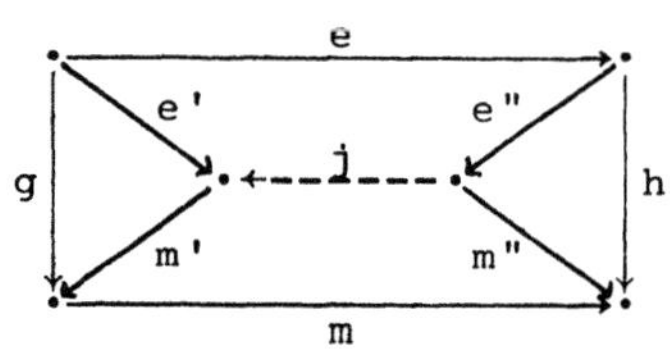

commutes. Hence, $k = m' \circ j \circ e''$ is the desired diagonal morphism.

(2) ⇒ (1): a) If $f = m \circ e = m' \circ e'$ are ($\mathcal{E}$,$\mathcal{M}$)-factorizations of $f \in \operatorname{Mor} \mathcal{C}$, then there exist $\mathcal{C}$-morphisms k and k' such that the diagrams

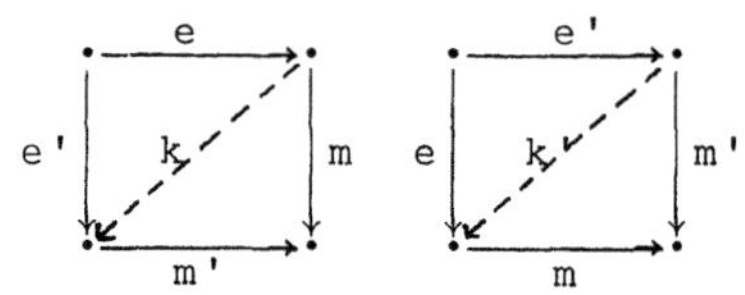

commute. Since e is an epimorphism, $1 \circ e = e = (k' \circ k) \circ e$ implies $1 = k' \circ k$ and since e' is an epimorphism $1 \circ e' = e' = (k \circ k') \circ e'$ implies $1 = k \circ k'$. Consequently, k is an isomorphism. Thus $\mathcal{C}$ is uniquely ($\mathcal{E}$,$\mathcal{M}$)-factorizable.

b) If $m_1, m_2 \in \mathcal{M}$ such that codomain(m_1) = domain(m_2) and $m_2 \circ m_1 = m \circ e$ is an ($\mathcal{E}$,$\mathcal{M}$)-factorization of $m_2 \circ m_1$, then there exists a $\mathcal{C}$-morphism k such that the diagram

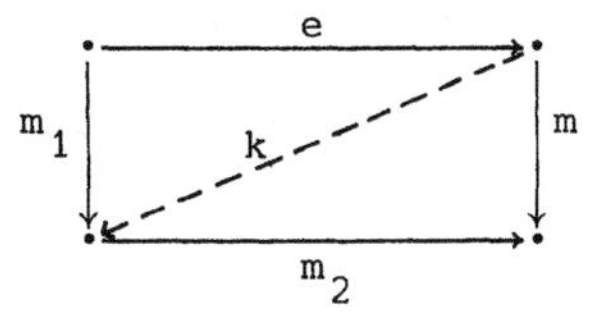

commutes. Furthermore, there exists some $k' \in \operatorname{Mor} \mathcal{C}$ such that the diagram

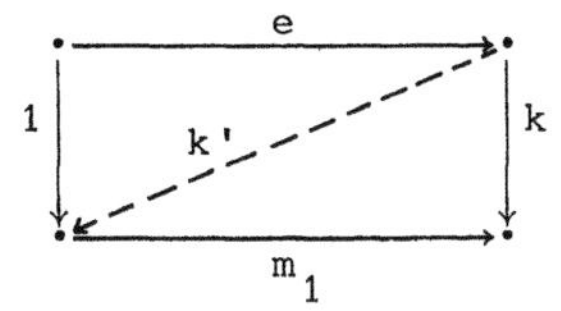

commutes. Since $k' \circ e = 1$ and e is an epimorphism, e is an isomorphism (Note that 1 is an extremal monomorphism!).
Hence, because of $m \in \mathcal{M}$, $m \circ e = m_2 \circ m_1 \in \mathcal{M}$.

c) Similarly to b) it follows that $\mathcal{E}$ is closed under composition.

1.3 Relative connectednesses and disconnectednesses in topological categories

1.3.1 In the following for a topological category $\mathcal{C}$ the category of pairs with respect to $\mathcal{C}$ is denoted by $\mathcal{C}_{(2)}$, i.e.

(1) *objects* of $\mathcal{C}_{(2)}$ are pairs $((X,\xi),\ (Y,\eta))$ where (X,ξ) is an object in $\mathcal{C}$, Y a subset of X and η the initial $\mathcal{C}$-structure with respect to $(Y,i,(X,\xi))$ where $i\colon Y \to X$ is the inclusion map.

(2) *morphisms* $f\colon ((X,\xi),\ (Y,\eta)) \to ((X',\xi'),\ (Y',\eta'))$ are morphisms $f\colon (X,\xi) \to (X',\xi')$ in $\mathcal{C}$ such that $f[Y] \subset Y'$.

For simplicity one often writes $(X,Y) \in |\mathcal{C}_{(2)}|$ instead of $((X,\xi),\ (Y,\eta)) \in |\mathcal{C}_{(2)}|$. If $(X,Y) \in |\mathcal{C}_{(2)}|$ and $f\colon X \to X'$ is a $\mathcal{C}$-morphism, then one writes usually $f|_Y$ instead of $f \circ i$, where $i\colon Y \to X$ is the inclusion map.

<u>1.3.2 Definition</u>. (1) Let $\mathcal{P}$ be a subclass of $|\mathcal{C}|$:

(a) Let $(X,Y) \in |\mathcal{C}_{(2)}|$. Y is called <u>$\mathcal{P}$-connected with respect to X</u> iff $f|_Y$ is constant for each $P \in \mathcal{P}$ and each $\mathcal{C}$-morphism $f: X \to P$.

(b) $C_{rel}\mathcal{P} = \{(X,Y) \in |\mathcal{C}_{(2)}|: Y$ is $\mathcal{P}$-connected with respect to $X\}$.

(c) $\mathcal{K} \subset |\mathcal{C}_{(2)}|$ is called a <u>relative connectedness</u> iff $\mathcal{K} = C_{rel}\mathcal{P}$ for some $\mathcal{P} \subset |\mathcal{C}|$.

(2) Let $\mathcal{K}$ be a subclass of $|\mathcal{C}_{(2)}|$:

(a) $D_{rel}\mathcal{K} = \{Z \in |\mathcal{C}|: f|_Y$ is constant for each $\mathcal{C}$-morphism $f: X \to Z$ and each $Y \subset X$ satisfying $(X,Y) \in \mathcal{K}\}$.

(b) $\mathcal{P} \subset |\mathcal{C}|$ is called a <u>relative disconnectedness</u> iff $\mathcal{P} = D_{rel}\mathcal{K}$ for some $\mathcal{K} \subset |\mathcal{C}_{(2)}|$.

<u>1.3.3 Corollaries</u>. 1. a) $\mathcal{P} \subset \mathcal{Q} \subset |\mathcal{C}|$ implies $C_{rel}\mathcal{P} \supset C_{rel}\mathcal{Q}$

b) $\mathcal{H} \subset \mathcal{K} \subset |\mathcal{C}_{(2)}|$ implies $D_{rel}\mathcal{H} \supset D_{rel}\mathcal{K}$

c) $\mathcal{P} \subset D_{rel}\, C_{rel}\mathcal{P}$ for each $\mathcal{P} \subset |\mathcal{C}|$

d) $\mathcal{K} \subset C_{rel}\, D_{rel}\mathcal{K}$ for each $\mathcal{K} \subset |\mathcal{C}_{(2)}|$

2. a) $C_{rel}\, D_{rel}\, C_{rel} = C_{rel}$

b) $D_{rel}\, C_{rel}\, D_{rel} = D_{rel}$

3. $P = C_{rel}\, D_{rel}$ and $Q = D_{rel}\, C_{rel}$ are <u>hull operators</u>, i.e. P and Q are extensive (cf. 1.(c) and (d)), isotonic ($\mathcal{H} \subset \mathcal{K} \subset |\mathcal{C}_{(2)}|$ implies $P\mathcal{H} \subset P\mathcal{K}$ and $\mathcal{P} \subset \mathcal{Q} \subset |\mathcal{C}|$ implies $Q\mathcal{P} \subset Q\mathcal{Q}$) and idempotent ($PP = P$ and $QQ = Q$).

<u>Proof</u>. 1. a)-d) follow immediately from the definitions.

2. a) α) $C_{rel}\mathcal{P} \subset C_{rel}\, D_{rel}\, C_{rel}\mathcal{P}$ (cf. 1. d))

β) $C_{rel}\mathcal{P} \supset C_{rel}\, D_{rel}\, C_{rel}\mathcal{P}$ follows from 1. c) and 1. a).

b) is proved analogously to a).

3. This follows immediately from 1. and 2.

<u>1.3.4 Definition</u>. A subclass $\mathcal{K}$ of $|\mathcal{C}_{(2)}|$ ($\mathcal{P}$ of $|\mathcal{C}|$) is called <u>P-closed</u> (<u>Q-closed</u>) iff $\mathcal{K} = P\mathcal{K}$ ($\mathcal{P} = Q\mathcal{P}$).

1.3.5 Proposition. a) A subclass K of $|C_{(2)}|$ is P-closed if and only if it is a relative connectedness.

b) A subclass P of $|C|$ is Q-closed if and only if it is a relative disconnectedness.

Proof. a) α) $K = PK$ implies $K = C_{rel}P$ where $P = D_{rel}K$.

β) Applying 1.3.3.3 and 1.3.3.2 a) one obtains from $K = C_{rel}P$ immediately $PK = C_{rel}\, D_{rel}K = C_{rel}\, D_{rel}\, C_{rel}P = C_{rel}P = K$.

b) is proved analogously to a).

1.3.6 Theorem. There exists a one-one-correspondence between the relative connectednesses of $|C_{(2)}|$ and the relative disconnectednesses of $|C|$ which converts the inclusion relation (Galois correspondence), and is obtained by the operators C_{rel} and D_{rel}.

Proof. By means of C_{rel} one obtains a one-one-correspondence which assigns to each relative disconnectedness, i.e. Q-closed subclass of $|C|$, a relative connectedness, i.e. P-closed subclass of $|C_{(2)}|$. The inverse correspondence is obtained by D_{rel}. From 1.3.3.1 a) and b) follows that the inclusion relation is converted.

1.3.7 Theorem. Let K be a subclass of $|C_{(2)}|$. Then the following are equivalent:

(1) K is a relative connectedness.

(2) $K = PK$.

(3) (a) $\{(X,Y) \in |C_{(2)}|: Y$ consists at most of a single element$\} \subset K$.

(b) Let $(X,Y) \in K$, and $f: (X,Y) \to (X',Y')$ be a $C_{(2)}$-morphism such that $f[Y] = Y'$. Then $(X',Y') \in K$.

(c) Let $(X,A_i) \in K$ for each i belonging to an index set I, and $\bigcap_{i\in I} A_i \neq \emptyset$. Then $(X, \bigcup_{i\in I} A_i) \in K$.

(d) Let $f: (X,Y) \to (X',Y')$ be a quotient map such that $f[Y] = Y'$. Further let $(X',Y') \in K$, and $(X,f^{-1}(x')) \in K$

for each $x' \in X'$. Then $(X,Y) \in \mathcal{K}$.

<u>Proof</u>. The equivalence of (1) and (2) was shown in 1.3.5.
(1) $\Rightarrow$ (3). Let $\mathcal{K} = C_{rel}\mathcal{P}$ for some $\mathcal{P} \subset |\mathcal{C}|$.

(a) is trivial.

(b) Let $P \in \mathcal{P}$ and $h \in [X',P]_{\mathcal{C}}$. If $Y' \neq \emptyset$ (the case $Y' = \emptyset$ is trivial) and $a',b' \in Y'$, then there exist $a,b \in Y$ such that $f(a) = a'$ and $f(b) = b'$. Since Y is $\mathcal{P}$-connected with respect to X , $(h \circ f)|_Y$ is constant. Consequently, $h(a') = h(f(a)) = h(f(b)) = h(b')$, i.e. $h|_{Y'}$ is constant.

(c) Let $f: X \to P$ be a $\mathcal{C}$-morphism for some $P \in \mathcal{P}$. If $a,b \in V = \bigcup_{i \in I} A_i$ (the case $V = \emptyset$ is trivial) and $c \in \bigcap_{i \in I} A_i$, then there exist $i_o, i_1 \in I$ such that $a \in A_{i_o}$ and $b \in A_{i_1}$. Since A_{i_o} and A_{i_1} are $\mathcal{K}$-connected with respect to X , $f|_{A_{i_o}}$ and $f|_{A_{i_1}}$ are constant maps. Consequently, $f(a) = f(c) = f(b)$.

(d) Let us define an equivalence relation $R_{\mathcal{P}}$ on X as follows:

$a \; R_{\mathcal{P}} \; b \Leftrightarrow a$ and b belong to a subset Y of X which is $\mathcal{P}$-connected with respect to X .

Let $\omega_{\mathcal{P}}: X \to X/_{R_{\mathcal{P}}}$ be the natural map. Now we endow $X/_{R_{\mathcal{P}}}$ with the final structure with respect to $((X,\xi),\omega_{\mathcal{P}},X/_{R_{\mathcal{P}}})$. Hence $\omega_{\mathcal{P}}$ is a $\mathcal{C}$-morphism. $h: X' \to X/_{R_{\mathcal{P}}}$ is defined by the property that it makes the diagram

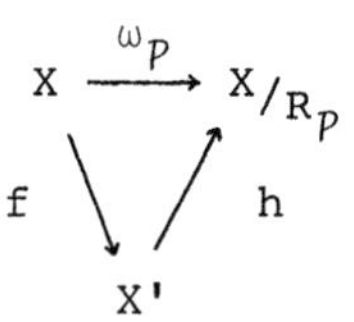

commutative. Thus h is well-defined (let $f(x) = f(y)$, then x and y belong to $f^{-1}(f(x))$ which is $\mathcal{P}$-connected with respect to X , and consequently, $\omega_{\mathcal{P}}(x) = \omega_{\mathcal{P}}(y)$, i.e. $h(f(x)) = h(f(y)))$ and it is a $\mathcal{C}$-morphism by the definition of the final structure on X'. Since $f[Y]$ is $\mathcal{P}$-connected with respect to X', it follows from (b) that $\omega_{\mathcal{P}}[Y] = h[f[Y]]$ is $\mathcal{P}$-connected with respect to $X/_{R_{\mathcal{P}}}$. Consequently, $\omega_{\mathcal{P}}[Y]$ consists at most of a single element (If $[x]$ and $[y]$ were distinct elements of $\omega_{\mathcal{P}}[Y]$, then y would not belong to $K_x = \bigcup \{V \subset X\colon V$ is $\mathcal{P}$-connected with respect to X and contains $x\}$. Thus there would exist some $P \in \mathcal{P}$ and some $\mathcal{C}$-morphism $g\colon X \to P$ such that $g(x) \neq g(y)$ and for the $\mathcal{C}$-morphism $\bar{g}\colon X/_{R_{\mathcal{P}}} \to P$ defined by $\bar{g}\circ\omega_{\mathcal{P}} = g$, $\bar{g}|_{\omega_{\mathcal{P}}[Y]}$ would not be constant which is impossible.). Therefore Y is $\mathcal{P}$-connected with respect to X.

(3) ⇒ (2). It suffices to show (cf. 1.3.3.1 d)):

$$P\mathcal{K} \subset \mathcal{K}\ .$$

Let $(X,Y) \in P\mathcal{K} = C_{rel}\, D_{rel}\, \mathcal{K}$. Since $\mathcal{K}$ satisfies (c) and (a), each $\mathcal{C}$-object (Z,ξ) may be decomposed into $\mathcal{K}$-quasicomponents which are defined as follows:

$$K_z = \bigcup_{\substack{z\in A\\ (Z,A)\in\mathcal{K}}} A \quad \text{for each } z \in Z$$

($\mathcal{K}$-quasicomponent of Z containing z).

Let R be the equivalence relation on X corresponding to the decomposition of X into $\mathcal{K}$-quasicomponents. We endow $X/_R$ with the final structure with respect to the natural map $\omega\colon X \to X/_R$. Since $\mathcal{K}$ satisfies (d), each of the $\mathcal{K}$-quasicomponents of $X/_R$ consists only of a single element (If $K \subset X/_R$ were a $\mathcal{K}$-quasicomponent of $X/_R$ containing at least two elements, then $\omega\colon (X,\omega^{-1}[K]) \to (X/_R,K)$ would satisfy (d) and therefore $(X,\omega^{-1}[K])$ would belong to $\mathcal{K}$. Thus $\omega^{-1}[K]$ would be a subset of a $\mathcal{K}$-quasicomponent of X and must be additionally the union of at least two $\mathcal{K}$-quasicomponents of X, which is impossible.). That means $X/_R \in D_{rel}\mathcal{K}$ (cf. 1.3.10). Thus since $(X,Y) \in C_{rel}\, D_{rel}\mathcal{K}$, $\omega|_Y$ is constant, i.e. $\omega[Y]$ consists at most of a single element. Since $\mathcal{K}$ satisfies (a), we obtain

$(X/_R, \omega[Y]) \in K$. Hence (d) may be applied to $\omega: (X,Y) \to (X/_R, \omega[Y])$. Consequently, $(X,Y) \in K$.

1.3.8 Remarks. (1) In order to prove 1.3.7 we have used the fact that for each subclass K of $|C_{(2)}|$ satisfying (a) and (c) of 1.3.7 (3) each C-object may be decomposed into maximal subsets M such that $(X,M) \in K$, the so-called K-quasicomponents. This generalizes the concept of P-quasicomponent in the sense of G. Preuß: Allgemeine Topologie (choose for C the category Top of topological spaces and continuous maps and let $K = C_{rel}P$) as well as the concept of quasicomponent in the sense of Hausdorff (choose for P the subclass of $|\text{Top}|$ consisting only of the two-point discrete space).

(2) One obtains the concept of relative connectedness in a natural way by looking for a concept of connectedness such that the decomposition into components yields the quasicomponents.

(3) Let C be a topological category, $P \subset |C|$ and $(X,X) \in C_{rel}P$. Then X is called P-connected. Let D_2 be a two-point set endowed with the discrete C-structure. Then $X \in |C|$ is called connected iff X is $\{D_2\}$-connected. This is the immediate translation of the classical concept of connectedness into the language of topological categories. Correspondingly Y is called connected with respect to X iff $(X,Y) \in C_{rel} \{D_2\}$. Maximal elements of the set of all subsets of X which are connected with respect to X are called quasicomponents instead of $C_{rel} \{D_2\}$-quasicomponents.

1.3.9 Proposition. Let K be a subclass of $|C_{(2)}|$ satisfying (3) (b) of 1.3.7. Then the following are equivalent for each $Z \in |C|$:

(1) $Z \in D_{rel}K$

(2) If $A \subset Z$, $A \neq \emptyset$ and $(Z,A) \in K$, then A consists only of a single element.

Proof. (1) $\Rightarrow$ (2). The identity map $1_Z: Z \to Z$ is a C-morphism and $A \subset Z$ such that $(Z,A) \in K$. Thus since $Z \in D_{rel}K$,

$1_Z|_A : A \to Z$ is constant, i.e. A consists only of a single element $(A \neq \emptyset!)$.

(2) ⇒ (1). Let $f: X \to Z$ be a $\mathcal{C}$-morphism and $Y \subset X$ such that $(X,Y) \in \mathcal{K}$. Since $\mathcal{K}$ satisfies (3) (b) of 1.3.7, we obtain $(X,f[Y]) \in \mathcal{K}$. Thus since (2) is valid, $f|_Y$ is constant.

<u>1.3.10 Corollary</u>. Let $\mathcal{K}$ be a subclass of $|\mathcal{C}_{(2)}|$ satisfying (3) (a), (b) and (c) of 1.3.7. Then $X \in |\mathcal{C}|$ belongs to $D_{rel}\mathcal{K}$ if and only if for each $x \in X$ the $\mathcal{K}$-quasicomponent of X containing x is a singleton.

<u>Proof</u>. Applying 1.3.9, the proof is obvious.

<u>1.3.11 Corollary</u>. Let $\mathcal{P}$ be a subclass of $|\mathcal{C}|$ and $X \in |\mathcal{C}|$. Then the following are equivalent:

(1) $X \in Q\mathcal{P}$

(2) For any two distinct elements $x,y \in X$ there exists an object $P \in \mathcal{P}$ and a $\mathcal{C}$-morphism $f: X \to P$ such that $f(x) \neq f(y)$.

<u>Proof</u>. This follows immediately from 1.3.10, if one chooses $\mathcal{K} = C_{rel}\mathcal{P}$.

<u>1.3.12 Examples for the category Top</u>

① a) Let $\mathcal{P} = \{S\}$, where S is the Sierpinski space $(\{0,1\},\{\emptyset,\{0\},\{0,1\}\})$. Then

$$Q\{S\} = \{T_0\text{-spaces}\}$$

(α) Let $X \in Q\{S\}$ and $x \neq y$. Then there exists a continuous map $f: X \to S$ such that $f(x) \neq f(y)$. Suppose $f(x) = 0$ and $f(y) = 1$. Then $y \notin f^{-1}[\{0\}] \in \overset{\circ}{\mathcal{U}}(x)$. Thus, X is a T_0-space.

β) Let X be a T_0-space and $x \neq y$. Then without loss of generality there exists $O_x \in \overset{\circ}{\mathcal{U}}(x)$ such that $y \notin O_x$. Hence $f: X \to S$ defined by

$$f(z) = \begin{cases} 0 & \text{for } z \in O_x \\ 1 & \text{for } z \in X \setminus O_x \end{cases}$$

is a continuous map satisfying $f(x) \neq f(y)$. Thus $X \in Q\{S\}$) .

b) Let $\mathcal{P} = \{T_o\text{-spaces}\}$: $Q\mathcal{P} = \{T_o\text{-spaces}\}$ (Apply a) and note $QQ = Q$!)

(2) a) $\mathcal{P}$ = {spaces with the cofinite topology}:
$Q\mathcal{P} = \{T_1\text{-spaces}\}$

(α) Let $X \in Q\mathcal{P}$ and $x \neq y$. Then there exist $P \in \mathcal{P}$ and a continuous map $f: X \to P$ such that $f(x) \neq f(y)$. Since P is a T_1-space, there exist $O_{f(x)} \in \overset{\circ}{\mathcal{U}}(f(x))$ and $O_{f(y)} \in \overset{\circ}{\mathcal{U}}(f(y))$ such that $f(x) \notin O_{f(y)}$ and $f(y) \notin O_{f(x)}$. Thus $x \notin f^{-1}[O_{f(y)}] \in \overset{\circ}{\mathcal{U}}(y)$ and $y \notin f^{-1}[O_{f(x)}] \in \overset{\circ}{\mathcal{U}}(x)$, i.e. X is a T_1-space.

β) Let $(X,\mathcal{X})$ be a T_1-space and $\mathcal{X}'$ be the cofinite topology on X . Then $\mathcal{X}' \subset \mathcal{X}$. Therefore $1_X: (X,\mathcal{X}) \to (X,\mathcal{X}')$ is continuous. If $x,y \in X$ such that $x \neq y$, then $1_X(x) = x \neq y = 1_X(y)$, i.e. $(X,\mathcal{X}) \in Q\mathcal{P}$.)

b) $\mathcal{P} = \{T_1\text{-spaces}\}$: $Q\mathcal{P} = \{T_1\text{-spaces}\}$ (Apply a) and note $QQ = Q$!).

(3) $\mathcal{P} = \{T_2\text{-spaces}\}$:
$Q\mathcal{P} = \{T_2\text{-spaces}\}$

(4) $\mathcal{P}$ = {Urysohn spaces}:
$Q\mathcal{P}$ = {Urysohn spaces}

(5) a) $\mathcal{P} = \{\mathbb{R}\}$
b) $\mathcal{P} = \{[0,1]\}$
: $Q\mathcal{P}$ = {completely Hausdorff spaces}

(6) $\mathcal{P} = \{D_2\}$: $Q\mathcal{P}$ = {totally separated spaces}[4)]

<u>1.3.13 Proposition</u>. Let $\mathcal{C}$ be a topological category and $\mathcal{P}$ an isomorphism-closed subclass of $|\mathcal{C}|$ (i.e. $X \in \mathcal{P}$ and $Y \in |\mathcal{C}|$ such that $Y \cong X$ imply $Y \in \mathcal{P}$) . Then the following are equivalent:

---

4) A topological space X is called <u>totally separated</u> provided its quasi-components consist at most of a singleton.

(1) $\mathcal{P} = Q\mathcal{P}$

(2) For every $X \in |\mathcal{C}|$ there exist $Y \in \mathcal{P}$ and an extremal epimorphism $e: X \to Y$ satisfying:

(A) For every $P \in \mathcal{P}$ and for every $\mathcal{C}$-morphism $f: X \to P$ there exists a unique $\mathcal{C}$-morphism $\bar{f}: Y \to P$ such that the diagram

$$\begin{array}{ccc} X & \xrightarrow{f} & P \\ & {\scriptstyle e}\searrow \quad \nearrow{\scriptstyle \bar{f}} & \\ & Y & \end{array}$$

commutes.

Proof. (1) $\Rightarrow$ (2). Let $\mathcal{K} = C_{rel}\mathcal{P}$ and $R_{\mathcal{P}}$ the equivalence relation on X corresponding to the decomposition of X into $\mathcal{K}$-quasicomponents. Let $X/_{R_{\mathcal{P}}}$ be endowed with the final $\mathcal{C}$-stucture with respect to the natural map $\omega_{\mathcal{P}}: X \to X/_{R_{\mathcal{P}}}$. It was shown in the proof of 1.3.7 (cf. "(3) $\Rightarrow$ (2)") that $X/_{R_{\mathcal{P}}} \in D_{rel}\mathcal{K} = D_{rel}\, C_{rel}\mathcal{P} = Q\mathcal{P}$. Thus since $\mathcal{P} = Q\mathcal{P}$, $X/_{R_{\mathcal{P}}} \in \mathcal{P}$. Since $\omega_{\mathcal{P}}$ is a quotient map, it is an extremal epimorphism. Let $P \in \mathcal{P}$ and $f: X \to P$ be a $\mathcal{C}$-morphism. Then $\bar{f}: X/_{R_{\mathcal{P}}} \to P$ defined by the commutative diagram

$$\begin{array}{ccc} X & \xrightarrow{f} & P \\ & {\scriptstyle \omega_{\mathcal{P}}}\searrow \quad \nearrow{\scriptstyle \bar{f}} & \\ & X/_{R_{\mathcal{P}}} & \end{array}$$

is a $\mathcal{C}$-morphism (note the final structure on $X/_{R_{\mathcal{P}}}$). Obviously, every $\mathcal{C}$-morphism $\underline{\underline{f}}: X/_{R_{\mathcal{P}}} \to P$ such that $\underline{\underline{f}} \circ \omega_{\mathcal{P}} = f$ coincides with $\bar{f}$.

(2) $\Rightarrow$ (1). It suffices to prove $Q\mathcal{P} \subset \mathcal{P}$ ($\mathcal{P} \subset Q\mathcal{P}$ is always true). If $X \in Q\mathcal{P}$, then there exist a $Y \in \mathcal{P}$ and an extremal epimorphism $e: X \to Y$ satisfying (A). Let $x,y \in X$ such that $x \neq y$. Then there exist a $P \in \mathcal{P}$ and a $\mathcal{C}$-morphism $f: X \to P$ such that $f(x) \neq f(y)$. By (A) there exist a unique

C-morphism $\bar{f}: Y \to P$ with $\bar{f} \circ e = f$. Hence $e(x) \neq e(y)$ (otherwise $f(x)$ would be equal to $f(y)$). Consequently, e is injective, i.e. a monomorphism. Thus e is an isomorphism. Since P is isomorphism-closed, $X \in P$.

1.3.14 Theorem. Let C be a topological category and P be an isomorphism-closed subclass of $|C|$. Then the following are equivalent:

(1) P is a relative disconnectedness

(2) $P = QP$

(3) P is closed under formation of weak subobjects[5] and products in C .

Proof. The equivalence of (1) and (2) was shown in 1.3.5 b). The equivalence of (2) and (3) will be shown in the following chapter in connection with general categorical investigations. (By 1.3.13 condition (2) means that P is an "extremal epireflective" subclass of $|C|$. These classes and others will be studied in the second chapter).

[5] $Y \in |C|$ is called a weak subobject of $X \in |C|$ iff there is a monomorphism $m: Y \to X$.

CHAPTER II

REFLECTIVE AND COREFLECTIVE SUBCATEGORIES

As well-known mathematical objects may be described by means of maps. There is an analogous description of categories via so-called functors. The classical definition of universal maps in the sense of N. Bourbaki corresponds to a categorical one with respect to a functor. The existence of all universal maps with respect to a given functor F is related to a pair of adjoint functors (G,F) where G (resp. F) is called a left adjoint (resp. a right adjoint). The relations between these functors are described by means of natural transformations u and v (which occur as "maps" between functors). Thus an adjoint situation (G,F,u,v) is obtained. In the first part of this chapter adjoint situations are studied together with some examples. In the second part an important special case of adjoint situations (G,F,u,v) is investigated, namely the case in which F is the inclusion functor I (the notion of inclusion functor corresponds to the notion of inclusion map in classical mathematics). Then G is called a reflector. If the morphisms belonging to all universal maps with respect to I are epimorphisms (resp. extremal epimorphisms) then G is called an epireflector (resp. extremal epireflector). The corresponding subcategory is called reflective (epireflective or extremal epireflective respectively) provided G is a reflector (epireflector or extremal epireflector respectively). The famous characterization theorem for epireflective (resp. extremal epireflective) subcategories is proved and several topological examples are added. Especially, the background of important constructions in topology like the Stone-Čech compactification or the Hausdorff completion of a uniform space is discovered. All concepts are dualized and the special features in topological categories are considered. Finally, epireflective hulls (dually: monocoreflective hulls) are investigated and it is shown that each reflector is the composite of two epireflectors. Thus epireflective subcategories are more important than reflective ones. Furthermore a categorical method is obtained for constructing the Hausdorff completion of a uniform space (this includes a categorical construction of the real numbers!).

2.1 Universal maps and adjoint functors

2.1.1 Definition. Let $\mathcal{C}$ and $\mathcal{D}$ be categories, let $F_1: |\mathcal{C}| \to |\mathcal{D}|$ and F_2: Mor $\mathcal{C} \to$ Mor $\mathcal{D}$ be maps. Instead of $F_1(A)$ we write $F(A)$ and instead of $F_2(f)$ we write $F(f)$. Then $F := (\mathcal{C},\mathcal{D},F_1,F_2)$ is called a functor from $\mathcal{C}$ to $\mathcal{D}$ or more exactly a covariant functor (denoted by $F: \mathcal{C} \to \mathcal{D}$) provided the following are satisfied:

F_1) $f \in [A,B]_{\mathcal{C}}$ implies $F(f) \in [F(A),F(B)]_{\mathcal{D}}$.

F_2) $F(f \circ g) = F(f) \circ F(g)$, provided $f \circ g$ is defined (i.e. the domain of f is equal to the codomain of g).

F_3) $F(1_A) = 1_{F(A)}$ $(A \in |A|)$.

If F_1) and F_2) are replaced by

F_1') $f \in [A,B]_{\mathcal{C}}$ implies $F(f) \in [F(B),F(A)]_{\mathcal{D}}$

and F_2') $F(f \circ g) = F(g) \circ F(f)$ (provided $f \circ g$ is defined in $\mathcal{C}$) respectively, then F is called a contravariant functor from $\mathcal{C}$ to $\mathcal{D}$ (which may also be defined as a covariant functor from $\mathcal{C}^*$ to $\mathcal{D}$).

2.1.2 Examples. (1) The identity functor $I: \mathcal{C} \to \mathcal{C}$ maps objects and morphisms identically to themselves (covariant functor!).

(2) Constant functors: Let $\mathcal{C}$ and $\mathcal{D}$ be arbitrary categories, let $X \in |\mathcal{D}|$. For every $A \in |\mathcal{C}|$ and every $f \in$ Mor $\mathcal{C}$, put $F(A) = X$ and $F(f) = 1_X$ (co- and contravariant functor!).

(3) Forgetful (or underlying) functors: Let $\mathcal{C}$ be a topological category, Set be the category of sets and maps and let $F: \mathcal{C} \to$ Set be defined by $F((X,\xi)) = X$ and $F(f) = f$ (= map between the underlying sets) (covariant functor!).

(4) The dualizing functor $F: \mathcal{C} \to \mathcal{C}^*$ is defined by $F(X) = X$ and $F(f) = f^*$ (contravariant functor!).

(5) <u>Inclusion functors</u>: Let $\mathcal{C}$ be a category, $\mathcal{A}$ be a <u>subcategory</u> of $\mathcal{C}$, i.e. $\mathcal{A}$ is a category such that

1. $|\mathcal{A}| \subset |\mathcal{C}|$.
2. $[A,B]_{\mathcal{A}} \subset [A,B]_{\mathcal{C}}$ for each $(A,B) \in |\mathcal{A}| \times |\mathcal{A}|$.
3. The composition of morphisms in $\mathcal{A}$ coincides with the composition of these morphisms in $\mathcal{C}$.
4. For each $A \in |\mathcal{A}|$ the identity 1_A is the same in $\mathcal{A}$ and $\mathcal{C}$.

(If $[A,B]_{\mathcal{A}} = [A,B]_{\mathcal{C}}$ is satisfied for each $(A,B) \in |\mathcal{A}| \times |\mathcal{A}|$ instead of 2., then $\mathcal{A}$ is called *full*).
The inclusion functor $\mathcal{F}_e: \mathcal{A} \to \mathcal{C}$ is defined by $\mathcal{F}_e(A) = A$ for each $A \in |\mathcal{A}|$ and $\mathcal{F}_e(f) = f$ for each $f \in \text{Mor}\,\mathcal{A}$ (covariant functor!).

<u>2.1.3 Remark</u>. Because of the property F_2) a functor preserves commutative diagrams and by F_2) and F_3) it preserves isomorphisms.

<u>2.1.4 Definition</u>. Let $\mathcal{A}$ and $\mathcal{B}$ be categories, $\mathcal{F}: \mathcal{A} \to \mathcal{B}$ be a functor and $B \in |\mathcal{B}|$. A pair (u,A) with $A \in |\mathcal{A}|$ and $u: B \to \mathcal{F}(A)$ is called a <u>universal map for B with respect to $\mathcal{F}$</u> provided that for each $A' \in |\mathcal{A}|$ and each $f: B \to \mathcal{F}(A')$ there exists a unique $\mathcal{A}$-morphism $\bar{f}: A \to A'$ such that the diagram

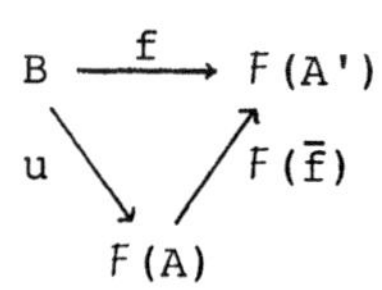

commutes.

<u>2.1.5 Examples</u>. (1) Let $\mathcal{C}$ be a topological category and $\mathcal{P} \subset |\mathcal{C}|$ be a relative disconnectedness. A full subcategory $\mathcal{A}$ of $\mathcal{C}$ is defined by $|\mathcal{A}| = \mathcal{P}$. If $\mathcal{F}_e$ is the inclusion functor from $\mathcal{A}$ to $\mathcal{C}$, then with the notation of 1.3.13(2), (e,Y) is a universal map for $X \in |\mathcal{C}|$ with respect to $\mathcal{F}_e$.

(2) Let <u>$C\ Reg_1$</u> be the category of completely regular T_1-spaces and continuous maps, let <u>$Comp\ T_2$</u> be the cate-

gory of compact Hausdorff spaces and continuous maps and F_e: $\underline{\text{Comp } T_2} \to \underline{\text{C Reg}_1}$ be the inclusion functor. If the Stone-Čech compactification of $X \in |\underline{\text{C Reg}_1}|$ is denoted by $\beta(X)$ and $\beta_X: X \to \beta(X)$ is the canonical map, then $(\beta_X, \beta(X))$ is a universal map for X with respect to F_e .

③ Let C be a topological category and F_u: $C \to \underline{\text{Set}}$ be the forgetful functor. If M is a set and ξ_D is the discrete C-structure on M , then $(1_M, (M, \xi_D))$ is a universal map for M with respect to F_u .

<u>2.1.6 Proposition</u>. Let each of (u,A) and (u',A') be a universal map for some $B \in |B|$ with respect to $F: A \to B$. Then there is an isomorphism $f: A \to A'$ such that the diagram

$$\begin{array}{ccc} B & \xrightarrow{u} & F(A) \\ & {\scriptstyle u'}\searrow \quad \swarrow{\scriptstyle F(f)} & \\ & F(A') & \end{array}$$

commutes.

<u>Proof</u>. Since (u,A) is a universal map, there is a unique morphism $f: A \to A'$ such that the diagram

$$(D_1) \qquad \begin{array}{ccc} B & \xrightarrow{u'} & F(A') \\ & {\scriptstyle u}\searrow \quad \nearrow{\scriptstyle F(f)} & \\ & F(A) & \end{array}$$

commutes. Since (u',A') is a universal map, there is a unique morphism $g: A' \to A$ such that the diagram

$$(D_2) \qquad \begin{array}{ccc} B & \xrightarrow{u} & F(A) \\ & {\scriptstyle u'}\searrow \quad \nearrow{\scriptstyle F(g)} & \\ & F(A') & \end{array}$$

commutes. The diagrams (D_1) and (D_2) form the following commutative diagram:

$$(D) \qquad \begin{array}{ccc} & \overset{u}{\nearrow} & F(A) \\ & & \uparrow F(g) \\ B & \xrightarrow{u'} & F(A') \\ & & \uparrow F(f) \\ & \underset{u}{\searrow} & F(A) \end{array}$$

There is a unique morphism from A to A whose image under F makes the outer triangle of (D) commutative. Since F is a functor, $g \circ f$ and 1_A are two morphisms satisfying this property. Thus, $g \circ f = 1_A$. Similarly one can show that $f \circ g = 1_{A'}$. Consequently, f is an isomorphism.

<u>2.1.7.</u> In the same way as we have defined functors for the description of categories, we define now natural transformations to describe functors and we explain when we do not distinguish between two functors, i.e. when they are isomorphic:

<u>Definitions</u>. Let C and $\mathcal{D}$ be categories and $F,G\colon C \to \mathcal{D}$ be functors:

1.) A family $\eta = (\eta_A)_{A \in |C|}$ such that $\eta_A \in [F(A),G(A)]_{\mathcal{D}}$ for each $A \in |C|$ is called a <u>natural transformation</u> (denoted by $\eta\colon F \to G$) provided that for each pair $(A,B) \in |C| \times |C|$ and each $f \in [A,B]_C$, the following diagram

$$\begin{array}{ccc} F(A) & \xrightarrow{\eta_A} & G(A) \\ {\scriptstyle F(f)}\downarrow & & \downarrow{\scriptstyle G(f)} \\ F(B) & \xrightarrow{\eta_B} & G(B) \end{array}$$

commutes.

2.) A natural transformation $\eta\colon F \to G$ is called a <u>natural equivalence</u> (or a natural isomorphism) provided that for each $A \in |C|$, η_A is an isomorphism.

3.) F and G are said to be <u>naturally equivalent</u> (or naturally isomorphic) [denoted by $F \approx G$] iff there exists a natural equivalence from F to G .

2.1.8. Let $F\colon \mathcal{A} \to \mathcal{B}$ and $G\colon \mathcal{B} \to \mathcal{C}$ be functors, i.e. $F = (\mathcal{A},\mathcal{B},F_1,F_2)$ and $G = (\mathcal{B},\mathcal{C},G_1,G_2)$. Then a functor $G\circ F\colon \mathcal{A} \to \mathcal{C}$ (the composition of F and G) is defined by $G\circ F = (\mathcal{A},\mathcal{C},G_1\circ F_1,G_2\circ F_2)$, i.e. $(G\circ F)(A) = G(F(A))$ for each $A \in |\mathcal{A}|$ and $(G\circ F)(f) = G(F(f))$ for each $f \in \operatorname{Mor}\mathcal{A}$. With this notation the following holds:

Theorem. Let $F\colon \mathcal{A} \to \mathcal{B}$ be a functor. If for each $B \in |\mathcal{B}|$ there exists a universal map (u_B,A_B) with respect to F then there is a unique functor $G\colon \mathcal{B} \to \mathcal{A}$ such that the following are satisfied:

(1) $G(B) = A_B$ for each $B \in |\mathcal{B}|$.

(2) $u = (u_B)\colon I_{\mathcal{B}} \to F\circ G$ is a natural transformation ($I_{\mathcal{B}}\colon \mathcal{B} \to \mathcal{B}$ is the identity functor) .

Corollary. There is a unique natural transformation $v = (v_A)\colon G\circ F \to I_{\mathcal{A}}$ ($I_{\mathcal{A}}\colon \mathcal{A} \to \mathcal{A}$ the identity functor) such that the following are valid:

(a) $F(v_A)\circ u_{F(A)} = 1_{F(A)}$ for each $A \in |\mathcal{A}|$

(b) $v_{G(B)}\circ G(u_B) = 1_{G(B)}$ for each $B \in |\mathcal{B}|$.

Proof of the theorem. By (1) a map is defined from the object class of $\mathcal{B}$ to the object class of $\mathcal{A}$. We try to find a map between the corresponding morphism classes. If $f\colon B \to B'$ is a $\mathcal{B}$-morphism, then there is a unique $\mathcal{A}$-morphism $\bar{f}\colon A_B \to A_{B'}$ such that the diagram

$$(D_1)\qquad \begin{array}{ccc} I_{\mathcal{B}}(B) = B & \xrightarrow{\ u_B\ } & F(A_B) = F(G(B)) \\ f\big\downarrow & & \big\downarrow F(\bar{f}) \\ I_{\mathcal{B}}(B') = B' & \xrightarrow{\ u_{B'}\ } & F(A_{B'}) = F(G(B')) \end{array}$$

commutes, because (u_B,A_B) is a universal map for B with respect to F . For each $f \in \operatorname{Mor}\mathcal{B}$, put $G(f) = \bar{f}$. Then G is a functor: If $f\colon B \to B'$ and $g\colon B' \to B''$ are $\mathcal{B}$-morphisms, then the following diagram

$$(D_2)\qquad \begin{array}{ccc} B & \xrightarrow{u_B} & F(A_B) \\ {\scriptstyle f}\downarrow & & \downarrow{\scriptstyle F(\bar{f})} \\ B' & \xrightarrow{u_{B'}} & F(A_{B'}) \\ {\scriptstyle g}\downarrow & & \downarrow{\scriptstyle F(\bar{g})} \\ B'' & \xrightarrow{u_{B''}} & F(A_{B''}) \end{array}$$

is commutative. Hence, $\bar{g}\circ\bar{f}$ is a morphism such that its image under F makes the outer square of (D_2) commutative (because $F(\bar{g}\circ\bar{f}) = F(\bar{g})\circ F(\bar{f})$) . Since there is a unique morphism of this kind, namely $\overline{g\circ f}$, it follows that $\overline{g\circ f} = \bar{g}\circ\bar{f}$, i.e. $G(g\circ f) = G(g)\circ G(f)$. For each $B \in |\mathcal{B}|$, $1_{A_B} : A_B \to A_B$ is a morphism such that its image under F makes the diagram

$$(D_3)\qquad \begin{array}{ccc} B & \xrightarrow{u_B} & F(A_B) \\ {\scriptstyle 1_B}\downarrow & & \downarrow{\scriptstyle F(1_{A_B})} \\ B & \xrightarrow{u_B} & F(A_B) \end{array}$$

commutative (because $F(1_{A_B}) = 1_{F(A_B)}$) . Since there is a unique morphism of this kind, namely $\overline{1_B}$, it follows that $\overline{1_B} = 1_{A_B}$, i.e. $G(1_B) = 1_{G(B)}$.
From the commutativity of (D_1) it follows that $u = (u_B): I_{\mathcal{B}} \to F\circ G$ is a natural transformation.
Let $G': \mathcal{B} \to \mathcal{A}$ be any functor satisfying (1) and (2). Then $G'(B) = A_B = G(B)$ by definition and $G'(f) = \bar{f} = G(f)$ by uniqueness of $\bar{f}$. Thus, $G = G'$.

<u>Proof of the corollary</u>. $F(A) \in |\mathcal{B}|$ for each $A \in |\mathcal{A}|$. Since $(u_{F(A)}\ , G(F(A)))$ is a universal map for $F(A)$ with respect to F , there is a unique morphism $v_A: G(F(A)) \to A$ such that the diagram

(D_1')

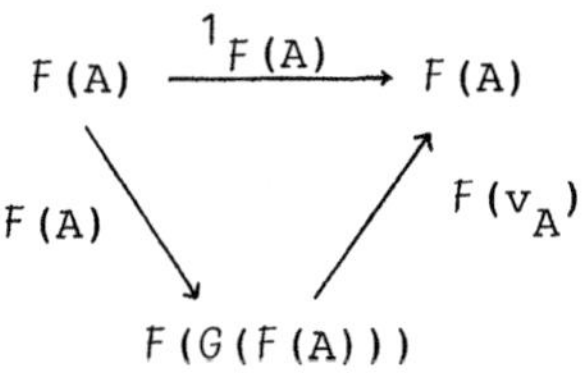

commutes, i.e. (a) is satisfied. Now it must be shown that (b) is also fulfilled, i.e. that the diagram

(D_2')

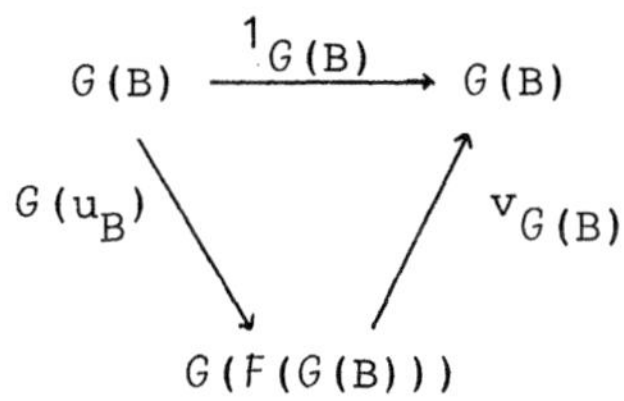

commutes. Since $(u_B, G(B))$ is a universal map for B with respect to F, there is a unique morphism $h\colon G(B) \to G(B)$ such that the diagram

(D_3')

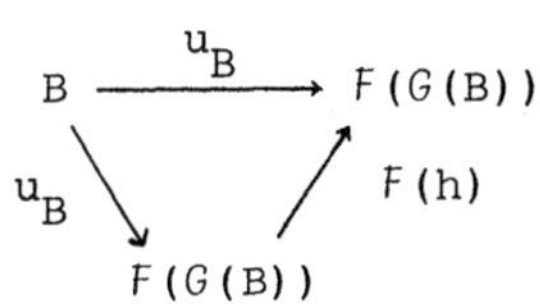

commutes. But since F is a functor, $F(1_{G(B)}) \circ u_B = 1_{F(G(B))} \circ u_B = u_B$. Further $u\colon I_{\mathcal{B}} \to F \circ G$ is a natural transformation, i.e. the diagram

(D_4')

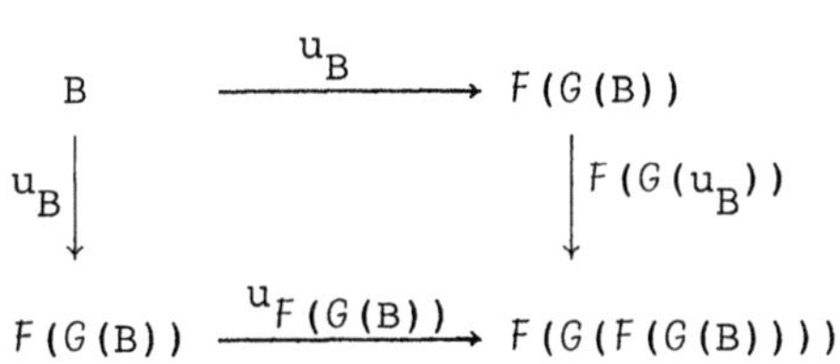

commutes and additionally (a) is valid. Thus,
$F(v_{G(B)} \circ G(u_B)) \circ u_B = F(v_{G(B)}) \circ F(G(u_B)) \circ u_B = F(v_{G(B)}) \circ u_{F(G(B))} \circ u_B =$
$= 1_{F(G(B))} \circ u_B = u_B$. Consequently, $h = v_{G(B)} \circ G(u_B) = 1_{G(B)}$,

i.e. (b) is fulfilled. It remains to show: $v = (v_A): \mathcal{G}\circ\mathcal{F} \to \mathcal{I}_{\mathcal{A}}$ is a natural transformation, i.e. for each pair $(A,A') \in |\mathcal{A}| \times |\mathcal{A}|$ and each $f \in [A,A']_{\mathcal{A}}$, the following diagram commutes:

$$(D_5')\qquad \begin{array}{ccc} \mathcal{G}(\mathcal{F}(A)) & \xrightarrow{v_A} & A \\ {\scriptstyle \mathcal{G}(\mathcal{F}(f))}\downarrow & & \downarrow{\scriptstyle f} \\ \mathcal{G}(\mathcal{F}(A')) & \xrightarrow{v_{A'}} & A' \end{array} .$$

Since $(u_{\mathcal{F}(A)}, \mathcal{G}(\mathcal{F}(A)))$ is a universal map for $\mathcal{F}(A)$ with respect to $\mathcal{F}$, there is a unique morphism $h: \mathcal{G}(\mathcal{F}(A)) \to A'$ such that the diagram

$$(D_6')\qquad \begin{array}{ccc} \mathcal{F}(A) & \xrightarrow{\mathcal{F}(f)} & \mathcal{F}(A') \\ & {\scriptstyle u_{\mathcal{F}(A)}}\searrow \quad \nearrow{\scriptstyle \mathcal{F}(h)} & \\ & \mathcal{F}(\mathcal{G}(\mathcal{F}(A))) & \end{array}$$

commutes. Since $\mathcal{F}$ is a functor and (a) is satisfied, $\mathcal{F}(f\circ v_A)\circ u_{\mathcal{F}(A)} = \mathcal{F}(f)\circ\mathcal{F}(v_A)\circ u_{\mathcal{F}(A)} = \mathcal{F}(f)\circ 1_{\mathcal{F}(A)} = \mathcal{F}(f)$. Further $u = (u_B): \mathcal{I}_{\mathcal{B}} \to \mathcal{F}\circ\mathcal{G}$ is a natural transformation, i.e. the diagram

$$(D_7')\qquad \begin{array}{ccc} \mathcal{F}(A) & \xrightarrow{u_{\mathcal{F}(A)}} & \mathcal{F}(\mathcal{G}(\mathcal{F}(A))) \\ {\scriptstyle \mathcal{F}(f)}\downarrow & & \downarrow{\scriptstyle \mathcal{F}(\mathcal{G}(\mathcal{F}(f)))} \\ \mathcal{F}(A') & \xrightarrow{u_{\mathcal{F}(A')}} & \mathcal{F}(\mathcal{G}(\mathcal{F}(A'))) \end{array}$$

commutes and additionally (a) is valid. Thus, $\mathcal{F}(v_{A'}\circ\mathcal{G}(\mathcal{F}(f)))\circ u_{\mathcal{F}(A)} = \mathcal{F}(v_{A'})\circ\mathcal{F}(\mathcal{G}(\mathcal{F}(f)))\circ u_{\mathcal{F}(A)} = \mathcal{F}(v_{A'})\circ u_{\mathcal{F}(A')}\circ\mathcal{F}(f) = 1_{\mathcal{F}(A')}\circ\mathcal{F}(f) = \mathcal{F}(f)$. Consequently, $h = f\circ v_A = v_{A'}\circ\mathcal{G}(\mathcal{F}(f))$, i.e. (D_5') is commutative. Therefore the corollary is proved.

2.1.9 Remark. If for every $B \in |\mathcal{B}|$ a universal map (u_B', A_B') with respect to $\mathcal{F}$ is chosen, then by the preceding theorem, there is a functor $\mathcal{G}': \mathcal{B} \to \mathcal{A}$ with properties corresponding to $\mathcal{G}$. Since (u_B, A_B) and (u_B', A_B') are isomorphic in the

sense of 2.1.6, G is naturally equivalent to G':
For every $B \in |\mathcal{B}|$, let $i_B: A_B = G(B) \to A'_B = G'(B)$ be the isomorphism existing by 2.1.6 with $F(i_B) \circ u_B = u'_B$. Then $i = (i_B): G \to G'$ is the desired natural equivalence, because for every $(B,B') \in |\mathcal{B}| \times |\mathcal{B}|$ and every $f \in [B,B']_{\mathcal{B}}$, the partial diagrams (I), (III), (IV) and the outer square of the diagram

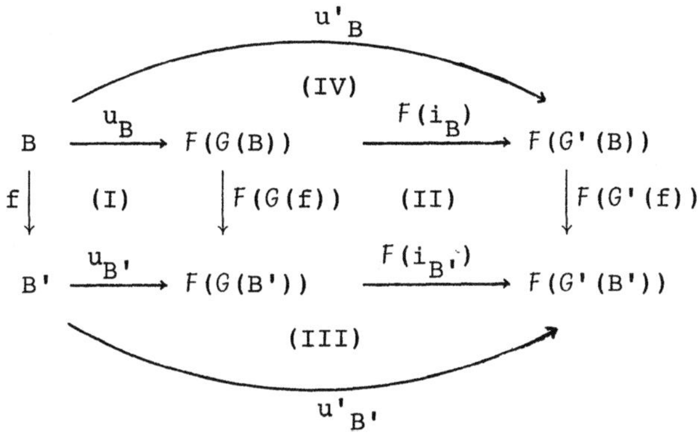

are commutative. Since there is a unique morphism $h: G(B) \to G'(B')$ such that the diagram

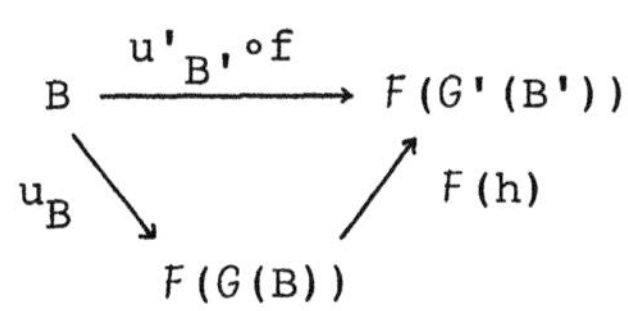

commutes $\big((u_B, G(B))$ is a universal map for B with respect to $F\big)$, then $h = G'(f) \circ i_B = i'_B \circ G(f)$ (so the whole diagram above is commutative, because F is a functor).

<u>2.1.10 Definition</u>. If $F: \mathcal{A} \to \mathcal{B}$ and $G: \mathcal{B} \to \mathcal{A}$ are functors and

$$v = (v_A): G \circ F \to I_{\mathcal{A}}$$

and $u = (u_B): I_{\mathcal{B}} \to F \circ G$

are natural transformations such that

(1) $F(v_A) \circ u_{F(A)} = 1_{F(A)}$ for each $A \in |\mathcal{A}|$

(2) $v_{G(B)} \circ G(u_B) = 1_{G(B)}$ for each $B \in |\mathcal{B}|$,

then G is said to be a left adjoint of F , F is said to be a right adjoint of G and (G,F) is called a pair of adjoint functors.

2.1.11 Remark. We have proved already that if $F: A \to B$ is a functor and each $B \in |B|$ has a universal map $(u_B, G(B))$ with respect to F , then there is a functor G which is a left adjoint of F (cf. the preceding theorem including the corollary). Now we show the converse.

2.1.12 Theorem. If $G: B \to A$ is a left adjoint of $F: A \to B$ and $u = (u_B): I_B \to F \circ G$ is a corresponding natural transformation, then for each $B \in |B|$, $(u_B, G(B))$ is a universal map with respect to F .

Proof. Let $B \in |B|$. It must be shown that for each $A \in |A|$ and each $f: B \to F(A)$, there is a unique $\bar{f}: G(B) \to A$ such that the diagram

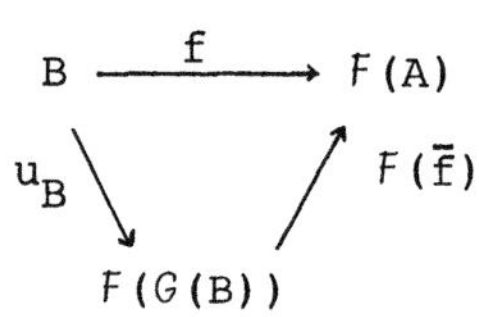

commutes. Since G is a left adjoint of F there is a natural transformation $v = (v_A): G \circ F \to I_A$ such that

$$(1) \quad F(v_A) \circ u_{F(A)} = 1_{F(A)} \qquad \text{for each } A \in |A|$$

and

$$(2) \quad v_{G(B)} \circ G(u_B) = 1_{G(B)} \qquad \text{for each } B \in |B| .$$

Put $\bar{f} = v_A \circ G(f)$. Then by (1), $F(\bar{f}) \circ u_B = F(v_A) \circ F(G(f)) \circ u_B = F(v_A) \circ u_{F(A)} \circ f = 1_{F(A)} \circ f = f$, because F is a functor and u is a natural transformation. Thus, $\bar{f}$ is the desired morphism provided its uniqueness can be shown: If $\bar{\bar{f}}: G(B) \to A$ is a morphism such that $F(\bar{\bar{f}}) \circ u_B = f$, then $G(F(\bar{\bar{f}})) \circ G(u_B) = G(f)$ and since $v: G \circ F \to I_A$ is a natural transformation, the diagram

$$\begin{array}{ccccc} & & G(F(G(B))) & \xrightarrow{v_{G(B)}} & G(B) \\ & \nearrow^{G(u_B)} & \Big\downarrow{\scriptstyle G(F(\bar{\bar{f}}))} & & \Big\downarrow{\scriptstyle \bar{\bar{f}}} \\ G(B) & \xrightarrow{G(f)} & G(F(A)) & \xrightarrow{v_A} & A \end{array}$$

is commutative. Thus, $\bar{\bar{f}} \circ v_{G(B)} \circ G(u_B) = v_A \circ G(f) = \bar{f}$. Consequently, by (2), $\bar{\bar{f}} = \bar{f}$.

<u>2.1.13 Remarks</u>. (1) For every functor $H\colon C \to D$, there is (in a natural way) an <u>opposite functor</u> $H^*\colon C^* \to D^*$. One obtains H^* by applying first the dualizing functor $C^* \to (C^*)^* = C$, then the functor $H\colon C \to D$ and finally the dualizing functor $D \to D^*$. It holds $(H^*)^* = H$. If $(G\colon B \to A, F\colon A \to B)$ is a pair of adjoint functors by means of $((u_B)\colon I_B \to F \circ G$, $(v_A)\colon G \circ F \to I_A)$, then obviously $(F^*\colon A^* \to B^*, G^*\colon B^* \to A^*)$ is a pair of adjoint functors by means of $((v_A^*)\colon I_{A^*} \to G^* \circ F^*$, $(u_B^*)\colon F^* \circ G^* \to I_{B^*})$. Especially, G is a left adjoint of F if and only if G^* is a right adjoint of F^* . (<u>Duality principle for adjoint functors</u>).

(2) By the preceding theorems a functor $F\colon A \to B$ has a left adjoint $G\colon B \to A$ if and only if every $B \in |B|$ has a universal map with respect to F . By the remark 2.1.9 for a given functor $F\colon A \to B$ a functor $G\colon B \to A$ is uniquely determined up to a natural equivalence by the property of being a left adjoint of F . Moreover, by the duality principle for adjoint functors, the functor F is also uniquely determined (up to a natural equivalence) by the property of being a right adjoint of a given functor G . Thus,

adjoint functors are uniquely determined by each other up to natural equivalence.

(3) Let F, G, H be functors from C to D and $s = (s_X)\colon F \to G$, $t = (t_X)\colon G \to H$ be natural transformations. Then a natural transformation from F to H is defined by $t \circ s = (t_X \circ s_X)_{X \in |C|}$. If $(G\colon B \to A, F\colon A \to B)$ is a pair of adjoint functors, then there are natural transformations

$$u = (u_B): I_B \to F\circ G$$

$$v = (v_A): G\circ F \to I_A$$

such that

(1) $F(v_A)\circ u_{F(A)} = 1_{F(A)}$ for each $A \in |A|$

and

(2) $v_{G(B)}\circ G(u_B) = 1_{G(B)}$ for each $B \in |B|$.

If $(u': I_B \to F\circ G,\ v': G\circ F \to I_A)$ is a second pair of natural transformations satisfying the relations corresponding to (1) and (2), then by the preceding theorem, $(u_B, G(B)), (u'_B, G(B))$ are universal maps with respect to F for each $B \in |B|$ and by 2.1.6 there is an isomorphism $i_B: G(B) \to G(B)$ for each $B \in |B|$ such that $F(i_B)\circ u_B = u'_B$. Then $j = (F(i_B))_{B\in|B|}: F\circ G \to F\circ G$ is a natural equivalence satisfying $j\circ u = u'$ (cf. the first diagram in 2.1.8 and put $G = G'$) , i.e. u and u' coincide up to a natural equivalence. Similarly one concludes that (v_A^*) coincides with $(v_A'^*)$ up to a natural equivalence and therefore (v_A) coincides with (v'_A) up to a natural equivalence. Thus, the transformations u,v are uniquely determined (up to natural equivalence) by (1) and (2). Consequently, it makes sense to speak of <u>the</u> natural transformations u,v belonging to a pair of adjoint functors. Then (G,F,u,v) is called an <u>adjoint situation</u>.

④ a) Especially the inclusion functor $F_e: A \to C$ from a full subcategory A of a topological category C into C , where $|A|$ is a relative disconnectedness, has a left adjoint.

b) The inclusion functor F_e: <u>Comp T_2</u> $\to$ <u>C Reg_1</u> (cf. 2.1.5 ②) has a left adjoint β: <u>C Reg_1</u> $\to$ <u>Comp T_2</u> . Thereby the meaning of the symbol "β" of the Stone-Čech compactification $\beta(X)$ of a completely regular Hausdorff space X is clarified.

c) The forgetful functor $F_u: C \to$ <u>Set</u> (C topological category) has a left adjoint D: <u>Set</u> $\to C$, which assigns to each set X the set X endowed with the discrete C-structure.

2.2 Definitions and characterization theorems of C-reflective and M-coreflective subcategories

2.2.1. In the following we restrict ourselves to the case that an inclusion functor has a left adjoint, or a right adjoint respectively.

2.2.2 Definitions. Let A be a subcategory of a category C and $F_e: A \to C$ be the inclusion functor. Then A is called

a) reflective in C iff one of the two following (equivalent) conditions is satisfied:

(1) F_e has a left adjoint R.

(2) Each $X \in |C|$ has a universal map with respect to F_e, i.e. for each $X \in |C|$, there exist an A-object X_A and a C-morphism $r_X: X \to X_A$ such that for each A-object Y and each C-morphism $f: X \to Y$, there is a unique A-morphism $\bar{f}: X_A \to Y$ such that $\bar{f} \circ r_X = f$

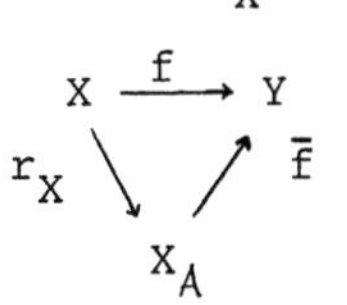

(The functor R is called a reflector).

a') coreflective in C iff A^* is reflective in C^*, i.e. iff one of the two following (equivalent) conditions is satisfied:

(1') F_e has a right adjoint R_c (i.e. F_e^* has a left ad-adjoint R_c^*).

(2') Each $X \in |C^*| = |C|$ has a universal map with respect to F_e^*, i.e. for each $X \in |C|$, there exist an A-object X_A and a C-morphism $m_X: X_A \to X$ such that for each A-object Y and each C-morphism $f: Y \to X$, there is a unique A-morphism $\bar{f}: Y \to X_A$ such that $m_X \circ \bar{f} = f$

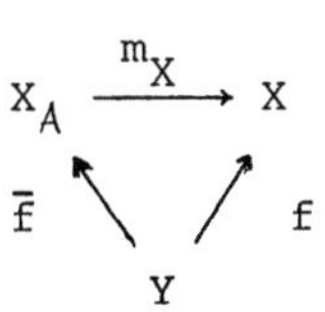

(The functor R_c is called a coreflector).

b) epireflective (resp. extremal epireflective) in C provided that A is reflective in C and for each $X \in |C|$, the C-morphisms $r_X: X \to X_A$ are epimorphisms (resp. extremal epimorphisms). (Then the functor R is called epireflector, resp. extremal epireflector).

b') monocoreflective (resp. extremal monocoreflective) in C provided that A is coreflective in C and for each $X \in |C|$, the C-morphisms $m_X: X_A \to X$ are monomorphisms (resp. extremal monomorphisms), i.e. iff A^* is epireflective (resp. extremal epireflective) in C^*. (Then the functor R_C is called monocoreflector, resp. extremal monocoreflector).

The morphisms

$r_X: X \to X_A$ are called reflections of $X \in |C|$ with respect to A in case a) and epireflections (resp. extremal epireflections) of $X \in |C|$ with respect to A in case b).

$m_X: X_A \to X$ are called coreflections of $X \in |C|$ with respect to A in case a') and monocoreflections (resp. extremal monocoreflections) of $X \in |C|$ with respect to A in case b').

2.2.3 Remarks. ① Let E (resp. M) be a class of epimorphisms (resp. monomorphisms) which is closed under composition with isomorphisms. Then one may introduce analogously to 2.2.2 b) (resp. 2.2.2 b')) the concept "E-reflective" (resp. "M-coreflective") subcategory by requiring that all reflections (resp. all coreflections) belong to E (resp. M).

② In the following we will often study subcategories A of a category C, which are

(1) full

and (2) isomorphism-closed, i.e. each $X \in |C|$ which is isomorphic to an $A \in |A|$ belongs to $|A|$.

Obviously, we obtain

a) Each subclass $|A|$ of the object class $|C|$ of a category C can be extended to a full subcategory A of C in a natural way (Put $[A,B]_A = [A,B]_C$ for each $(A,B) \in |A|\times|A|$).

b) Full subcategories A of a category C defined by

$$|A| = \{X \in |C|: X \text{ has the property } P\}$$

for each property P of C-objects which is a C-invariant are obviously isomorphism-closed (e.g. for C = Top the property "connected" is a Top-invariant, i.e. a topological invariant).

2.2.4. A full subcategory A of a category C is called closed under formation of products (resp. subobjects, weak subobjects) in C provided the product of a family of A-objects formed in C belongs always to A (resp. the subobject (weak subobject) of an A-object formed in C belongs always to A). With this manner of speaking the following is valid:

Theorem. If A is a full and isomorphism-closed subcategory of a

A) co-well-powered, (epi, extremal mono)-factorizable category C that has products, then the following are equivalent:

(1) A is epireflective in C.

(2) A is closed under formation of products and subobjects in C.

A') well-powered, (extremal epi, mono)-factorizable category C that has coproducts, then the following are equivalent:

(1) A is monocoreflective in C.

(2) A is closed under formation of coproducts and quotient objects in C.

B) co-well-powered, (extremal epi, mono)-factorizable category that has products, then the following are equivalent:

(1) A is extremal epireflective in C.

(2) A is closed under formation of products and weak subobjects in C.

B') well-powered (epi, extremal mono)-factorizable category that has coproducts, then the following are equivalent:

(1) A is extremal monocoreflective.

(2) A is closed under formation of coproducts and weak quotient objects in C.

Proof. Because of duality it suffices to prove the left part of the above theorem.

A) (2) ⇒ (1) .	B) (2) ⇒ (1) .

Let $X \in |C|$ and $\{e_i : X \to A_i\}$ be a representative set of

C-epimorphisms	extremal C-epimorphisms

which have objects of A as codomain. If $(P,(p_i))$ is the product of the family (A_i) , then $P \in |A|$ by assumption. By the definition of a product there is a unique $f: X \to P$ such that all diagrams

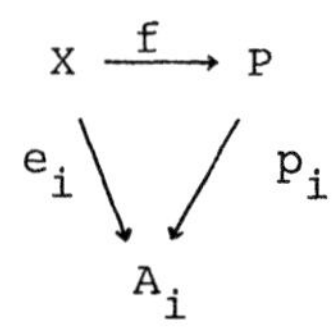

commute. Let

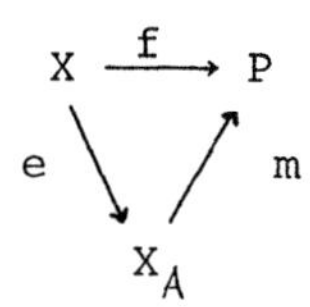

be an

(epi, extremal mono)-factorization	(extremal epi, mono)-factorization

of f . By (2), $X_A \in |A|$, because X_A is a

subobject	weak subobject

of P . It remains to show that e is a reflection. Let $g \in [X,Y]_C$ such that $Y \in |A|$. Let $X \xrightarrow{e'} A \xrightarrow{m'} Y$ be an

(epi, extremal mono)- factorization	(extremal epi, mono)-factorization

of g . Thus, $A \in |A|$. Without loss of generality it may be assumed that $e' = e_i$ and $A = A_i$ for some $i \in I$. Then the following diagram

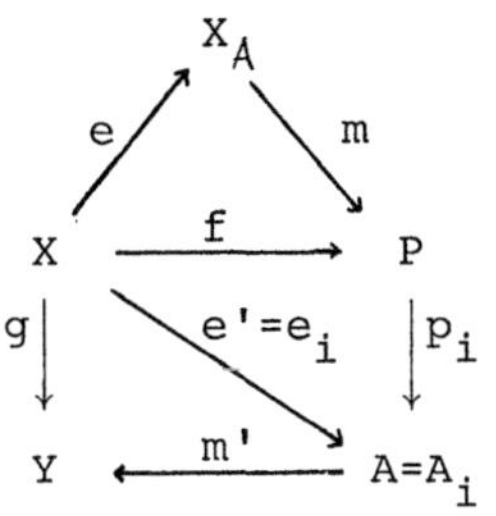

is commutative. Put $\bar{g} = m' \circ p_i \circ m$. Then $\bar{g} \circ e = g$ and since e is an epimorphism, $\bar{g}$ is the unique morphism with this property. Thus, everything is shown.

A) (1) ⇒ (2) . | B) (1) ⇒ (2) .

a) Let $(A_i)_{i \in I}$ be a family of objects of A and $(P,(p_i))$ the product of this family in C . If the

epireflection | extremal epireflection

of P with respect to A is denoted by $r_P\colon P \to P_A$, then there is a unique $\overline{p_i}\colon P_A \to A_i$ such that $\overline{p_i} \circ r_P = p_i$ for each $i \in I$. By definition of a product there is a unique $s_P\colon P_A \to P$ such that $\overline{p_i} = p_i \circ s_P$ for each $i \in I$. Hence, $p_i \circ 1_P = p_i = \overline{p_i} \circ r_P = p_i \circ (s_P \circ r_P)$ for each $i \in I$. Thus, $s_P \circ r_P = 1_P$, because, by definition of the product, there is a unique $h\colon P \to P$ such that $p_i \circ h = p_i$ for each $i \in I$. Since 1_P is an extremal monomorphism and r_P is an

epimorphism, | extremal epimorphism,

r_P is an isomorphism. Consequently, $P \in |A|$, because A is isomorphism-closed.

b) Let $f \in [X,Y]_C$ be

an extremal monomorphism | a monomorphism

and let $Y \in |A|$. If the

epireflection | extremal epireflection

of X with respect to A is denoted by $r_X\colon X \to X_A$, then, since $Y \in |A|$, there is a unique $\bar{f}\colon X_A \to Y$ such that $\bar{f} \circ r_X = f$.

Since r_X is an epimorphism and f is an extremal monomorphism, r_X has to be an isomorphism. | Since r_X is an extremal epi- and additionally a monomorphism, (because $\bar{f} \circ r_X$ is a monomorphism), r_X is an isomorphism.

Thus, since $\mathcal{A}$ is isomorphism-closed, $X \in |\mathcal{A}|$.

<u>2.2.5 Remarks</u>. (1) Obviously, the preceding theorem can be generalized as follows: Let $\mathcal{E}$ (resp. $\mathcal{M}$) be a class of epimorphisms (resp. monomorphisms) which is closed under composition with isomorphisms.

(a) $\mathcal{A}$ is closed under formation of products and $\mathcal{M}$-subobjects[6] in $\mathcal{C}$

implies

(b) $\mathcal{A}$ is $\mathcal{E}$-reflective in $\mathcal{C}$

provided that $\mathcal{C}$ is $\mathcal{E}$-co-well-powered (i.e. for every $X \in |\mathcal{C}|$ there is a representative set of $\mathcal{E}$-quotient objects) and $(\mathcal{E},\mathcal{M})$-factorizable and $\mathcal{C}$ has products. If $\mathcal{C}$ is also an $(\mathcal{E},\mathcal{M})$-category, then (a) and (b) are equivalent.

Proof. (a) ⇒ (b) is proved analogously to "(2) ⇒ (1)" of 2.2.4 A) (resp. B)). Conversely, let (b) be satisfied and $m: X \to A$ be a morphism of $\mathcal{M}$ such that $A \in |\mathcal{A}|$. If $r_X: X \to X_{\mathcal{A}}$ is the $\mathcal{E}$-reflection of X with respect to $\mathcal{A}$, then there exists an f such that $f \circ r_X = m$. By 1.2.3.5, since $\mathcal{C}$ is an $(\mathcal{E},\mathcal{M})$-category, $\mathcal{C}$ satisfies the $(\mathcal{E},\mathcal{M})$-diagonalization property. Therefore there exists an e such that the diagram

$$\begin{array}{ccc} X & \xrightarrow{r_X} & X_{\mathcal{A}} \\ {\scriptstyle 1_X}\downarrow & \swarrow{\scriptstyle e} & \downarrow{\scriptstyle f} \\ X & \xrightarrow{m} & A \end{array}$$

commutes. Especially, $e \circ r_X = 1_X$. Thus r_X is an isomorphism, because it is an epimorphism and 1_X is an extremal monomorphism. Consequently $X \in |\mathcal{A}|$, because $\mathcal{A}$ is isomorphism-closed. (The fact that $\mathcal{A}$ is closed under formation of products in $\mathcal{C}$ is proved as above.)

---

6) $Y \in |\mathcal{C}|$ is called an <u>$\mathcal{M}$-subobject</u> of $X \in |\mathcal{C}|$ provided that there is a $\mathcal{C}$-morphism $f: Y \to X$ such that $f \in \mathcal{M}$. Similarly an $\mathcal{E}$-quotient object is defined.

② A topological category C satisfies the conditions of 2.2.4 (cf. 1.2.2.9, 1.2.3.3, 1.2.1.10). If $P \subset |C|$ is isomorphism-closed and a full subcategory A of C is defined by $|A| = P$, then the following are equivalent:

(1) A is extremal epireflective in C.

(2) P is a relative disconnectedness.

(3) P is closed under formation of products and weak subobjects in C.

This follows immediately from 2.2.4 B) and 1.3.13 together with 1.3.5 b). Thus 1.3.14 is proved.

③ The category Haus of Hausdorff spaces (and continuous maps) satisfies the conditions of 2.2.4:

(1) (a) Haus is well-powered.

(b) Haus is co-well-powered.

(2) (a) Haus is (epi, extremal mono)-factorizable.

(b) Haus is (extremal epi, mono)-factorizable.

(3) Haus has products and coproducts.

Proof. (1) By part A) of the proof of 1.2.2.9 we obtain: If K is a cardinal number, then there is a set $\mathcal{Q}$ of topological spaces such that every space $(Y,\mathcal{Y})$ satisfying $|Y| \leq K$ is homeomorphic to a space of $\mathcal{Q}$.

(a) If $(X,\mathcal{X}) \in |\text{Haus}|$ and $f\colon (Y,\mathcal{Y}) \to (X,\mathcal{X})$ is a monomorphism in Haus (i.e. an injective continuous map), then $|Y| \leq |X| = K$. By the previous remark, there is a representative set of monomorphisms with codomain $(X,\mathcal{X})$.

(b) If $f\colon (X,\mathcal{X}) \to (Y,\mathcal{Y})$ is an epimorphism in Haus, then f is continuous and dense (i.e. $\overline{f[X]} = Y$). Thus for each $y \in Y$, the trace of the neighbourhood filter $\mathcal{U}(y)$ on $f[X]$ is a filter on $f[X]$, i.e. $i^{-1}(\mathcal{U}(y))$ exists ($i\colon f[X] \to Y$ is the inclusion map). For each $y \in Y$, a map $g\colon Y \to \mathcal{P}(\mathcal{P}(f[X]))$ is defined by $g(y) = i^{-1}(\mathcal{U}(y))$ which is obviously injective. (Let $g(y_1) = g(y_2)$ for $y_1, y_2 \in Y$, i.e. $i^{-1}(\mathcal{U}(y_1)) = i^{-1}(\mathcal{U}(y_2))$. Then $\mathcal{U}(y_1) \subset i(i^{-1}(\mathcal{U}(y_1))) = i(i^{-1}(\mathcal{U}(y_2))) \supset \mathcal{U}(y_2)$. Thus a filter on Y is obtained which converges to y_1 and to y_2. Consequently, since Y is a Hausdorff space, $y_1 = y_2$).

Since there is an injective map from $f[X]$ to X (which assigns to each $y \in f[X]$ a unique $x \in f^{-1}(y)$), there is an injective map $h: \mathcal{P}(\mathcal{P}(f[X])) \to \mathcal{P}(\mathcal{P}(X))$ (For every injective map j from a set M to a set N, an injective map $j^*: \mathcal{P}(M) \to \mathcal{P}(N)$ can be defined by $j^*(A) = j[A]$ for each $A \in \mathcal{P}(M)$). Hence the map

$$h \circ g: Y \to \mathcal{P}(\mathcal{P}(X))$$

is injective, i.e. $|Y| \leq K = |\mathcal{P}(\mathcal{P}(X))| = 2^{2^{|X|}}$.
Consequently, together with the previous remark, there is a representative set of epimorphisms with domain $(X,\mathcal{X})$.

(2) (a) If $f: (X,\mathcal{X}) \to (Y,\mathcal{Y})$ is a morphism in Haus, then $f = i \circ \hat{f}$ (where $\hat{f}: X \to f[X]$ is defined by $\hat{f}(x) = f(x)$ for each $x \in X$ and $i: f[X] \to Y$ is the inclusion map) is the desired (epi, extremal mono)-factorization of f.

(b) If $f: (X,\mathcal{X}) \to (Y,\mathcal{Y})$ is a morphism in Haus, then the desired (extremal epi, mono)-factorization of f is given by the commutative diagram

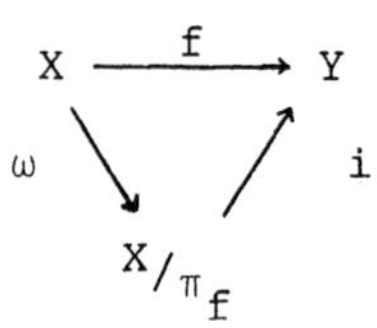

(Since $(Y,\mathcal{Y})$ is a Hausdorff space and i is injective and continuous, $X/_{\pi_f}$ endowed with the quotient topology with respect to ω is also a Hausdorff space).

(3) The products and coproducts of objects of Haus formed in Top are exactly the products and coproducts of these objects formed in Haus, because Haus is full and closed under formation of products and coproducts in Top.

Since the extremal monomorphisms in Haus are exactly the closed embeddings, the subobjects in Haus of objects of Haus are precisely the closed subspaces. *Especially a full and isomor-*

phism-closed subcategory A *of* Haus *is epireflective in* Haus *iff* A *is closed under formation of closed subspaces and products in* Haus .

Example. The full and isomorphism-closed subcategory A of Haus defined by $|A|$ = {compact Hausdorff spaces} is epireflective in Haus .

2.2.6 Further examples. (1) The full and isomorphism-closed subcategories A of Top which are defined in the following are epireflective in Top:

a) $|A|$ is an (isomorphism-closed) relative disconnectedness.
b) $|A|$ = {regular spaces} resp. $|A|$ = {regular T_1-spaces} .
c) $|A|$ = {completely regular spaces} resp. $|A|$ = {completely regular T_1-spaces} .
d) $|A|$ = {zero-dimensional spaces}[7] .

(2) The full and isomorphism-closed subcategory A of Top defined by $|A|$ = {locally connected spaces} (resp. $|A|$ = {locally path connected spaces}) is monocoreflective in Top (even bicoreflective in Top [cf. 2.2.11 (1)]).

2.2.7 Definition. An object G of a category C is called a separator provided that for each pair of distinct morphisms $f,g: A \to B$ with the same domain and the same codomain, there is a morphism $h: G \to A$ such that $f \circ h \neq g \circ h$.

2.2.8 Example. Each object (X,ξ) of a topological category C such that $X \neq \emptyset$ is a separator.

2.2.9 Theorem. Let G be a separator of a category C . Then each coreflective subcategory A of C such that $G \in |A|$ is epicoreflective, i.e. the coreflections are epimorphisms.

---

[7] A topological space $(X,\mathcal{X})$ is called zero-dimensional provided that the open-closed subsets of X form a base of $\mathcal{X}$.

Proof. Let $m: A \to X$ be the coreflection of $X \in |\mathcal{C}|$ with respect to $\mathcal{A}$ and let $\alpha, \beta: X \to Y$ be $\mathcal{C}$-morphisms such that $\alpha \circ m = \beta \circ m$. Then for each $\mathcal{C}$-morphism $h: G \to X$, there is a unique $\mathcal{A}$-morphism $\bar{h}: G \to A$ satisfying $m \circ \bar{h} = h$. Thus $\alpha \circ h = \alpha \circ m \circ \bar{h} = \beta \circ m \circ \bar{h} = \beta \circ h$. Since G is a separator, $\alpha = \beta$. Consequently, m is an epimorphism.

2.2.10 Theorem. Every epicoreflective full subcategory $\mathcal{A}$ of a category $\mathcal{C}$ is bicoreflective, i.e. the coreflections are bimorphisms (Dual: Every monoreflective full subcategory $\mathcal{A}$ of a category $\mathcal{C}$ is bireflective, i.e. the reflections are bimorphisms).

Corollary. Let G be a separator of a category $\mathcal{C}$. Then every coreflective full subcategory $\mathcal{A}$ of $\mathcal{C}$ such that $G \in |\mathcal{A}|$ is bicoreflective.

Proof. Let $e_X: A_X \to X$ be the epicoreflection of $X \in |\mathcal{C}|$ with respect to $\mathcal{A}$ and let $\alpha, \beta: Y \to A_X$ be $\mathcal{C}$-morphisms such that $e_X \circ \alpha = e_X \circ \beta$. If $e_Y: A_Y \to Y$ is the epicoreflection of Y with respect to $\mathcal{A}$, then

$$(*) \qquad e_X \circ \alpha \circ e_Y = e_X \circ \beta \circ e_Y .$$

Since $\mathcal{A}$ is full, $\alpha \circ e_Y$ and $\beta \circ e_Y$ are $\mathcal{A}$-morphisms. Thus, applying $(*)$, $\alpha \circ e_Y = \beta \circ e_Y$, because e_X is a coreflection. Since e_Y is an epimorphism, $\alpha = \beta$. Consequently, e_X is a monomorphism. Therefore the theorem is proved. The corollary is an application of this theorem in connection with 2.2.9.

2.2.11 Remarks. (1) By the preceding considerations every coreflective, full and isomorphism-closed subcategory $\mathcal{A}$ of a topological category $\mathcal{C}$ is bicoreflective, if $|\mathcal{A}|$ contains at least one object with a non-empty underlying set. In this case the coreflection $m_X: (Y_A, \eta_A) \to (X, \xi)$ of $(X, \xi) \in |\mathcal{C}|$ with respect to $\mathcal{A}$ is bijective. By 1.2.2.7 there is a $\mathcal{C}$-structure ξ_A on X such that $(Y_A, \eta_A) \cong (X, \xi_A)$ (by means of m_X !). *Obviously* ξ_A *is the coarsest of all C-structures* ξ' *which are finer than* ξ *and for which* $(X,\xi') \in |A|$ ($(Y_A,\eta_A) \xrightarrow{m_X} (X,\xi_A) \xrightarrow{1_X} (X,\xi)$ is a C-morphism; furthermore $m_X^{-1}: (X,\xi_A) \to (Y_A,\eta_A)$ is a C-morphism (even a C-isomorphism). Thus $1_X = (1_X \circ m_X) \circ m_X^{-1}$ is a C-morphism since the composition of two C-morphisms is a C-morphism. Consequently,

$$\xi_A \leq \xi \text{ , i.e. } \xi_A \text{ is finer than } \xi \text{ .}$$

If ξ' is a C-structure on X such that $\xi' \leq \xi$ and $(X,\xi') \in |A|$, then $1_X: (X,\xi') \to (X,\xi)$ is a C-morphism and there exists a unique C-morphism $\overline{1_X}: (X,\xi') \to (Y_A,\eta_A)$ such that the diagram

$$(*) \qquad \begin{array}{ccc} (Y_A,\eta_A) & \xrightarrow{\;\; m_X \;\;} & (X,\xi) \\ & {\scriptstyle \overline{1}_X} \nwarrow \quad \nearrow {\scriptstyle 1_X} & \\ & (X,\xi') & \end{array}$$

is commutative. $(X,\xi') \xrightarrow{\overline{1}_X} (Y_A,\eta_A) \underset{\cong}{\xrightarrow{m_X}} (X,\xi_A)$ is a C-morphism (as a composition of two C-morphisms) whose underlying map is the identity map on X because the diagram (*) commutes. Consequently, $\xi' \leq \xi_A$) .

Therefore $1_X: (X,\xi_A) \to (X,\xi)$ *is the coreflection of* (X,ξ) *with respect to* A , *i.e. one obtains the coreflection of a C-object* (X,ξ) *with respect to* A *(up to an isomorphism) by a modification of the C-structure* ξ *on* X .

Moreover A *contains obviously all discrete objects of* C because for each discrete C-object (X,ξ) , the coreflection $1_X: (X,\xi_A) \to (X,\xi)$ is an isomorphism.

(2) By dualization of the concept "separator" one obtains the concept "coseparator". Obviously every indiscrete object in a topological category whose underlying set consists at least of two elements is a coseparator.

By applying the dual assertion of 2.2.10 Cor. one obtains the following theorem.

Theorem. Let C be a topological category and let A be a full and isomorphism-closed subcategory of C. Then the following are equivalent:

(1) A is bireflective in C.

(2) A is reflective in C and contains all indiscrete C-objects.

If $r_X\colon (X,\xi) \to (Y_A,\eta_A)$ is the bireflection of $(X,\xi) \in |C|$ with respect to A, then by 1.2.2.7 there is a unique C-structure ξ_A on X such that $r_X^{-1}\colon (Y_A,\eta_A) \to (X,\xi_A)$ is a C-isomorphism. *Especially* $1_X\colon (X,\xi) \to (X,\xi_A)$ *is (up to an isomorphism) the bireflection of* (X,ξ) *with respect to* A *and* ξ_A *is the finest of all* C*-structures* ξ' *on* X *which are coarser than* ξ *and for which* $(X,\xi') \in |A|$. (This is proved analogously to the corresponding assertion of (1) with respect to bicoreflections.)

2.2.12 Theorem. Every bicoreflective and every bireflective, full and isomorphism-closed subcategory of a topological category C is topological.

Proof. 1) Let A be a full and isomorphism-closed bicoreflective subcategory of C:

Cat top_1) Let X be a set, $((X_i,\xi_i))_{i\in I}$ a family of A-objects and $f_i\colon X \to X_i$ maps for each $i \in I$. Let ξ be the initial C-structure on X with respect to $(X,f_i,(X_i,\xi_i),I)$ and let $1_X\colon (X,\xi_A) \to (X,\xi)$ be the coreflection of (X,ξ) with respect to A. Then ξ_A is the initial A-structure on X with respect to $(X,f_i,(X_i,\xi_i),I)$.

a) Let $g\colon (Y_A,\eta_A) \to (X,\xi_A)$ be an A-morphism (which is also a C-morphism). Since $f_i\colon (X,\xi) \to (X_i,\xi_i)$ is a C-morphism for each $i \in I$, $f_i = f_i \circ 1_X\colon (X,\xi_A) \to (X_i,\xi_i)$ is a C-morphism between A-objects. Thus it is an A-morphism for each $i \in I$. Consequently, $f_i \circ g\colon (Y_A,\eta_A) \to (X_i,\xi_i)$ is an A-morphism for each $i \in I$.

b) Let $f_i \circ g\colon (Y_A,\eta_A) \to (X_i,\xi_i)$ be an A-morphism for each $i \in I$. $1_X \circ g\colon (Y_A,\eta_A) \to (X,\xi)$ is a C-morphism, because $f_i \circ g = f_i \circ 1_X \circ g\colon (Y_A,\eta_A) \to (X_i,\xi_i)$ is an A-morphism (= C-morphism) and ξ is an initial C-structure. Since $1_X\colon (X,\xi_A) \to (X,\xi)$ is the coreflection of (X,ξ) with respect to A , there exists a unique A-morphism $h\colon (Y_A,\eta_A) \to (X,\xi_A)$ such that the diagram

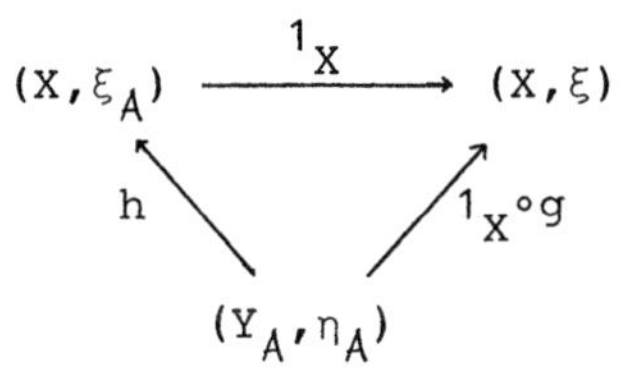

commutes, i.e. $1_X \circ h = 1_X \circ g$. Thus h and g coincide as maps between the underlying sets. Consequently, $g\colon (Y_A,\eta_A) \to (X,\xi_A)$ is an A-morphism. If ξ_A' is also an initial A-structure on X with respect to $(X,f_i,(X_i,\xi_i),I)$, then, by the definition of initial structures, both $1_X\colon (X,\xi_A) \to (X,\xi_A')$ and $1_X\colon (X,\xi_A') \to (X,\xi_A)$ are A-morphisms and therefore C-morphisms. Thus $\xi_A \le \xi_A'$ and $\xi_A' \le \xi_A$ in C . Consequently, by 1.1.3 ②, $\xi_A = \xi_A'$ because C is topological.

Cat top$_2$) is obviously satisfied. Otherwise it would not be valid in C .

Cat top$_3$) On every set X with cardinality one there is the indiscrete A-structure. Since every A-structure is a C-structure and Cat top$_3$) is valid in C , it is the unique A-structure on X .

2) The case that A is a full and isomorphism-closed bireflective subcategory of C is proved analogously to 1).

<u>2.2.13 Remarks</u>. ① Let C be a topological category and A be a (full and isomorphism-closed) subcategory which is bicoreflective in C . Further let $((X_i,\xi_i))_{i\in I}$ be a family of A-objects and $f_i\colon X_i \to X$ maps for each $i \in I$. By 2.2.12 there exists the final A-structure ξ on X with respect to $((X_i,\xi_i),f_i,X,I)$. ξ coincides with the final C-structure on

X with respect to $((X_i,\xi_i),f_i,X,I)$. (If ξ is the final C-structure on X with respect to $((X_i,\xi_i),f_i,X,I)$ and the bicoreflection of (X,ξ) with respect to A is denoted by $1_X\colon (X,\xi_A) \to (X,\xi)$, then $1_X\colon (X,\xi_A) \to (X,\xi)$ is a C-isomorphism, i.e. $(X,\xi) \in |A|$ [$1_X\colon (X,\xi) \to (X,\xi_A)$ is a C-morphism if and only if $1_X \circ f_i = f_i\colon (X_i,\xi_i) \to (X,\xi_A)$ is a C-morphism for each $i \in I$. Obviously for every $i \in I$, there exists a unique C-morphism $g_i\colon (X_i,\xi_i) \to (X,\xi_A)$ such that the diagram

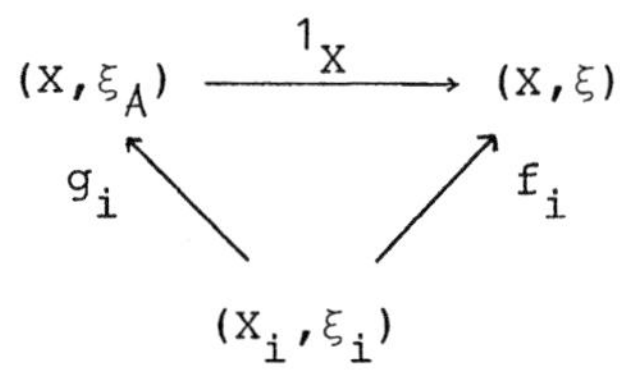

commutes, i.e. $g_i = f_i$ (as maps between the underlying sets). Therefore the condition is satisfied]) .

(2) Let A be a bireflective (full and isomorphism-closed) subcategory of a topological category C . Then the initial structures in A are formed as in C , whereas the final structures in A arise from the final structures in C by applying the bireflector.

2.3 E-reflective and M-coreflective hulls

2.3.1. Let A,B be full subcategories of a category C . Then A is called smaller than B (resp. B larger than A) provided that $|A| \subset |B|$. One often says "B contains A" instead of "A is smaller than B" (resp. "B is larger than A") and writes $A \subset B$.

Now the problem is investigated whether for every full subcategory A of a category C , there is a smallest (full and isomorphism-closed) epireflective (extremal epireflective) [resp. monocoreflective (extremal monocoreflective)] subcategory of C containing A , i.e. an epireflective (extremal epireflective) [resp. monocoreflective (extremal monocoreflective)] "hull" and how it may be characterized if it exists.

2.3.2 Theorem. Let A be a subcategory of a

A) co-well-powered, (epi, extremal mono)-factorizable category C that has products. Then there is a smallest epireflective, full and isomorphism-closed subcategory $R_C A$ of C containing A. If C is additionally an (epi, extremal mono)-category, then $|R_C A|$ consists exactly of all $X \in |C|$ which are subobjects of products of A-objects (The subobjects and the products are formed in C!).

A') well-powered, (extremal epi, mono)-factorizable category C that has coproducts. Then there is a smallest monocoreflective, full and isomorphism-closed subcategory $R_C^{co} A$ of C containing A. If C is additionally an (extremal epi, mono)-category, then $|R_C^{co} A|$ consists exactly of all $X \in |C|$ which are quotient objects of coproducts of A-objects (The quotient objects and the coproducts are formed in C!).

B) co-well-powered (extremal epi, mono)-factorizable category C that has products. Then there is a smallest extremal epireflective, full and isomorphism-closed subcategory $Q_C A$ of C containing A. Especially $|Q_C A|$ consists exactly of all $X \in |C|$ which are weak subobjects of products of A-objects (The weak subobjects and the products are formed in C!)

B') well-powered, (epi, extremal mono)-factorizable category C that has coproducts. Then there is a smallest extremal monocoreflective, full and isomorphism-closed subcategory $Q_C^{co} A$ of C containing A. Especially $|Q_C^{co} A|$ consists exactly of all $X \in |C|$ which are weak quotient objects of coproducts of A-objects. (The weak quotient objects and the coproducts are formed in C!).

Proof. By duality it suffices to prove the left part of the above theorem.

A) a) A full subcategory $R_{\mathcal{C}}\mathcal{A}$ of $\mathcal{C}$ is defined by $|R_{\mathcal{C}}\mathcal{A}| = \{X \in |\mathcal{C}|: X \in |\mathcal{B}|$ for each full and isomorphism-closed epireflective subcategory $\mathcal{B}$ of $\mathcal{C}$ such that $|\mathcal{A}| \subset |\mathcal{B}|\}$. Obviously, $R_{\mathcal{C}}\mathcal{A}$ is isomorphism-closed and epireflective in $\mathcal{C}$ by the characterization theorem 2.2.4. Moreover, $|\mathcal{A}| \subset |R_{\mathcal{C}}\mathcal{A}|$ and $|R_{\mathcal{C}}\mathcal{A}| \subset |\mathcal{B}|$ for each full and isomorphism-closed epireflective subcategory $\mathcal{B}$ of $\mathcal{C}$ with $|\mathcal{A}| \subset |\mathcal{B}|$ (by construction of $|R_{\mathcal{C}}\mathcal{A}|$!).

b) Additionally let $\mathcal{C}$ be an (epi, extremal mono)-category. A full subcategory $\mathcal{D}$ of $\mathcal{C}$ is defined by $|\mathcal{D}| = \{X \in |\mathcal{C}|: X$ is a subobject of a product of $\mathcal{A}$-objects$\}$. Obviously, $\mathcal{D}$ is isomorphism-closed. It must be shown that $\mathcal{D} = R_{\mathcal{C}}\mathcal{A}$.

b_1) $|\mathcal{A}| \subset |\mathcal{D}|$: Since each $X \in |\mathcal{A}| \subset |\mathcal{C}|$ is a product of itself (1_X is the corresponding projection) and $1_X: X \to X$ is an extremal monomorphism (even an isomorphism!), this assertion is trivial.

b_2) $|\mathcal{D}| \subset |R_{\mathcal{C}}\mathcal{A}|$ is valid, because each $X \in |\mathcal{D}|$ belongs to each (full and isomorphism-closed) subcategory $\mathcal{B}$ of $\mathcal{C}$ which is closed under formation of subobjects and products in $\mathcal{C}$ (i.e. epireflective by 2.2.4) and which contains $\mathcal{A}$.

b_3) Because of b_1), $|R_{\mathcal{C}}\mathcal{A}| \subset |\mathcal{D}|$ is satisfied provided that $\mathcal{D}$ is epireflective in $\mathcal{C}$. We show this fact by applying the characterization theorem 2.2.4:

α) Let $X \in |\mathcal{D}|$ and $Y \in |\mathcal{C}|$ be a subobject of X. Then since the composition of two extremal monomorphisms is an extremal monomorphism, Y is a subobject of a product of $\mathcal{A}$-objects.

β) Let $(X_i)_{i\in I}$ be a family of objects of $\mathcal{D}$, i.e. every X_i is a subobject of a product P_i of $\mathcal{A}$-objects. For each $i \in I$, let $P_i = \prod_{k\in K_i} A_k$. It may be assumed that $K_i \cap K_j = \emptyset$ for each $(i,j) \in I\times I$ such that $i \neq j$ (otherwise replace K_i by $K_i\times\{i\}$ for each $i \in I$). Put $K = \bigcup_{i\in I} K_i$. Then obviously $\prod_{i\in I} P_i = \prod_{k\in K} A_k$ (up to isomorphism) [One obtains the corresponding projections $q_k: \prod_{i\in I} P_i \to A_k$ by determining for each

$k \in K$ the corresponding $i \in I$ such that $k \in K_i$ and putting $p'_k \circ p_i = q_k$, where $p_i\colon \prod_{i\in I} P_i \to P_i$ and $p'_k\colon P_i \to A_k$ are the projections]. Furthermore, for each $i \in I$, there exists an extremal monomorphism $m_i\colon X_i \to P_i$. If $\left(\prod_{i\in I} X_i, (\tilde{p}_i)\right)$ is the product of $(X_i)_{i\in I}$, then there is a unique $f\colon \prod_{i\in I} X_i \to \prod_{i\in I} P_i = \prod_{k\in K} A_k$ such that the diagram

$$\begin{array}{ccc} \prod\limits_{i\in I} X_i & \xrightarrow{\quad f \quad} & \prod\limits_{i\in I} P_i = \prod\limits_{k\in K} A_k \\ {\scriptstyle \tilde{p}_i}\big\downarrow & & \big\downarrow{\scriptstyle p_i} \\ X_i & \xrightarrow{\quad m_i \quad} & P_i \end{array} \qquad \text{(D)}$$

commutes for each $i \in I$, because $\left(\prod_{i\in I} P_i, (p_i)\right)$ is a product. If it can be shown that f is an extremal monomorphism, then $\prod_{i\in I} X_i \in |\mathcal{D}|$:

β_1) If $\alpha, \beta\colon Y \to \prod_{i\in I} X_i$ are $\mathcal{C}$-morphisms such that $f\circ\alpha = f\circ\beta$, then $p_i\circ f\circ\alpha = p_i\circ f\circ\beta$ for each $i \in I$ and since (D) is commutative, $m_i\circ\tilde{p}_i\circ\alpha = m_i\circ\tilde{p}_i\circ\beta$. Thus $\tilde{p}_i\circ\alpha = \tilde{p}_i\circ\beta$ for each $i \in I$, because all m_i are monomorphisms. Consequently, since $\left(\prod_{i\in I} X_i, (\tilde{p}_i)\right)$ is a product, $\alpha = \beta$.

β_2) Let $f = h\circ e$, where e is an epimorphism. Since $\mathcal{C}$ is an (epi, extremal mono)-category, $\mathcal{C}$ satisfies the (epi, extremal mono)-diagonalization property and thus there exist $\mathcal{C}$-morphisms $h_i\colon Z \to X_i$ such that the diagram

$$\begin{array}{ccccc} \prod\limits_{i\in I} X_i & & \xrightarrow{\quad f \quad} & & \prod\limits_{i\in I} P_i = \prod\limits_{k\in K} A_k \\ & {\scriptstyle e}\searrow & & \nearrow{\scriptstyle h} & \\ {\scriptstyle \tilde{p}_i}\big\downarrow & & Z & & \big\downarrow{\scriptstyle p_i} \\ & \swarrow{\scriptstyle h_i} & & & \\ X_i & & \xrightarrow{\quad m_i \quad} & & P_i \end{array}$$

commutes for each $i \in I$. Since $(\prod_{i\in I} X_i, (\tilde{p}_i))$ is a product, there is a unique $\mathcal{C}$-morphism $e': Z \to \prod_{i\in I} X_i$ such that $\tilde{p}_i \circ e' = h_i$ for each $i \in I$ and since $\tilde{p}_i \circ 1_{\prod X_i} = \tilde{p}_i = h_i \circ e = \tilde{p}_i \circ e' \circ e$ for each $i \in I$, $e' \circ e = 1_{\prod X_i}$. Consequently, e is an isomorphism because e is an epimorphism.
$|\mathcal{D}| = |R_{\mathcal{C}}\mathcal{A}|$ follows from b_2) and b_3) and thus $\mathcal{D} = R_{\mathcal{C}}\mathcal{A}$.
B) is proved analogously to A) by replacing "epireflective" by "extremal epireflective" (resp. "extremal monomorphism" by "monomorphism" [i.e. "subobject" by "weak subobject"]). Since the class of all monomorphisms is already closed under composition and there is no analogue to β_2), the assumption "(extremal epi, mono)-category" is omitted for the analogous assertion to b).

2.3.3 Definition. The (full and isomorphism-closed) subcategory $R_{\mathcal{C}}\mathcal{A}$ (resp. $Q_{\mathcal{C}}\mathcal{A}$) of $\mathcal{C}$ explained in 2.3.2 is called the epireflective (resp. extremal epireflective) hull of $\mathcal{A}$ in $\mathcal{C}$ (Correspondingly $R_{\mathcal{C}}^{co}\mathcal{A}$ (resp. $Q_{\mathcal{C}}^{co}\mathcal{A}$) is called the monocoreflective (resp. extremal monocoreflective) hull of $\mathcal{A}$ in $\mathcal{C}$).

2.3.4 Corollary. Let $\mathcal{A}$ be a full and isomorphism-closed subcategory of a

co-well-powered (epi, extremal mono)-factorizable category $\mathcal{C}$ that has products. Then the following is satisfied:

(1) $\mathcal{A}$ is epireflective in $\mathcal{C}$ if and only if $\mathcal{A} = R_{\mathcal{C}}\mathcal{A}$.

(2) $\mathcal{A}$ is extremal epireflective if and only if $\mathcal{A} = Q_{\mathcal{C}}\mathcal{A}$.

well-powered (epi, extremal mono)-factorizable category $\mathcal{C}$ that has coproducts. Then the following is satisfied:

(1') $\mathcal{A}$ is monocoreflective in $\mathcal{C}$ if and only if $\mathcal{A} = R_{\mathcal{C}}^{co}\mathcal{A}$.

(2') $\mathcal{A}$ is extremal monocoreflective in $\mathcal{C}$ if and only if $\mathcal{A} = Q_{\mathcal{C}}^{co}\mathcal{A}$.

Proof. (1): If $\mathcal{A}$ is epireflective in $\mathcal{C}$, then automatically $R_{\mathcal{C}}\mathcal{A} \subset \mathcal{A}$ by the construction of $R_{\mathcal{C}}\mathcal{A}$. Consequently, $\mathcal{A} = R_{\mathcal{C}}\mathcal{A}$ since $\mathcal{A} \subset R_{\mathcal{C}}\mathcal{A}$. The converse is trivial.

(2): Analogously to (1) .

2.3.5 Corollary. Let $\mathcal{A},\mathcal{B}$ be full subcategories of a co-well-powered (epi, extremal mono)-factorizable category $\mathcal{C}$ that has products. Then the following are satisfied:

A) (1) $\mathcal{A} \subset R_{\mathcal{C}}\mathcal{A}$.

(2) $\mathcal{A} \subset \mathcal{B}$ implies $R_{\mathcal{C}}\mathcal{A} \subset R_{\mathcal{C}}\mathcal{B}$.

(3) $R_{\mathcal{C}}R_{\mathcal{C}}\mathcal{A} = R_{\mathcal{C}}\mathcal{A}$.

B) (1) $\mathcal{A} \subset Q_{\mathcal{C}}\mathcal{A}$.

(2) $\mathcal{A} \subset \mathcal{B}$ implies $Q_{\mathcal{C}}\mathcal{A} \subset Q_{\mathcal{C}}\mathcal{B}$.

(3) $Q_{\mathcal{C}}Q_{\mathcal{C}}\mathcal{A} = Q_{\mathcal{C}}\mathcal{A}$.

well-powered (extremal epi, mono)-factorizable category that has coproducts. Then the following are satisfied:

A') (1') $\mathcal{A} \subset R^{co}_{\mathcal{C}}\mathcal{A}$.

(2') $\mathcal{A} \subset \mathcal{B}$ implies $R^{co}_{\mathcal{C}}\mathcal{A} \subset R^{co}_{\mathcal{C}}\mathcal{B}$.

(3') $R^{co}_{\mathcal{C}}R^{co}_{\mathcal{C}}\mathcal{A} = R^{co}_{\mathcal{C}}\mathcal{A}$.

B') (1') $\mathcal{A} \subset Q^{co}_{\mathcal{C}}\mathcal{A}$.

(2') $\mathcal{A} \subset \mathcal{B}$ implies $Q^{co}_{\mathcal{C}}\mathcal{A} \subset Q^{co}_{\mathcal{C}}\mathcal{B}$.

(3') $Q^{co}_{\mathcal{C}}Q^{co}_{\mathcal{C}}\mathcal{A} = Q^{co}_{\mathcal{C}}\mathcal{A}$.

Proof. A) (1) is satisfied by definition.

(2) follows from $\mathcal{A} \subset \mathcal{B} \subset R_{\mathcal{C}}\mathcal{B}$ and the definition of $R_{\mathcal{C}}\mathcal{A}$.

(3) follows from 2.3.4.

B) analogously to A) .

2.3.6 Remark. Obviously the theorem 2.3.2 can be generalized as follows provided E (resp. M) is a class of epimorphisms (resp. monomorphisms) which is closed under composition with isomorphisms:

Theorem. Let $\mathcal{C}$ be an E-co-well-powered (E,M) - category that has products and let $\mathcal{A}$ be a full subcategory of $\mathcal{C}$. Then there is a smallest full and isomorphism-closed E-reflective subcategory $E(\mathcal{A})$ of $\mathcal{C}$ containing $\mathcal{A}$ whose objects are precisely the M-subobjects of products of A-objects in C, i.e. the <u>E-reflective hull of A in C</u> [8].

2.3.7 Examples. (1) Let C be a topological category and $P \subset |C|$. A full subcategory A of C is defined by $|A| = P$. Then the following holds:

$$|Q_C A| = QP \qquad \text{(cf. 1.3)} .$$

(Since $Q(QP) = QP$, QP is a relative disconnectedness, i.e. the object class of a full and isomorphism-closed extremal epireflective subcategory of C containing $|A| = P$. Thus $|Q_C A| \subset QP$. Conversely, if $X \in QP$ and $r_X: X \to X_{Q_C A}$ is the extremal epireflection of X with respect to $Q_C A$ then r_X is also a monomorphism [If $x,y \in X$ such that $x \neq y$, then there exist a $P \in P$ and a morphism $f: X \to P$ satisfying $f(x) \neq f(y)$. For this f, there is a unique $\bar{f}: X_{Q_C A} \to P$ such that $\bar{f} \circ r_X = f$. Thus $r_X(x) \neq r_X(y)$], i.e. an isomorphism. Consequently, $X \in |Q_C A|$.)

(2) Let P be a subclass of $|\underline{Top}|$. A full subcategory A of $\underline{Top}$ is defined by $|A| = P$. Then the object class $|R_C A|$ of the epireflective hull $R_C A$ of A consists precisely of all topological spaces (X,X) which are subobjects of products of P-objects, i.e. which are homeomorphic to subspaces of product spaces of spaces of P (the subobjects in $\underline{Top}$ are subspaces up to homeomorphism). We write briefly RP instead of $|R_{\underline{Top}} A|$:

a) $P = \{[0,1]\}$: $RP = \{$completely regular T_1-spaces$\}$
b) $P = \{\mathbb{R}\}$: $RP = \{$completely regular T_1-spaces$\}$
c) $P = \{D_2\}$: $RP = \{$zero-dimensional T_1-spaces$\}$.

Definition (Mrowka, Engelking, Herrlich). Let P be a class of Hausdorff spaces. A topological space (X,X) is called

---

8) Dually the <u>M-coreflective hull of A in C</u> can be introduced.

P-regular provided that $(X,\mathcal{X})$ is homeomorphic to a subspace of a product of spaces of P, i.e. $(X,\mathcal{X}) \in RP$.

Remarks. 1) Every P-regular space is a Hausdorff space by definition.

2) Especially it follows from 2.3.2 and the characterization theorem of epireflective subcategories that a product of P-regular spaces is again P-regular and a subspace of a P-regular space is also P-regular.

③ Let P be a subclass of $|\underline{\text{Haus}}|$. A full subcategory $\mathcal{A}$ of Haus is defined by $|\mathcal{A}| = P$. Then the object class $|R_{\underline{\text{Haus}}}\mathcal{A}|$ of the epireflective hull $R_{\underline{\text{Haus}}}\mathcal{A}$ consists precisely of all those Hausdorff spaces which are subobjects (in Haus) of products of $\mathcal{A}$-objects, i.e. which are homeomorphic to closed subspaces of product spaces of spaces of P (the subobjects in Haus are just the closed subspaces in Top [up to homeomorphism]!). We write CP instead of $|R_{\underline{\text{Haus}}}\mathcal{A}|$:

a) $P = \{[0,1]\}$: $CP = \{$compact Hausdorff spaces$\}$

b) If $P = \{\mathbb{R}\}$, then $CP \supset \{$compact Hausdorff spaces$\}$ but the equality is not true because $\mathbb{R} \in CP$ is not compact. Obviously, CP consists of all Hausdorff spaces X such that X is homeomorphic to a closed subspace of $\mathbb{R}^I$ for some I. These spaces are called real-compact. Thus by definition:

$P = \{\mathbb{R}\}$: $CP = \{$real-compact spaces$\}$.

c) $P = \{D_2\}$: $CP = \{$zero-dimensional compact Hausdorff spaces$\}$.

Definition (Mrowka, Herrlich). Let P be a class of Hausdorff spaces. Then a topological space $(X,\mathcal{X})$ is called P-compact provided that $(X,\mathcal{X})$ is homeomorphic to a closed subspace of a product of spaces of P, i.e. $(X,\mathcal{X}) \in CP$.

Remarks. 1) By definition every P-compact space is P-regular and thus a Hausdorff space.

2) Especially it follows from 2.3.2 and the characterization theorem of epireflective subcategories that a product of P-compact spaces is again P-compact and each closed subspace of a P-compact space is also P-compact.

④ Let A be a full subcategory of Top. Then the object class $|R^{co}_{\underline{Top}}A|$ of the monocoreflective hull $R^{co}_{\underline{Top}}A$ of A consists precisely of all those topological spaces which are homeomorphic to quotient spaces of sums of spaces of $|A|$ (the quotient objects in Top are just the quotient spaces [up to homeomorphism]!).

a) $|A|$ = {locally connected spaces} ∩ {connected spaces}:
$|R^{co}_{\underline{Top}}A|$ = {locally connected spaces} .

b) $|A|$ = {locally path-connected spaces} ∩ {path-connected spaces}:
$|R^{co}_{\underline{Top}}A|$ = {locally path-connected spaces} .

⑤ Let A be a full subcategory of Haus. Then the object class $|R^{co}_{\underline{Haus}}A|$ of the monocoreflective hull $R^{co}_{\underline{Haus}}A$ of A consists precisely of all those Hausdorff spaces which are homeomorphic to quotient spaces of sums of spaces of $|A|$ (the quotient objects in Haus are just the quotient spaces of Hausdorff spaces which are Hausdorff spaces [up to homeomorphism]!):
$|A|$ = {compact Hausdorff spaces}: $|R^{co}_{\underline{Haus}}A|$ = {k-spaces} .

2.3.8 Theorem. Let A be a subcategory of a co-well-powered (epi, extremal mono)- and (extremal epi, mono)-factorizable category C that has products. Then the following is satisfied:

$$A \subset R_C A \subset Q_C A \ .$$

Moreover, if C is balanced, then

$$R_C A = Q_C A \ .$$

Proof: trivial.

2.3.9 Remark. There is an interesting topological interpretation of the fact that $R_C A$ and $Q_C A$ coincide in balanced "nice" categories. The category Comp T_2 of compact Hausdorff spaces (and continuous maps) is balanced. The epimorphisms in Comp T_2 are exactly the surjective continuous maps (a) Let $(X,\mathcal{X})$, $(Y,\mathcal{Y}) \in |\text{Comp } T_2|$. If $f: (X,\mathcal{X}) \to (Y,\mathcal{Y})$ is surjective and continuous, then for each $y \in Y$, there is an $x \in X$ such that $f(x) = y$. If $\alpha \circ f = \beta \circ f$, then $\alpha(y) = \alpha(f(x)) = \beta(f(x)) = \beta(y)$ for each $y \in Y$, i.e. $\alpha = \beta$. Consequently, f is an epimorphism. b) If $f \in [(X,\mathcal{X}),(Y,\mathcal{Y})]_{\underline{\text{Comp } T_2}}$ is not surjective, then there is a $y \in Y \setminus f[X]$. Since $f[X]$ is closed ($f[X]$ is a compact subset of a Hausdorff space) and $(Y,\mathcal{Y})$ is a completely regular Hausdorff space ($(Y,\mathcal{Y}) \in |\underline{\text{Comp } T_2}|$!), there is a continuous map $h: Y \to [0,1]$ ($[0,1]$ is compact!) such that $h(y) = 0$ and $h[f[X]] = \{1\}$. $k: Y \to [0,1]$ defined by $k(z) = 1$ for each $z \in Y$ is continuous and $k \circ f = h \circ f$ but $k \neq h$. Thus $f: (X,\mathcal{X}) \to (Y,\mathcal{Y})$ is not an epimorphism). The monomorphisms in $\underline{\text{Comp } T_2}$ are the injective continuous maps. (This is proved analogously to part a) of the proof of 1.2.2.4). The extremal epimorphisms in $\underline{\text{Comp } T_2}$ are the quotient maps (analogous to part b) of the proof of 1.2.2.5). The extremal monomorphisms in $\underline{\text{Comp } T_2}$ are the embeddings (analogous to part a) of the proof of 1.2.2.5). Corresponding to the proof for topological categories one easily verifies the following:

1. $\underline{\text{Comp } T_2}$ is (epi, extremal mono)-factorizable.
2. $\underline{\text{Comp } T_2}$ is (extremal epi, mono)-factorizable.
3. $\underline{\text{Comp } T_2}$ is co-well-powered .

If one chooses $|\mathcal{A}| = \{D_2\}$, then $|R_{\underline{\text{Comp } T_2}}\mathcal{A}|$ consists precisely of all those compact Hausdorff spaces which are isomorphic to a subspace of D_2^I for a suitable I, i.e. of all zero-dimensional compact Hausdorff spaces, and $|Q_{\underline{\text{Comp } T_2}}\mathcal{A}|$ consists of all those compact Hausdorff spaces which can be mapped onto a subspace of D_2^I for a suitable I by a continuous bijection, i.e. of all totally separated compact Hausdorff spaces. Therefore 2.3.8 contains the following special case:

Theorem. A compact Hausdorff space is totally separated if and only if it is zero-dimensional.

2.4 Reflectors as compositions of epireflectors

2.4.1. The category Comp T_2 of compact Hausdorff spaces (and continuous maps) is epireflective in the category C Reg_1 of completely regular T_1-spaces (and continuous maps). (Let $X \in |C\ Reg_1|$ and $\gamma, \delta \in [\beta(X), Z]_{C\ Reg_1}$ such that $\gamma \circ \beta_X = \delta \circ \beta_X$ and put $h = \beta_Z \circ \gamma \circ \beta_X = \beta_Z \circ \delta \circ \beta_X$. Then there is a unique $\bar{h}: \beta(X) \to \beta(Z)$ such that $\bar{h} \circ \beta_X = h$. Hence $\bar{h} = \beta_Z \circ \gamma = \beta_Z \circ \delta$ so that since β_Z is a monomorphism in Top, $\gamma = \delta$. Consequently, $\beta_X: X \to \beta(X)$ is an epimorphism for each $X \in |C\ Reg_1|$.)
β: C Reg_1 $\to$ Comp T_2 is the corresponding epireflector, the epireflection of $X \in |C\ Reg_1|$ with respect to Comp T_2 is denoted by β_X. It follows from the investigations on epireflective subcategories that C Reg_1 is epireflective in Top. Let α: Top $\to$ C Reg_1 be the corresponding epireflector and let the epireflection of $X \in |Top|$ with respect to C Reg_1 be denoted by α_X. Then the composition $\gamma = \beta \circ \alpha$: Top $\to$ Comp T_2 is a reflector but not an epireflector (in general the maps $\gamma_X = \beta_{\alpha(X)} \circ \alpha_X: X \to \beta(\alpha(X)) = \gamma(X)$ for $X \in |Top|$ are not surjective!). Hence Comp T_2 is only reflective in Top. The factorization of γ by means of two epireflectors is not unique because Comp T_2 is also epireflective in Haus and Haus is epireflective in Top.
However, C Reg_1 is obviously the smallest of those (full and isomorphism-closed) subcategories $\mathcal{D}$ of Top for which Comp $T_2 \subset \mathcal{D} \subset$ Top such that Comp T_2 is epireflective in $\mathcal{D}$ and $\mathcal{D}$ is epireflective in Top, because each of these $\mathcal{D}$'s contains Comp T_2 and is closed under formation of subobjects in Top, i.e. it contains C Reg_1 (C Reg_1 consists exactly of all subobjects [in Top] of Comp T_2-objects). Now the problem is to be solved whether every reflector $R: \mathcal{C} \to \mathcal{A}$ can be obtained as a composition of two epireflectors by factorization through a smallest intermediate category.

<u>2.4.2 Theorem</u> (Kennison, Baron). Let $\mathcal{C}$ be an (epi, extremal mono)-category and $\mathcal{A}$ a (full and isomorphism-closed) reflective subcategory of $\mathcal{C}$. If a (full and isomorphism-closed) subcategory $\mathcal{B}$ of $\mathcal{C}$ is defined by

$$|\mathcal{B}| = \{X \in |\mathcal{C}|\colon \text{there is an extremal monomorphism } m\colon X \to A \text{ in } \mathcal{C} \text{ such that } A \in |\mathcal{A}|\}\ ,$$

then $\mathcal{B}$ is the smallest of those (full and isomorphism-closed) subcategories $\mathcal{D}$ for which $\mathcal{A} \subset \mathcal{D} \subset \mathcal{C}$ such that $\mathcal{A}$ is epireflective in $\mathcal{D}$ and $\mathcal{D}$ is epireflective in $\mathcal{C}$.

<u>Proof</u>. 1) $\mathcal{A} \subset \mathcal{B}$ is trivially satisfied since $1_A\colon A \to A$ is an extremal monomorphism (even an isomorphism) in $\mathcal{C}$ for each $A \in |\mathcal{A}|$.

2) $\mathcal{A}$ is epireflective in $\mathcal{B}$: Since $\mathcal{A}$ is reflective in $\mathcal{C}$, $\mathcal{A}$ is also reflective in $\mathcal{B}$. Let $X \in |\mathcal{B}|$ and let $r_X\colon X \to X_{\mathcal{A}}$ be the reflection of X with respect to $\mathcal{A}$ and $m\colon X \to A$ with $A \in |\mathcal{A}|$ an extremal monomorphism in $\mathcal{C}$ existing by the definition of $\mathcal{B}$. Then there is a unique $\bar{m}\colon X_{\mathcal{A}} \to A$ satisfying $\bar{m} \circ r_X = m$. Since m is also a monomorphism in $\mathcal{B}$, r_X is a monomorphism in $\mathcal{B}$. Hence $\mathcal{A}$ is monoreflective in $\mathcal{B}$ and consequently, epireflective in $\mathcal{B}$ (cf. 2.2.10).

3) $\mathcal{B}$ is epireflective in $\mathcal{C}$: Let $r_X\colon X \to X_{\mathcal{A}}$ be the reflection of $X \in |\mathcal{C}|$ with respect to $\mathcal{A}$. The (epi, extremal mono)-factorization of r_X is given by the commutative diagram

$$\begin{array}{ccc} & X & \\ {}^{e_X}\swarrow & & \searrow^{r_X} \\ X_{\mathcal{B}} & \xrightarrow[m_X]{} & X_{\mathcal{A}} \end{array}\ .$$

Hence by the definition of $\mathcal{B}$, $X_{\mathcal{B}} \in |\mathcal{B}|$, since m_X is an extremal monomorphism and $X_{\mathcal{A}} \in |\mathcal{A}|$. Then $e_X\colon X \to X_{\mathcal{B}}$ is the desired epireflection of X with respect to $\mathcal{B}$: Let $f\colon X \to B$ be a $\mathcal{C}$-morphism such that $B \in |\mathcal{B}|$. By the definition of $\mathcal{B}$ there is an extremal monomorphism $m\colon B \to A$ with $A \in |\mathcal{A}|$. Since r_X is the reflection of X with respect to

$\mathcal{A}$, there is a unique $f'\colon X_{\mathcal{A}} \to A$ such that $f'\circ r_X = f'\circ m_X \circ e_X = m\circ f$. By the (epi, extremal mono)-diagonalization property there exists an $\bar{f}\colon X_{\mathcal{B}} \to B$ such that the diagram

$$\begin{array}{ccc} X & \xrightarrow{e_X} & X_{\mathcal{B}} \\ {\scriptstyle f}\downarrow & \overset{\bar{f}}{\swarrow} & \downarrow{\scriptstyle f'\circ m_X} \\ B & \xrightarrow{m} & A \end{array}$$

commutes. Especially, $\bar{f}\circ e_X = f$. Given an $\bar{\bar{f}}\colon X_{\mathcal{B}} \to B$ satisfying $\bar{\bar{f}}\circ e_X = f$. Then since e_X is an epimorphism, $\bar{\bar{f}} = \bar{f}$.

4) Let $\mathcal{D}$ be a full and isomorphism-closed subcategory of $\mathcal{C}$ containing $\mathcal{A}$ such that $\mathcal{A}$ is epireflective in $\mathcal{D}$ and $\mathcal{D}$ is epireflective in $\mathcal{C}$. Since $\mathcal{D}$ is closed under formation of subobjects in $\mathcal{C}$ (the implication "(1) $\Rightarrow$ (2)" of the characterization theorem 2.2.4 remains true without any assumptions for $\mathcal{C}$ because they are not needed for the proof) and contains $\mathcal{A}$, it follows that $\mathcal{B} \subset \mathcal{D}$.

2.4.3 Remarks. (1) By the theorem of Kennison and Baron one may conclude that epireflections are more important than reflections.

(2) If the category of uniform spaces (and uniformly continuous maps) is denoted by Unif , then for each $X \in |\text{Unif}|$, there is a complete separated uniform space $\hat{X}$, namely the Hausdorff completion of X , and a canonical map $r_X\colon X \to \hat{X}$. Especially $r_X\colon X \to \hat{X}$ is the reflection of X with respect to the category C Sep (i.e. the full and isomorphism-closed subcategory of complete separated uniform spaces). By the theorem of Kennison and Baron the reflector $R\colon$ Unif $\to$ C Sep can be considered as a composition of two epireflectors (Unif is a topological category and so it satisfies the conditions of 2.4.2). Since for any separated uniform space X , the map $r_X\colon X \to \hat{X}$ is an embedding, i.e. an extremal monomorphism in Unif , and every (uniform) subspace of a separated uniform space is again separated, *the*

category U Sep *of separated uniform spaces (and uniformly continuous maps) is the smallest intermediate category which is epireflective in* Unif *and which has* C Sep *as an epireflective subcategory* (C Sep ⊂ U Sep ⊂ Unif) .

(3) The characterization theorem of epireflective subcategories yields an alternative construction of the Hausdorff completion $\hat{X}$ of a uniform space X . For this purpose we have to show (by applying this theorem) that C Sep is epireflective in U Sep (alternative construction of the complete hull!) and U Sep is epireflective in Unif . First we have to verify the assumptions in order to apply the theorem. It must be shown that Unif and U Sep are co-well-powered (epi, extremal mono)-factorizable categories that have products. Since Unif is a topological category, these conditions are satisfied. To show the assumptions for U Sep we first determine the epimorphisms. It holds the following

Proposition. f ∈ Mor U Sep is an epimorphism in U Sep if and only if f is dense (as a morphism in Top) .

Proof. 1) "⇐" If $f \in [(X,\mathcal{W}),(Y,\mathcal{R})]_{\text{U Sep}}$ is dense and α,β: $(Y,\mathcal{R}) \to (Z,\mathcal{S})$ are uniformly continuous maps into a separated uniform space $(Z,\mathcal{S})$ satisfying $\alpha\circ f = \beta\circ f$, i.e. α and β coincide on the dense subset $f[X]$ of Y , then $\alpha = \beta$. Consequently, f is an epimorphism in U Sep .

2) "⇒" (indirectly): If $f \in [(X,\mathcal{W}),(Y,\mathcal{R})]_{\text{U Sep}}$ is not dense, then there are $y_o \in Y$ and $V = V^{-1} \in \mathcal{R}$ such that $V(y_o) \cap f[X] = \emptyset$. For V , there is a pseudometric d_V on Y . Put $A = f[X]$; then a uniformly continuous map g_V: $(Y,d_V) \to ([0,1],d)$[9] is defined by $g_V(y) = d_V(y,A) = \inf\{d_V(y,z)\colon z \in A\}$ for each $y \in Y$ (note that $|d_V(y',A) - d_V(y'',A)| \leq d_V(y',y'')$ for each $(y',y'') \in Y\times Y)$. g_V has the property $g_V[A] = \{0\}$. Since $\mathcal{D}_V \subset \mathcal{R}$ for the uniformity $\mathcal{D}_V$ generated by d_V , the identical map

[9] d is the natural metric, i.e. the metric induced by the Euclidean metric.

$i\colon (Y,\mathcal{R}) \to (Y,\mathcal{D}_V)$ is uniformly continuous. Thus $\alpha = g_V \circ i$ is also uniformly continuous. $\beta\colon Y \to [0,1]$ defined by $\beta(y) = 0$ for each $y \in Y$ is a uniformly continuous map differing from α: since $\frac{1}{2} \leq d_V(y_o,z) \leq 1$ for each $z \in A$ (note that $\frac{1}{2} h_V(y_o,z) \leq d_V(y_o,z) \leq h_V(y_o,z)$ and since $A \cap V(y_o) = \emptyset$, $(y_o,z) \in CV$, i.e. $z \notin V(y_o)$, so that $h_V(y_o,z) = 1$), $\alpha(y_o) = g_V(y_o) = \inf\{d_V(y_o,z)\colon z \in A\} \neq 0$, i.e. $\alpha(y_o) \neq \beta(y_o) = 0$. Since the uniformity induced by the Euclidean metric on $[0,1]$ is separated, α and β are U Sep-morphisms such that, since $\alpha[A] = \beta[A] = \{0\}$, $\alpha \circ f = \beta \circ f$. Consequently f is not an epimorphism in U Sep.

The monomorphisms in U Sep are the injective uniformly continuous maps (this is proved analogously to the corresponding fact in topological categories). $f \in [X,Y]_{\underline{\text{U Sep}}}$ is an extremal monomorphism if and only if $f[X] = \overline{f[X]}$ and $f'\colon X \to f[X]$ defined by $f'(x) = f(x)$ for each $x \in X$ is an isomorphism (a) "⇒" Since f is an extremal monomorphism and there is the factorization $f = h \circ g$ where $g\colon X \to \overline{f[X]}$ is defined by $g(x) = f(x)$ for each $x \in X$ and $h\colon \overline{f[X]} \to Y$ is the inclusion map, the epimorphism g is an isomorphism and consequently, $f[X] = g[X] = \overline{f[X]}$.

b) "⇐" If $i\colon f[X] \to Y$ is the inclusion map, then $f = i \circ f'$ is a monomorphism as a composite of two monomorphisms. Let $f = h \circ g$ be a factorization where $g\colon X \to Z$ is an epimorphism. Since $h[Z] = h[\overline{g[X]}] \subset \overline{(h \circ g)[X]} = \overline{f[X]} = f[X]$ [continuity of h!], a morphism $h'\colon Z \to f[X]$ is defined by $i \circ h' = h$. Obviously $f' = h' \circ g$. Thus since f' is an extremal monomorphism, g has to be an isomorphism. Consequently, f is an extremal monomorphism.). Let $f \in [X,Y]_{\underline{\text{U Sep}}}$ and let $g\colon X \to \overline{f[X]}$ be defined by $g(x) = f(x)$ for each $x \in X$ and let the inclusion map be denoted by $h\colon \overline{f[X]} \to Y$. Then $f = h \circ g$ is the desired (epi, extremal mono)-factorization of f. The fact that U Sep is co-well-powered is proved analogously to 2.2.5 (3) (1) (b).

U Sep has also products (because the Hausdorff property is productive!).

Therefore the assumptions of the characterization theorem of epireflective subcategories are satisfied for $\mathcal{C}$ = Unif and $\mathcal{C}$ = U Sep . Since closed subspaces of complete spaces are complete and products of complete spaces are complete as well as subspaces and products of Hausdorff spaces are Hausdorff spaces, C Sep is closed under formation of subobjects and products in U Sep (the subobjects in U Sep are just the closed subspaces), i.e. epireflective in U Sep . On the other hand U Sep is closed under formation of subobjects and products in Unif , i.e. epireflective in Unif . If the epiflection of $X \in |\text{Unif}|$ with respect to U Sep is denoted by $s_X: X \to X'$ and the epireflection of $X' \in |\text{U Sep}|$ with respect to C Sep is denoted by $r'_{X'}: X' \to \hat{X}$ (i.e. $\hat{X}$ is the complete hull of X'), then $r_X = r'_{X'} \circ s_X: X \to \hat{X}$ is the desired reflection of X with respect to C Sep (i.e. $\hat{X}$ is just the Hausdorff completion of X).

Thus simultaneously a categorical method has been found for constructing the reals from the rationals.

CHAPTER III

RELATIONS BETWEEN SPECIAL TOPOLOGICAL CATEGORIES

Besides the categorical approach of chapter I there is another approach in order to handle problems of a "topological" nature namely the conceptual one. The aim of this approach is to find a basic topological concept by means of which any topological concept or idea can be expressed. A fundamental requirement is the following: By means of such a concept one should be able to explain "nearness". Axiomatizing the concept of nearness between a point x and a set A (usually denoted by $x \in \bar{A}$, i.e. x belongs to the closure of A) one can obtain topological spaces. But there are other types of spaces: Proximity spaces for instance are obtained by an axiomatization of the concept of nearness between two sets, and by means of contiguity spaces one can even explain axiomatically nearness between a finite collection of sets. Thus, H. Herrlich filled a gap by defining nearness spaces as an axiomatization of the concept of nearness between arbitrary collections of sets. Though there is a difference of a "topological" nature between removing a point from the usual topological space $\mathbb{R}$ of real numbers and removing a closed interval of length one respectively the obtained topological spaces are homeomorphic. But if we do the same with respect to the usual uniformities (resp. proximities) we obtain non-isomorphic uniform spaces (resp. proximity spaces). The reason why uniform (proximity) spaces behave "well" and topological spaces behave "badly" with respect to the formation of subobjects becomes clear in the realm of nearness structures: A subspace of a uniform (proximal) nearness space is uniform (proximal), but a subspace of a topological nearness space is not topological.

In this chapter we use the definition of nearness spaces by means of uniform covers instead of near collections because this definition is easier to handle. Thus the open covers of a topological space form a base for the set of uniform covers of the corresponding (topological) nearness space. It is shown that the category Near of nearness spaces and uniformly continuous maps contains the category Top of topological spaces and continuous maps (provided a certain symmetry condition is fulfilled) as well as the categories Unif of uniform spaces and uniformly continuous maps, Prox of proximity spaces and δ-maps and Cont of contiguity spaces and contiguity maps as nicely embedded subcategories. Thus in the realm of nearness spaces it is possible to consider those spaces which are simultaneously topological and uniform. It turns out that these spaces are precisely the fully normal topological spaces which differ from paracompact spaces only by the Hausdorff axiom. If paracompactness is considered as a nearness concept instead of a topological one it has a better structural behaviour (e.g. products of paracompact spaces are paracompact and each subspace of a paracompact space is paracompact).

There may be defined supercategories of Near by omitting some of the nearness axioms namely the categories S-Near of seminearness spaces (and uniformly continuous maps) and P-Near of prenearness spaces (and uniformly continuous maps) respectively. The category S-Near can also be obtained by omitting the star refinement axiom in Tukey's definition of a uniform space (in this chapter is also proved the equivalence between Weil's and Tukey's definition). A seminearness space is nothing else but a merotopic space in the sense of Katětov who axiomatized the concept of (what we now call) Cauchy systems, i.e. collections of sets containing arbitrary small members. It turns out later on (cf. the following chapter) that the category S-Near is of fundamental interest since it contains not only Near and its subcategories but also categories which are defined by means of "convergence" structures. The latter ones occur as subcategories of the category Grill of grill-determined prenearness spaces (introduced in the last part of this chapter) which is a subcategory of S-Near and which behaves very well with respect to function space structures considered in the next chapter. Grill contains addi-

tionally the categories Cont and Prox.
The relations between all these categories mentioned above may be described by means of bireflections and bicoreflections respectively.

3.1 The category Near and its subcategories

3.1.1 Topological spaces

3.1.1.1 Definitions. 1) A topological space $(X,\mathcal{X})$ is called an R_o-space provided that $x \in \overline{\{y\}}$ implies $y \in \overline{\{x\}}$ for each pair $(x,y) \in X\times X$.

2) A nearness space (X,μ) is called topological iff the following is satisfied:

(T) $X = \bigcup\{\mathrm{int}_\mu A\colon A \in \mathcal{A}\}$ implies $\mathcal{A} \in \mu$.

3.1.1.2 Remarks. ① Every T_1-space is an R_o-space (trivial!); the converse is not true (counterexample: $(\{0,1\},\{\emptyset,\{0,1\}\})$) .

② By N_3) the converse of (T) is always true for any nearness space (X,μ) , i.e. (X,μ) is topological if and only if $\mathcal{A} \in \mu$ is equivalent to $X = \bigcup\{\mathrm{int}_\mu A\colon A \in \mathcal{A}\}$.

③ In category theory we do not distinguish between two categories $\mathcal{A}$ and $\mathcal{B}$ if they are *isomorphic*, i.e. if there are functors $F\colon \mathcal{A} \to \mathcal{B}$ and $G\colon \mathcal{B} \to \mathcal{A}$ such that

$$G\circ F = I_{\mathcal{A}} \quad \text{and} \quad F\circ G = I_{\mathcal{B}} .$$

In this sense the following theorem has to be understood.

3.1.1.3 Theorem. The full subcategory T-Near of Near whose objects are the topological nearness spaces is

(1) bicoreflective in Near

and (2) isomorphic to the category R_o-Top of topological R_o-spaces and continuous maps.

<u>Proof</u>. (2) (a) *Let* $(X,\mathcal{X})$ *be an* R_0*-space and* $\mu_X = \{\mathcal{A} \subset P(X): X = \bigcup_{A \in \mathcal{A}} A^o\}$. *Then* (X, μ_X) *is a topological nearness space such that* $\text{int}_{\mu_X} A = A^o$ *for each* $A \in P(X)$.

Since $\{X\} \in \mu_X$, μ_X is a non-empty collection of non-empty covers.

N_1) Let $\mathcal{A} < \mathcal{B}$ and $\mathcal{A} \in \mu_X$. Hence there exists $B_A \in \mathcal{B}$ with $A \subset B_A$ for each $A \in \mathcal{A}$. Thus $X = \bigcup_{A \in \mathcal{A}} A^o = \bigcup_{A \in \mathcal{A}} B_A^o = \bigcup_{B \in \mathcal{B}} B^o$, i.e. $\mathcal{B} \in \mu_X$.

N_2) Let $\mathcal{A} \in \mu_X$ and $\mathcal{B} \in \mu_X$. Then also $\mathcal{A} \wedge \mathcal{B} = \{A \cap B: A \in \mathcal{A} \text{ and } B \in \mathcal{B}\} \in \mu_X$: If $x \in X$, then there exist $A \in \mathcal{A}$ and $B \in \mathcal{B}$ such that $x \in A^o$ and $x \in B^o$. Hence $x \in A^o \cap B^o = (A \cap B)^o \subset \bigcup_{\substack{A \in \mathcal{A} \\ B \in \mathcal{B}}} (A \cap B)^o$. Consequently,

$$X = \bigcup_{\substack{A \in \mathcal{A} \\ B \in \mathcal{B}}} (A \cap B)^o .$$

N_3) For the proof of N_3) it suffices to show:

(*) $\text{int}_{\mu_X} A = A^o$ for each $A \subset X$

(then $\mathcal{A} \in \mu_X$, i.e. $\bigcup_{A \in \mathcal{A}} A^o = X$, implies $\{\text{int}_{\mu_X} A: A \in \mathcal{A}\} \in \mu_X$, since $\bigcup_{A \in \mathcal{A}} (\text{int}_{\mu_X} A)^o = \bigcup_{A \in \mathcal{A}} A^o = X$).

α) $x \in \text{int}_{\mu_X} A = \{z \in X: \{A, X\setminus\{z\}\} \in \mu_X\}$, i.e. $X = A^o \cup (X\setminus\{x\})^o$ implies $x \in A^o$.

β) $x \in A^o$ implies $x \in \text{int}_{\mu_X} A$, i.e. $\{A, X\setminus\{x\}\} \in \mu_X$ because it can be shown that $X = A^o \cup (X\setminus\{x\})^o$: $z \in X$ implies either

1. $z \in \overline{\{x\}}$ and thus $x \in \overline{\{z\}}$ ($(X,\mathcal{X})$ is an R_0-space!) and consequently $z \in A^o \in \overset{o}{\mathcal{U}}(x)$

or

2. $z \notin \overline{\{x\}}$ and thus $z \in X\setminus\overline{\{x\}} = (X\setminus\{x\})^o$,

i.e. in each case $z \in A^o \cup (X\setminus\{x\})^o$.

(T) Applying (*), (T) is satisfied by the definition of μ_X.

(b) *If* $(X,\mu) \in |\underline{\text{Near}}|$, *then an* R_0*-topology on* X *is defined by* $\mathcal{X}_\mu = \{A \subset X: \text{int}_\mu A = A\}$ *such that* $A^o = \text{int}_\mu A$ *for each* $A \subset X$.

In order to prove (b) it suffices to show that $\text{int}_\mu: P(X) \to P(X)$ defines an interior operator:

$K_1)$ $\text{int}_\mu X = X$: Let $\mathcal{U} \in \mu$. Then $\mathcal{U} < \{X\} < \{X, X \setminus \{x\}\}$ for each $x \in X$. Thus by $N_1)$, $\{X, X\setminus\{x\}\} \in \mu$, i.e. $x \in \text{int}_\mu X$. Consequently, $X \subset \text{int}_\mu X$. The converse holds by definition.

$K_2)$ $\text{int}_\mu A \subset A$ for each $A \in P(X)$: If $x \in \text{int}_\mu A = \{x \in X: \{A, X\setminus\{x\}\} \in \mu\}$, then since $X = A \cup (X\setminus\{x\})$, x belongs to A.

$K_3)$ $\text{int}_\mu A \subset \text{int}_\mu(\text{int}_\mu A)$ for each $A \in P(X)$: Let $x \in \text{int}_\mu A$, i.e. $\{A, X\setminus\{x\}\} \in \mu$. Then by $N_3)$, $\{\text{int}_\mu A, \text{int}_\mu(X\setminus\{x\})\} \in \mu$ and since $\{\text{int}_\mu A, \text{int}_\mu(X\setminus\{x\})\} < \{\text{int}_\mu A, X\setminus\{x\}\}$ (note $K_2)$), $\{\text{int}_\mu A, X\setminus\{x\}\} \in \mu$ by $N_1)$; i.e. $x \in \text{int}_\mu(\text{int}_\mu A)$.

$K_4)$ $\text{int}_\mu(A \cap B) = \text{int}_\mu A \cap \text{int}_\mu B$ for each $(A,B) \in P(X)\times P(X)$: Since obviously for each $(C,D) \in P(X)\times P(X)$

$$C \subset D \text{ implies } \text{int}_\mu C \subset \text{int}_\mu D$$

it follows immediately from $A \cap B \subset A$ and $A \cap B \subset B$ that $\text{int}_\mu(A \cap B) \subset \text{int}_\mu A \cap \text{int}_\mu B$. Conversely, let $x \in \text{int}_\mu A \cap \text{int}_\mu B$. Then $\{A, X\setminus\{x\}\} \in \mu$ and $\{B, X\setminus\{x\}\} \in \mu$. Thus by $N_2)$, $\{A \cap B, A \cap (X\setminus\{x\}), B \cap (X\setminus\{x\}), X\setminus\{x\}\} \in \mu$ and consequently by $N_1)$, $\{A \cap B, X\setminus\{x\}\} \in \mu$, i.e. $x \in \text{int}_\mu(A \cap B)$.

It is shown by $K_1)$-$K_4)$ that $\mathcal{X}_\mu$ is a topology on X such that $A^o = \text{int}_\mu A$ for each $A \subset X$. Moreover, $\mathcal{X}_\mu$ is an R_o-topology: If $x,y \in X$ such that $x \in \overline{\{y\}}$, then $x \notin (X\setminus\{y\})^o = \text{int}_\mu(X\setminus\{y\})$, i.e. $\{X\setminus\{y\}, X\setminus\{x\}\} = \{X\setminus\{x\}, X\setminus\{y\}\} \notin \mu$, thus $y \notin \text{int}_\mu(X\setminus\{x\}) = (X\setminus\{x\})^o$. Consequently, $y \in \overline{\{x\}}$.

(c) *If* $(X,\mathcal{X})$ *and* $(X',\mathcal{X}')$ *are* R_o*-spaces and* $f: (X,\mathcal{X}) \to (X',\mathcal{X}')$ *is a continuous map, then* $f: (X,\mu_{\mathcal{X}}) \to (X',\mu_{\mathcal{X}'})$ *is a uniformly continuous map.*

Proof. If $\mathcal{A} \in \mu_{\mathcal{X}'}$ then $\mathcal{B} = \{\text{int}_{\mu_{\mathcal{X}'}} A = A^o: A \in \mathcal{A}\} \in \mu_{\mathcal{X}'}$ and $\mathcal{B} < \mathcal{A}$. Since f is continuous, $f^{-1}\mathcal{B} = \{f^{-1}[A^o]: A \in \mathcal{A}\}$ is an open cover of X. Thus $f^{-1}\mathcal{B}$ belongs to $\mu_{\mathcal{X}}$. Since $f^{-1}\mathcal{B} < f^{-1}\mathcal{A}$, $f^{-1}\mathcal{A}$ also belongs to $\mu_{\mathcal{X}}$.

(d) *If* (X,μ) *and* (X',μ') *are nearness spaces and* $f: (X,\mu) \to (X',\mu')$ *is a uniformly continuous map, then* $f: (X,\mathcal{X}_\mu) \to (X',\mathcal{X}_{\mu'})$ *is a continuous map.*

Proof. Let $A \subset X'$ and $x \in f^{-1}[A^{o}]$, i.e. $f(x) \in A^{o} = \operatorname{int}_{\mu'} A$. Thus $\{A, X' \setminus \{f(x)\}\} \in \mu'$. Since f is uniformly continuous, $\{f^{-1}[A], f^{-1}[X' \setminus \{f(x)\}]\} \in \mu$. Then by N_1), $\{f^{-1}[A], X \setminus \{x\}\} \in \mu$, i.e. $x \in \operatorname{int}_{\mu} f^{-1}[A] = (f^{-1}[A])^{o}$. Thus $f^{-1}[A^{o}] \subset (f^{-1}[A])^{o}$. Consequently, f is continuous.

(e) Put $\mathcal{F}((X,\mathcal{X})) = (X,\mu_{\mathcal{X}})$ for each R_0-topological space $(X,\mathcal{X})$, and for each continuous map f between R_0-spaces, let $\mathcal{F}(f)$ be the corresponding uniformly continuous map (cf. c)). Then a functor $\mathcal{F}$: <u>R_0-Top</u> $\to$ <u>T-Near</u> is defined. Put $\mathcal{G}((X,\mu)) = (X,\mathcal{X}_{\mu})$ for each topological nearness space (X,μ), and for each uniformly continuous map f between topological nearness spaces, let $\mathcal{G}(f)$ be the corresponding continuous map (cf. d)). Then a functor $\mathcal{G}$: <u>T-Near</u> $\to$ <u>R_0-Top</u> is defined. Especially $\mathcal{G} \circ \mathcal{F} = \mathcal{I}_{\underline{R_0\text{-Top}}}$ (note: $\mathcal{X}_{\mu_{\mathcal{X}}} = \mathcal{X}$ for each R_0-topology $\mathcal{X}$) and $\mathcal{F} \circ \mathcal{G} = \mathcal{I}_{\underline{\text{T-Near}}}$ (note: $\mu_{\mathcal{X}_{\mu}} = \mu$ for each <u>T-Near</u>-structure μ).

(1) Let $(X,\mu) \in |\underline{\text{Near}}|$. Then $(X,\mathcal{X}_{\mu})$ is an R_0-space (cf. (2)(b)) and $(X,\mu_{\mathcal{X}_{\mu}})$ is a topological nearness space such that $\operatorname{int}_{\mu_{\mathcal{X}_{\mu}}} A = \operatorname{int}_{\mu} A$ (cf. (2)(a)). Put $\mu_{\mathcal{X}_{\mu}} = \mu_t$. Then

$$\mu_t = \{\mathcal{A} \subset \mathcal{P}(X) : \bigcup\{\operatorname{int}_{\mu} A : A \in \mathcal{A}\} = X\} \supset \mu$$

and thus $1_X: (X,\mu_t) \to (X,\mu)$ is a uniformly continuous map. If $(X',\mu') \in |\underline{\text{T-Near}}|$ and $f: (X',\mu') \to (X,\mu)$ is a uniformly continuous map then by (2)(d), $f: (X',\mathcal{X}_{\mu'}) \to (X,\mathcal{X}_{\mu})$ is a continuous map and since $\mathcal{X}_{\mu} = \mathcal{X}_{\mu_t}$ (note $\operatorname{int}_{\mu} A = \operatorname{int}_{\mu_t} A$!), $f: (X',\mathcal{X}_{\mu'}) \to (X,\mathcal{X}_{\mu_t})$ is also continuous. By (2)(c) this means that $f: (X',\mu') \to (X,\mu_t)$ is a uniformly continuous map (note that μ' and μ_t are <u>T-Near</u>-structures!). Thus there is a unique uniformly continuous map $g: (X',\mu') \to (X,\mu_t)$ such that the diagram

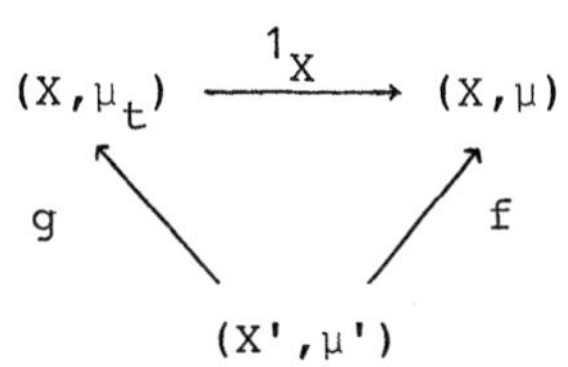

commutes, namely $g = f$. Consequently, $1_X: (X,\mu_t) \to (X,\mu)$ is the desired bicoreflection of $(X,\mu) \in |\underline{\text{Near}}|$ with respect to T-Near .

3.1.1.4 Remarks. (1) Since T-Near and R_0-Top are isomorphic, one can identify a topological nearness space (X,μ) with the corresponding R_0-space (X,X_μ) (the topology X_μ is generated by $\text{int}_\mu: P(X) \to P(X)$) . If P is a topological invariant, then we say that a topological nearness space (X,μ) has the property P if and only if (X,X_μ) has this property.

(2) If the bicoreflector is denoted by T: Near → T-Near , then $T((X,\mu)) = (X,\mu_t)$ (resp. (X,X_{μ_t})) is called the *underlying topological space of the nearness space* (X,μ) .

(3) Since Near is a topological category, the characterization theorem of monocoreflective subcategories (2.2.4 A')) is applicable and thus T-Near is closed under the formation of coproducts and quotient objects in Near . Moreover T-Near contains all discrete objects of Near (cf. the last sentence of 2.2.11 (1)) [obviously a Near-object (X,μ) is discrete if and only if each non-empty cover of X is a uniform cover] . A Near-object (X,μ) is indiscrete if and only if $\mu' = \{\{X\}\}$ is a base for μ . Obviously every indiscrete Near-object is topological. But T-Near is not epireflective in Near (and thus by 2.2.11 (2) also not reflective in Near) because the following example shows that T-Near is not closed under formation of products in Near:

$\mathcal{B} = \{[a,b): a,b \in \mathbb{R}\}$ is a base of a topology $\mathcal{R}$ on $\mathbb{R}$. $(\mathbb{R},\mathcal{R})$ is a Hausdorff space[10)] and thus

$\mu_{\mathcal{R}} = \{\mathcal{A} \subset P(\mathbb{R}) : \bigcup_{A\in\mathcal{A}} A^o = \mathbb{R}\}$ is a T-Near-structure on $\mathbb{R}$.

Let $(\mathbb{R}\times\mathbb{R},\mu_{\mathcal{R}}\times\mu_{\mathcal{R}})$ be the product of $(\mathbb{R},\mu_{\mathcal{R}})$ with itself in Near, i.e. $\{\mathcal{A}_1\times\mathcal{A}_2: \mathcal{A}_1,\mathcal{A}_2 \in \mu_{\mathcal{R}}\}$ is a base for $\mu_{\mathcal{R}}\times\mu_{\mathcal{R}}$ $(\mathcal{A}_1\times\mathcal{A}_2 := \{A_1\times A_2: A_1 \in \mathcal{A}_1, A_2 \in \mathcal{A}_2\})$. If $(\mathbb{R}^2, \mu_{\mathcal{R}}\times\mu_{\mathcal{R}}) \in |\underline{\text{T-Near}}|$, then the bicoreflection $1_{\mathbb{R}^2}: (\mathbb{R}^2,(\mu_{\mathcal{R}}\times\mu_{\mathcal{R}})_t) \to (\mathbb{R}^2,\mu_{\mathcal{R}}\times\mu_{\mathcal{R}})$ must be an isomorphism, i.e. $(\mu_{\mathcal{R}}\times\mu_{\mathcal{R}})_t = \mu_{\mathcal{R}}\times\mu_{\mathcal{R}}$. If one chooses

10) This space is usually called "Sorgenfrey line" .

$A = \{(x,-x): x \in \mathbb{Q}\}$ and $B = \{(x,-x): x \in \mathbb{R} \setminus \mathbb{Q}\}$, then it can be shown that

$$\{ \mathbb{R}^2 \setminus A, \mathbb{R}^2 \setminus B \} \notin \mu_R \times \mu_R$$

but

$$\{ \mathbb{R}^2 \setminus A, \mathbb{R}^2 \setminus B \} \in (\mu_R \times \mu_R)_t$$

(exercise!) [cf. 3.1.2.8 (2)] .

(4) It is an unsolved problem to find a "nice" characterization of the epireflective hull $R_{\underline{\text{Near}}}$ T-Near of T-Near in Near .

3.1.2 Uniform spaces

3.1.2.1 Definitions. 1) Let X be a set, $\mathcal{U}$ a cover of X and $A \subset X$. Then

$$St(A,\mathcal{U}) = \bigcup\{U \in \mathcal{U} : A \cap U \neq \emptyset\}$$

is called the star of A with respect to $\mathcal{U}$.

2) If $\mathcal{U}$ and $\mathcal{V}$ are covers of the set X , then $\mathcal{U}$ is called a star-refinement of $\mathcal{V}$ (denoted by $\mathcal{U} *< \mathcal{V}$) provided that for each $U \in \mathcal{U}$ there is some $V \in \mathcal{V}$ such that $St(U,\mathcal{U}) \subset V$.

3) A nearness space (X,μ) is called uniform provided that the following is satisfied:

(U) For each $\mathcal{A} \in \mu$ there exists some $\mathcal{B} \in \mu$ such that $\mathcal{B} *< \mathcal{A}$.

3.1.2.2 Theorem. The full subcategory U-Near of Near whose objects are the uniform nearness spaces is bireflective in Near . If (X,μ) is a nearness space and μ_u is the set of all $\mathcal{A} \in \mu$ for which there exists a sequence $(\mathcal{A}_n)_{n \in \mathbb{N}}$ in μ such that $\mathcal{A}_0 = \mathcal{A}$ and $\mathcal{A}_{n+1} *< \mathcal{A}_n$ for each $n \in \mathbb{N}$, then $1_X: (X,\mu) \to (X,\mu_u)$ is the bireflection of (X,μ) with respect to U-Near .

Proof. Since $\{X\} \in \mu$, $\{X\} *\!\!< \{X\}$ and thus $\{X\} \in \mu_u$, i.e. μ_u is a non-empty set of non-empty covers.

N_1) Let $\mathcal{A} \in \mu_u$ and $\mathcal{A} < \mathcal{B}$. Put $\mathcal{B}_o = \mathcal{B}$ and $\mathcal{B}_n = \mathcal{A}_n$ for $n > 0$. Then a sequence $(\mathcal{B}_n)_{n\in\mathbb{N}}$ in μ with the desired property has been found, i.e. $\mathcal{B} \in \mu_u$.

N_2) Let $\mathcal{A},\mathcal{B} \in \mu_u$. Then there exist sequences $(\mathcal{A}_n)_{n\in\mathbb{N}}$ and $(\mathcal{B}_n)_{n\in\mathbb{N}}$ in μ such that $\mathcal{A}_o = \mathcal{A}$, $\mathcal{B}_o = \mathcal{B}$, $\mathcal{A}_{n+1} *\!\!< \mathcal{A}_n$ and $\mathcal{B}_{n+1} *\!\!< \mathcal{B}_n$ for each $n \in \mathbb{N}$. Thus $(\mathcal{A}_n \wedge \mathcal{B}_n)_{n\in\mathbb{N}}$ is a sequence in μ such that $\mathcal{A}_o \wedge \mathcal{B}_o = \mathcal{A} \wedge \mathcal{B}$ and $\mathcal{A}_{n+1} \wedge \mathcal{B}_{n+1} *\!\!< \mathcal{A}_n \wedge \mathcal{B}_n$ for each $n \in \mathbb{N}$ (If $A_{n+1} \cap B_{n+1}$ belongs to $\mathcal{A}_{n+1} \wedge \mathcal{B}_{n+1}$, then there exist $A_n \in \mathcal{A}_n$ and $B_n \in \mathcal{B}_n$ such that $St(A_{n+1},\mathcal{A}_{n+1}) \subset A_n$ and $St(B_{n+1},\mathcal{B}_{n+1}) \subset B_n$. Thus, $St(A_{n+1} \cap B_{n+1},\mathcal{A}_{n+1} \wedge \mathcal{B}_{n+1}) \subset A_n \cap B_n$. [If $x \in St(A_{n+1} \cap B_{n+1},\mathcal{A}_{n+1} \wedge \mathcal{B}_{n+1})$, then there exist $A'_{n+1} \in \mathcal{A}_{n+1}$ and $B'_{n+1} \in \mathcal{B}_{n+1}$ such that $x \in A'_{n+1} \cap B'_{n+1}$ and $A'_{n+1} \cap$ $\cap B'_{n+1} \cap A_{n+1} \cap B_{n+1} \neq \emptyset$. Thus $A'_{n+1} \cap A_{n+1} \neq \emptyset$, i.e. $A'_{n+1} \subset St(A_{n+1},\mathcal{A}_{n+1}) \subset A_n$, and $B'_{n+1} \cap B_{n+1} \neq \emptyset$, i.e. $B'_{n+1} \subset St(B_{n+1},\mathcal{B}_{n+1}) \subset B_n$. Hence $x \in A_n \cap B_n$.]). Consequently, $\mathcal{A} \wedge \mathcal{B} \in \mu_u$.

N_3) Let $\mathcal{A} \in \mu_u$. By the definition of μ_u there exists some $\mathcal{B} \in \mu_u$ with $\mathcal{B} *\!\!< \mathcal{A}$. Then $\mathcal{B} < \{\text{int}_{\mu_u} A: A \in \mathcal{A}\}$; for if $B \in \mathcal{B}$, then there exists $A \in \mathcal{A}$ such that $St(B,\mathcal{B}) \subset A$ and thus $B \subset \text{int}_{\mu_u} A$ (namely $x \in B$ implies $St(\{x\},\mathcal{B}) \subset$ $\subset St(B,\mathcal{B}) \subset A$. Hence $\mathcal{B} < \{A,X \setminus \{x\}\}$[11], so that $\{A,X \setminus \{x\}\} \in \mu_u$, i.e. $x \in \text{int}_{\mu_u} A$). Consequently by N_1), $\{\text{int}_{\mu_u} A: A \in \mathcal{A}\} \in \mu_u$.

By the construction of μ_u, $\mu_u \subset \mu$. Hence $1_X: (X,\mu) \to (X,\mu_u)$ is a uniformly continuous map. (X,μ_u) is a nearness space (by N_1)-N_3)) which is obviously uniform. If (Y,ν) is a uniform nearness space and $f: (X,\mu) \to (Y,\nu)$ is a uniformly continuous map, then $f: (X,\mu_u) \to (Y,\nu)$ is a uniformly continuous map;

---

[11] If $B' \in \mathcal{B}$, then either $x \in B'$ and thus $\{x\} \cap B' \neq \emptyset$, i.e. $B' \subset St(\{x\},\mathcal{B}) \subset A$ or $x \notin B'$, i.e. $B' \subset X \setminus \{x\}$.

for if $A \in \nu$, then since ν is a U-Near-structure, there exists a sequence $(A_n)_{n\in\mathbb{N}}$ in ν such that $A_o = A$ and $A_{n+1} *\!< A_n$ for each $n \in \mathbb{N}$ (this sequence is constructed by applying successively the condition (U)). Hence $(f^{-1}A_n)_{n\in\mathbb{N}}$ is a sequence in μ ($f: (X,\mu) \to (Y,\nu)$ is a uniformly continuous map!) such that $f^{-1}A_o = f^{-1}A$ and

$$f^{-1}A_{n+1} *\!< f^{-1}A_n \quad \text{for each } n \in \mathbb{N}\,.$$

i.e. $f^{-1}A \in \mu_u$ (Namely if $f^{-1}[A_{n+1}] \in f^{-1}A_{n+1}$, then there exists some $A_n \in A_n$ such that $St(A_{n+1},A_{n+1}) \subset A_n$. Thus $St(f^{-1}[A_{n+1}], f^{-1}A_{n+1}) \subset f^{-1}[A_n] \in f^{-1}A_n$ [If $x \in St(f^{-1}[A_{n+1}], f^{-1}A_{n+1})$, then there exists some $A'_{n+1} \in A_{n+1}$ such that $x \in f^{-1}[A'_{n+1}]$ and $f^{-1}[A_{n+1}] \cap f^{-1}[A'_{n+1}] = f^{-1}[A_{n+1} \cap A'_{n+1}] \neq \emptyset$. Hence $A'_{n+1} \cap A_{n+1} \neq \emptyset$, so that $A'_{n+1} \subset St(A_{n+1},A_{n+1}) \subset A_n$. Consequently, $x \in f^{-1}[A'_{n+1}] \subset f^{-1}[A_n]]$). Therefore $1_X: (X,\mu) \to (X,\mu_u)$ is the desired bireflection of (X,μ) with respect to U-Near.

3.1.2.3 Remark. If a non-empty set μ of non-empty covers of a set X only satisfies the conditions N_1) and N_2), then (X,μ) is called a seminearness space. *If a seminearness space satisfies the condition* (U), *then it is already a uniform nearness space, i.e.* N_3) *is automatically fulfilled* (this is a consequence of the proof of 3.1.2.2 N_3), if there μ_u is formally replaced by μ; by the way one can see that N_1) and (U) already imply N_3).).

3.1.2.4 Theorem. The category U-Near is isomorphic to the category Unif of uniform spaces and uniformly continuous maps.

Proof. (a) *Let* (X,μ) *be a uniform nearness space. Then* $\mathcal{B}_\mu = \{\bigcup_{A\in A} A\times A : A \in \mu\}$ *is a base for a uniformity* $\mathcal{W}_\mu$ *on* X: Since μ is a non-empty set of non-empty covers, $\mathcal{B}_\mu \subset P(X\times X)$ is non-empty.

BU_1) If $B_\mu \in \mathcal{B}_\mu$, then there exists some $\mathcal{A} \in \mu$ such that $B_\mu = \bigcup_{A\in\mathcal{A}} A\times A$. Since $\mathcal{A}$ is a cover of X, $(x,x) \in \Delta$ implies the existence of some $A \in \mathcal{A}$ with $x \in A$, so that $(x,x) \in A\times A \subset B_\mu$. Thus $\Delta \subset B_\mu$.

BU_2) Let $B_\mu \in \mathcal{B}_\mu$, i.e. $B_\mu = \bigcup_{A\in\mathcal{A}} A\times A$ for some $\mathcal{A} \in \mu$. Obviously, $\left(\bigcup_{A\in\mathcal{A}} A\times A\right)^{-1} = \bigcup_{A\in\mathcal{A}} A\times A$.

BU_3) Let $B_\mu \in \mathcal{B}_\mu$. Then there exists some $\mathcal{A} \in \mu$ such that $B_\mu = \bigcup_{A\in\mathcal{A}} A\times A$. Since (U) is satisfied, there is some $\mathcal{B} \in \mu$ such that $\mathcal{B} \ast\!< \mathcal{A}$. Then $B'_\mu = \bigcup_{B\in\mathcal{B}} B\times B$ belongs to $\mathcal{B}_\mu$ and $B'_\mu \circ B'_\mu \subset B_\mu$ (If $(x,y) \in B'_\mu \circ B'_\mu$, then there exists some $z \in X$ such that $(x,z) \in B'_\mu$ and $(z,y) \in B'_\mu$. Hence there are $B_1, B_2 \in \mathcal{B}$ such that $(x,z) \in B_1\times B_1$ and $(z,y) \in B_2\times B_2$ and thus $\{x,y\} \subset St(B_1,\mathcal{B})$. Further, there exists some $A \in \mathcal{A}$ satisfying $St(B_1,\mathcal{B}) \subset A$. Therefore $(x,y) \in A\times A \subset$ $\subset \bigcup_{A'\in\mathcal{A}} A'\times A' = B_\mu$.).

BU_4) Let $B_\mu, B'_\mu \in \mathcal{B}_\mu$. Then there exist $\mathcal{A},\mathcal{B} \in \mu$ such that $B_\mu = \bigcup_{A\in\mathcal{A}} A\times A$ and $B'_\mu = \bigcup_{B\in\mathcal{B}} B\times B$. Thus by N_2), $\mathcal{A}\wedge\mathcal{B} \in \mu$. Consequently, $B''_\mu = \bigcup_{C\in\mathcal{A}\wedge\mathcal{B}} C\times C$ belongs to $\mathcal{B}_\mu$ and $B''_\mu \subset B_\mu \cap B'_\mu$.

(b) *Let* $(X,\mathcal{W})$ *be a uniform space. Then there exists a unique* U-Near-*structure* $\mu_{\mathcal{W}}$ *on* X *such that* $\mathcal{W}_{\mu_{\mathcal{W}}} = \mathcal{W}$.

α) For every $V \in \mathcal{W}$, let $\mathcal{A}_V = \{V(x) : x \in X\}$. Then the set $\mu_{\mathcal{W}}$ of all covers $\mathcal{A}$ of X for which there exists some $V \in \mathcal{W}$ such that $\mathcal{A}_V < \mathcal{A}$ is a U-Near-structure on X: Obviously $\mu_{\mathcal{W}}$ is a non-empty set of non-empty covers which satisfies

N_1) by definition.

N_2) Let $\mathcal{A}_1, \mathcal{A}_2 \in \mu_{\mathcal{W}}$. Then there exist $V,W \in \mathcal{W}$ such that $\mathcal{A}_V < \mathcal{A}_1$ and $\mathcal{A}_W < \mathcal{A}_2$. Hence
$\mathcal{A}_{V\cap W} = \{(V\cap W)(x) : x \in X\} = \{V(x)\cap W(x) : x \in X\} < \mathcal{A}_V \wedge \mathcal{A}_W < \mathcal{A}_1 \wedge \mathcal{A}_2$
Since $V \cap W \in \mathcal{W}$, $\mathcal{A}_1 \wedge \mathcal{A}_2 \in \mu_{\mathcal{W}}$.

(U) Let $\mathcal{A} \in \mu_{\mathcal{W}}$. Then there exists some $V \in \mathcal{W}$ such that $\mathcal{A}_V < \mathcal{A}$. For each V, there is a symmetric $V' \in \mathcal{W}$ such that $V'^2 \subset V$.

Then for each $x \in X$, $St(\{x\},A_{V'}) \subset V(x)$ $\big($If $y \in St(\{x\},A_{V'}) = = \bigcup_{\substack{z\in X \\ x\in V'(z)}} V'(z)$, then there exists some $z \in X$ with $x,y \in V'(z)$, i.e. $(z,x) \in V'$ and $(z,y) \in V'$. Thus by the symmetry of V', $(x,y) \in V'^2 \subset V$, i.e. $y \in V(x)$.$\big)$. Hence $A_{V'}$ is a barycentric refinement[12] of A_V and therefore of A. Thereby everything has already been shown because it can be easily checked that the following holds for a set X and covers $\mathcal{U}$, $\mathcal{V}$, $\mathcal{W}$ of X:

$$\mathcal{U} \Delta \mathcal{V} \text{ and } \mathcal{V} \Delta \mathcal{W} \text{ imply } \mathcal{U} \mathrel{*<} \mathcal{W}.$$

Consequently by 3.1.2.3, $\mu_{\mathcal{W}}$ is a U-Near-structure on X.

β) $\mathcal{W}_{\mu_{\mathcal{W}}} = \mathcal{W}$:

(1) If $W \in \mathcal{W}_{\mu_{\mathcal{W}}}$, then there exists $A \in \mu_{\mathcal{W}}$ and $V \in \mathcal{W}$ such that $\{V(x): x \in X\} = A_V < A$ and $\bigcup_{A\in A} A\times A \subset W$. Thus $V \subset W$ $\big((x,y) \in V$ implies $y \in V(x) \subset A$ for a suitable $A \in A$, i.e. $(x,y) \in A\times A \subset W\big)$ and consequently $W \in \mathcal{W}$.

(2) If $W \in \mathcal{W}$, then there exists a symmetric $V \in \mathcal{W}$ such that $V^2 \subset W$. Thus $\bigcup_{A\in A_V} A\times A \subset W$ $\big($If $(x,y) \in \bigcup_{A\in A_V} A\times A$, then there is some $z \in X$ with $(x,y) \in V(z)\times V(z)$. Hence by the symmetry of V, $(x,y) \in V^2 \subset W\big)$, i.e. $W \in \mathcal{W}_{\mu_{\mathcal{W}}}$.

γ) Let μ be a U-Near-structure on X such that $\mathcal{W}_\mu = \mathcal{W}$.

(1) If $A \in \mu_{\mathcal{W}}$, then there exists some $V \in \mathcal{W}$ such that $\{V(x): x \in X\} = A_V < A$. Since V also belongs to $\mathcal{W}_\mu$, there is some $\mathcal{B} \in \mu$ such that $\bigcup_{B\in\mathcal{B}} B\times B \subset V$. Thus $\mathcal{B} < A_V$ $\big($If $B \in \mathcal{B}$ and $x \in B$, then for every $y \in B$, the pair (x,y) belongs to $B\times B$ and hence to V, i.e. $B \subset V(x)\big)$ and consequently (by N_1)), $A \in \mu$.

(2) Let $A \in \mu$. Then there is some $\mathcal{B} \in \mu$ such that $\mathcal{B} \mathrel{*<} A$. Hence $V = \bigcup_{B\in\mathcal{B}} B\times B \in \mathcal{B}_\mu \subset \mathcal{W}_\mu = \mathcal{W}$ and $A_V < A$, i.e.

[12] If $\mathcal{U}$ and $\mathcal{V}$ are covers of a set X, then $\mathcal{U}$ is called a barycentric refinement of $\mathcal{V}$ (denoted by $\mathcal{U} \Delta \mathcal{V}$) provided that $\{St(\{x\},\mathcal{U}): x \in X\} < \mathcal{V}$.

$A \in \mu_{\mathcal{W}}$ (If $x \in X$, then $V(x) = \left(\bigcup_{B \in \mathcal{B}} B\times B\right)(x) = \bigcup_{B \in \mathcal{B}} (B\times B)(x) = \mathrm{St}(\{x\},\mathcal{B})$ [1. $y \in \bigcup_{B\in\mathcal{B}} (B\times B)(x)$ implies the existence of some $B \in \mathcal{B}$ with $(x,y) \in B\times B$. Thus $y \in B \subset \mathrm{St}(\{x\},\mathcal{B})$.
2. $y \in \mathrm{St}(\{x\},\mathcal{B})$ implies the existence of some $B \in \mathcal{B}$ with $x,y \in B$, hence $(x,y) \in B\times B$, i.e. $y \in (B\times B)(x) \subset V(x)$] and since $\mathcal{B}$ is a cover of X, there exists some $B \in \mathcal{B}$ with $x \in B$ and by $\mathcal{B} *\!< \mathcal{A}$, there is some $A \in \mathcal{A}$ such that $V(x) = \mathrm{St}(\{x\},\mathcal{B}) \subset \mathrm{St}(B,\mathcal{B}) \subset A$).

From (1) and (2) follows $\mu = \mu_{\mathcal{W}}$.

(c) *Let* (X,μ), (Y,ν) *be uniform nearness spaces and* $f\colon X \to Y$ *a map. Then the following are equivalent:*

(1) $f\colon (X,\mu) \to (Y,\nu)$ *is uniformly continuous.*

(2) $f\colon (X,\mathcal{W}_\mu) \to (Y,\mathcal{W}_\nu)$ *is uniformly continuous.*

Proof. "(1) ⇒ (2)": Let $B_\nu \in \mathcal{B}_\nu$, i.e. there exists some $\mathcal{A} \in \nu$ with $B_\nu = \bigcup_{A\in\mathcal{A}} A\times A$. Then by assumption, $f^{-1}\mathcal{A} \in \mu$ and hence $B_\mu = \bigcup_{A\in\mathcal{A}} (f^{-1}[A]\times f^{-1}[A]) \in \mathcal{B}_\mu$. Thus $(x,y) \in B_\mu$ implies the existence of some $A \in \mathcal{A}$ such that $(x,y) \in f^{-1}[A]\times f^{-1}[A]$, i.e. $(f(x),f(y)) \in A\times A \subset B_\nu$. Therefore (2) has been proved.

"(2) ⇒ (1)": Let $\mathcal{A} \in \nu$. Then there exists some $\mathcal{B} \in \nu$ such that $\mathcal{B} *\!< \mathcal{A}$. Since $B_\nu = \bigcup_{B\in\mathcal{B}} B\times B \in \mathcal{B}_\nu$, $(f\times f)^{-1}[B_\nu] \in \mathcal{W}_\mu$ by (2), i.e. there exists some $\mathcal{C} \in \mu$ such that $\bigcup_{C\in\mathcal{C}} C\times C \subset (f\times f)^{-1}[B_\nu]$. Then $\mathcal{C} < f^{-1}\mathcal{A}$ and thus $f^{-1}\mathcal{A} \in \mu$ (Let $C \in \mathcal{C}$ and $x \in C$. Then there is some $B \in \mathcal{B}$ with $f(x) \in B$ and some $A \in \mathcal{A}$ such that $\mathrm{St}(\{f(x)\},\mathcal{B}) \subset \mathrm{St}(B,\mathcal{B}) \subset A$. Hence $C \subset f^{-1}[A]$ [$y \in C$ implies $(x,y) \in C\times C \subset (f\times f)^{-1}[B_\nu]$. Thus $(f(x),f(y)) \in B_\nu$, so that there is some $B' \in \mathcal{B}$ with $f(x),f(y) \in B'$. Consequently, $f(y) \in \mathrm{St}(\{f(x)\},\mathcal{B}) \subset A$, i.e. $y \in f^{-1}[A]$]).

(d) Put $F((X,\mu)) = (X,\mathcal{W}_\mu)$ for every uniform nearness space (X,μ), and for each uniformly continuous map $f\colon (X,\mu) \to (Y,\nu)$, let $F(f)$ be the corresponding uniformly continuous map $f\colon (X,\mathcal{W}_\mu) \to (Y,\mathcal{W}_\nu)$ (cf. (c)). Thereby a functor F: U-Near → Unif has been defined. Put $G((X,\mathcal{W})) = (X,\mu_{\mathcal{W}})$ for each uniform space $(X,\mathcal{W})$, and for each uniformly continuous map $f\colon (X,\mathcal{W}) \to (Y,\mathcal{R})$, let $G(f)$ be the corresponding uniformly continuous map $f\colon (X,\mu_{\mathcal{W}}) \to (Y,\mu_{\mathcal{R}})$ [note (c) and $\mathcal{W}_{\mu_{\mathcal{W}}} = \mathcal{W}$ (resp. $\mathcal{R}_{\mu_{\mathcal{R}}} = \mathcal{R}$)]. Thus a functor G: Unif $\to$ U-Near has been defined and the following hold:

(1) $G \circ F = I_{\text{U-Near}}$ (note $\mu_{\mathcal{W}_\mu} = \mu$ for every U-Near-structure μ)

and (2) $F \circ G = I_{\text{Unif}}$ (note $\mathcal{W}_{\mu_{\mathcal{W}}} = \mathcal{W}$ for every uniformity $\mathcal{W}$).

Consequently it has been shown that U-Near and Unif are isomorphic.

3.1.2.5 Remarks. ① The isomorphism between U-Near and Unif means nothing else but the equivalence between the definition of uniform spaces by means of covers (in the sense of J.W. Tukey (1940)) and the definition of uniform spaces by means of entourages (in the sense of A. Weil (1937)). The definition of Weil is more common.

② By 3.1.2.3 one can identify each uniform space $(X,\mathcal{W})$ with the corresponding uniform nearness space $(X,\mu_{\mathcal{W}})$. If P is a uniform invariant, then $(X,\mu_{\mathcal{W}})$ is said to have the property P iff $(X,\mathcal{W})$ has the property P.

③ If U: Near $\to$ U-Near denotes the bireflector, then $U((X,\mu)) = (X,\mu_u)$ (cf. 3.1.2.2) [resp. $(X,\mathcal{W}_{\mu_u})$] is called *the underlying uniform space of the nearness space* (X,μ).

④ It follows from 3.1.2.2 and the characterization theorem of epireflective subcategories that the product (in Near) of any family of uniform nearness spaces is again a uniform nearness space and that any subspace (in Near) of a uniform nearness space is also a uniform nearness space. Moreover one obtains by applying 2.2.11 ② that every indiscrete Near-object is uniform. But this is trivial.

⑤ α) A nearness space (X,μ) is called pseudometrizable provided that there is a pseudometric d on X such that $\{A_\varepsilon\colon \varepsilon > 0\}$ is a base for μ, where $A_\varepsilon = \{U(x,\varepsilon)\colon x \in X\}$ and $U(x,\varepsilon) = \{y \in X\colon d(x,y) < \varepsilon\}$. We obtain the following

Proposition. A nearness space (X,μ) is pseudometrizable if and only if the following conditions are satisfied:

(1) μ is a U-Near-structure on X .

(2) μ has a countable base.

Proof. a) "⇒" . $A_{\frac{\varepsilon}{3}} *\!< A_\varepsilon$ for each $\varepsilon > 0$ and $\{A_{\frac{1}{n}}: n \in \mathbb{N} \setminus \{0\}\}$ is a countable base for μ .

b) "⇐" . If $\mu' \subset \mu$ is a countable base for μ , then $\{\bigcup_{B \in \mathcal{B}} B\times B: \mathcal{B} \in \mu'\}$ is a countable base for $\mathcal{W}_\mu$, i.e. $\mathcal{W}_\mu$ is pseudometrizable. Hence there is a pseudometric d on X such that for each $V \in \mathcal{W}_\mu$, there exists some $\varepsilon > 0$ with $V_\varepsilon = \{(x,y): d(x,y) < \varepsilon\} \subset V$. Then by $V_\varepsilon(x) = U(x,\varepsilon)$ (for every $x \in X$) ,

$$A_{V_\varepsilon} = A_\varepsilon < A_V .$$

Thus $(X,\mu_{\mathcal{W}_\mu})$ is pseudometrizable and since (1) is valid, $\mu = \mu_{\mathcal{W}_\mu}$.

β) Let Ps Near be the category of pseudometrizable nearness spaces (and uniformly continuous maps) and Ps Unif the category of pseudometrizable uniform spaces (and uniformly continuous maps). Then the isomorphism between U-Near and Unif yields an isomorphism between Ps Near and Ps Unif (If (X,μ) is a pseudometrizable nearness space, then by the preceding proposition, μ is a U-Near-structure having a countable base. Hence $\mathcal{W}_\mu$ is a uniformity having a countable base and thus $(X,\mathcal{W}_\mu)$ is pseudometrizable. Conversely, if $(X,\mathcal{W})$ is a pseudometrizable uniform space, then $\mathcal{W}$ has a countable base $\mathcal{B}$ and consequently, $\mu_\mathcal{W}$ is a U-Near-structure with the countable base $\{A_B: B \in \mathcal{B}\}$.).

γ) It is a well-known fact that each uniform space $(X,\mathcal{W})$ is (up to isomorphism) a subspace of a product of pseudometrizable uniform spaces (the subspaces and the products are formed in Unif) [For each $V \in \mathcal{W}$, there exists a pseudometric d_V such that for the induced uniformities $\mathcal{D}_V$,

$\mathcal{W} = \sup\{\mathcal{D}_V : V \in \mathcal{W}\}$, i.e. $\mathcal{W}$ is the initial uniformity on X with respect to $(1_X^V : X \to (X,\mathcal{D}_V))_{V\in\mathcal{W}}$, where $1_X^V : X \to X$ is the identity map. Let $\prod_{V\in\mathcal{W}} (X,\mathcal{D}_V)$ be the product of the family $((X,\mathcal{D}_V))_{V\in\mathcal{W}}$ in Unif and let $p_V : \prod_{V\in\mathcal{W}} (X,\mathcal{D}_V) \to (X,\mathcal{D}_V)$ be the projection maps for each $V \in \mathcal{W}$. Then there exists a unique (uniformly continuous) map $e: (X,\mathcal{W}) \to \prod_{V\in\mathcal{W}} (X,\mathcal{D}_V)$ such that $p_V \circ e = 1_X^V$ for each $V \in \mathcal{W}$. Since e is injective and $\mathcal{W}$ is the initial uniformity on X with respect to e , e is an embedding.]. But then by β), each uniform nearness space (X,μ) is also a subspace of a product of pseudometrizable nearness spaces (the subspaces and the products are formed in U-Near) . Since U-Near is bireflective in Near , the subspaces and products are formed in U-Near as in Near (cf. 2.2.13 ②).

Thus:

$$R_{\underline{\text{Near}}}\ \underline{\text{Ps Near}} = \underline{\text{U-Near}}$$

⑥ Obviously $|R^{co}_{\underline{\text{Near}}}\ \underline{\text{U-Near}}|$ consists of those nearness spaces which are quotient objects of coproducts of uniform nearness spaces. One can easily show that U-Near is closed under formation of coproducts. Therefore $|R^{co}_{\underline{\text{Near}}}\ \underline{\text{U-Near}}|$ consists precisely of all nearness spaces which are quotient objects of uniform nearness spaces. It is a nontrivial result of M. Katětov (1967) that $R^{co}_{\underline{\text{Near}}}\ \underline{\text{U-Near}} = \underline{\text{Near}}$.

3.1.2.6 Definitions. 1) A nearness space (X,μ) is called an N_1-space provided that $T((X,\mu))$ is a T_1-space .

2) A topological space $(X,\mathcal{X})$ is called fully normal provided that every open cover of X has an open star-refinement.

3) A topological space $(X,\mathcal{X})$ is called paracompact provided that $(X,\mathcal{X})$ is fully normal and a T_1-space.

3.1.2.7 Theorem. If (X,μ) is a topological nearness space, then the following are satisfied:

(1) (X,μ) is a uniform nearness space if and only if (X,μ) is fully normal.

(2) (X,μ) is a uniform N_1-space if and only if (X,μ) is paracompact.

Proof. (1) a) "⇒" . If $\mathcal{U}$ is an open cover of (X,X_μ) , then since (X,μ) is topological, $\mathcal{U} \in \mu$. Since (X,μ) is uniform, there exists some $\mathcal{V} \in \mu$ such that $\mathcal{V} *< \mathcal{U}$. Moreover, there is an open cover $\mathcal{W}$ of (X,X_μ) with $\mathcal{W} < \mathcal{V}$. Hence $\mathcal{W} *< \mathcal{U}$ (for every $W \in \mathcal{W}$ there is some $V \in \mathcal{V}$ such that $W \subset V$ and since $\mathcal{V} *< \mathcal{U}$, there exists some $U \in \mathcal{U}$ with $St(V,\mathcal{V}) \subset U$. Furthermore, $St(W,\mathcal{W}) \subset St(V,\mathcal{V})$ [$y \in St(W,\mathcal{W})$ implies the existence of some $W' \in \mathcal{W}$ such that $y \in W'$ and $W \cap W' \neq \emptyset$. Since $\mathcal{W} < \mathcal{V}$, there exists some $V' \in \mathcal{V}$ with $W' \subset V'$. Then $W \cap W' \subset V \cap V'$, hence $V \cap V' \neq \emptyset$ and $y \in V'$, i.e. $y \in St(V,\mathcal{V})$] . Consequently $St(W,\mathcal{W}) \subset U$).

b) "⇐" . If $\mathcal{A} \in \mu$, then $\mathcal{B} = \{int_\mu A: A \in \mathcal{A}\} \in \mu$. Since (X,X_μ) is fully normal and $\mathcal{B}$ is an open cover, there is an open cover $\mathcal{C}$ with $\mathcal{C} *< \mathcal{B}$. Since (X,μ) is topological, $\mathcal{C} \in \mu$.

(2) is trivial if (1) has been proved.

3.1.2.8 Remarks. (1) The category U-Near$_1$ of uniform N_1-spaces (and uniformly continuous maps) is epireflective in Near , because U-Near$_1$ is isomorphic to the category U Sep of separated uniform spaces (and uniformly continuous maps), which is closed under formation of subspaces and products in Unif , i.e. epireflective in Unif, and Unif is isomorphic to the bireflective subcategory U-Near of Near (the composition of the two reflectors yields the desired epireflector).

(2) As well-known a topological space $(X,\mathcal{X})$ is paracompact if and only if $(X,\mathcal{X})$ is a Hausdorff space and each open cover $\mathcal{U}$ of X has an open locally finite refinement $\mathcal{V}$ (i.e., for every $x \in X$ there is a neighbourhood U_x of x which intersects only finitely many $V \in \mathcal{V}$) . Although the paracompact spaces play an important role in topology, they behave badly with respect to standard constructions like the formation of products or subspaces. Not even the product of two paracompact spaces is paracompact (the most known counterexample is the pro-

duct of the Sorgenfrey line [cf. 3.1.1.4 (3)] with itself).
3.1.2.7 (2) gives rise to a definition of paracompactness for nearness spaces: A nearness space (X,μ) is called paracompact provided that (X,μ) is a uniform N_1-space. If (X,μ) is a topological nearness space, then this concept coincides with the classical one and it turns out that *the paracompact topological spaces are precisely those* N_1*-spaces, which are simultaneously topological and uniform* - a completely new insight!
Now by (1) the product of any family of paracompact nearness spaces formed in Near is a paracompact nearness space and every subobject (in Near) of a paracompact nearness space is a paracompact nearness space. These preserving properties are also true for paracompact topological spaces provided that the products and subobjects are formed in Near (naturally the result is no topological nearness space).
Therefore Top is not the right category to handle paracompactness.

(3) A topological space $(X,\mathcal{X})$ is uniformizable if and only if $(X,\mathcal{X})$ is completely regular (especially this characterization may be applied to R_o-topological spaces). A topological nearness space (X,μ) is called uniformizable provided that there exists some U-Near-structure μ' on X such that $(\mu')_t = \mu$. Correspondingly, a uniform nearness space (X,μ) is called topologizable provided that there exists some T-Near-structure μ' on X such that $(\mu')_u = \mu$. *A uniform nearness space* (X,μ) *is topologizable if and only if* (X,μ) *is a fine uniform space, i.e. if* μ *is the finest of those* U-Near-*structures* μ^* *on* X *for which* $(\mu^*)_t = \mu_t$. [1) "⇒". If μ^* is a U-Near-structure on X with $(\mu^*)_t = \mu_t$, i.e. $\mathcal{X}_{\mu^*} = \mathcal{X}_\mu$, which means $\mathrm{int}_{\mu^*} = \mathrm{int}_\mu$, and if $\mathcal{A} \in \mu^*$, then $\{\mathrm{int}_{\mu^*} A : A \in \mathcal{A}\} \in \mu^*$. Thus $X = \bigcup_{A\in\mathcal{A}} \mathrm{int}_{\mu^*} A = \bigcup_{A\in\mathcal{A}} \mathrm{int}_\mu A$ and additionally $X = \bigcup_{A\in\mathcal{A}} \mathrm{int}_{\mu'} A$, because $\mathrm{int}_\mu A \subset \mathrm{int}_{\mu'} A$ $\left(x \in \mathrm{int}_\mu A \text{ implies } \{A, X\setminus\{x\}\} \in \mu = (\mu')_u \subset \mu', \text{ i.e. } x \in \mathrm{int}_{\mu'} A\right)$. Since μ' is a T-Near-structure, $\mathcal{A} \in \mu'$. Furthermore there exists some $\mathcal{B} \in \mu^*$ with $\mathcal{B} *\!< \mathcal{A}$. Corresponding to the above arguments we get $\mathcal{B} \in \mu'$ and so on. Thus $\mathcal{A} \in (\mu')_u = \mu$, i.e. $\mu^* \subset \mu$.

2) "⇐". We will show that $\mu = (\mu_t)_u$.

a) If $A \in \mu \subset \mu_t$, then since μ is a U-Near-structure, $A \in (\mu_t)_u$.

b) If $A \in (\mu_t)_u$, then there exists a sequence $(A_n)_{n\in\mathbb{N}}$ in μ_t such that $A_o = A$ and $A_{n+1} *< A_n$ for each $n \in \mathbb{N}$. $\mu \cup \{A_o, A_1, \ldots\}$ is a subbase for a U-Near-structure μ^* with $(\mu^*)_t = \mu_t$. Since μ is the finest structure of this kind, $\mu \cup \{A_o, A_1, \ldots\} \subset \mu^* \subset \mu$, i.e. $A_o = A \in \mu$.]

3.1.3 Contigual spaces

3.1.3.1 Definition. A nearness space (X,μ) is called *contigual*[13)] provided that

(C) For every $A \in \mu$ there exists some finite subset $B \subset A$ with $B \in \mu$.

3.1.3.2 Remarks. ① A nearness space (X,μ) is contigual if and only if for every $A \in \mu$ there is some finite $B \in \mu$ with $B < A$.

② The full subcategory C-Near of Near whose objects are the contigual nearness spaces is isomorphic to the category Cont of "contiguity spaces" and "contiguity maps" in the sense of V.M. Ivanova and A.A. Ivanov (1959).

3.1.3.3 Theorem. The category C-Near is bireflective in Near. Especially if (X,μ) is a nearness space and μ_c is the set of all covers of X which are refined by some finite element of μ, then $1_X\colon (X,\mu) \to (X,\mu_c)$ is the bireflection of (X,μ) with respect to C-Near.

Proof. 1) (X,μ_c) is a contigual nearness space:
Obviously μ_c is a non-empty set of non-empty covers.
N_1) If $A \in \mu_c$ and $A < B$, then there exists some finite $C \in \mu$

---

13) Sometimes "totally bounded" is used instead of "contigual".

such that $\mathcal{C} < \mathcal{A} < \mathcal{B}$. Hence $\mathcal{B} \in \mu_c$.

N_2) If $\mathcal{A}_1, \mathcal{A}_2 \in \mu_c$, then there exist finite $\mathcal{B}_1, \mathcal{B}_2 \in \mu$ with $\mathcal{B}_i < \mathcal{A}_i$ (i=1,2) . Thus $\mathcal{B}_1 \wedge \mathcal{B}_2 \in \mu$ is a finite refinement of $\mathcal{A}_1 \wedge \mathcal{A}_2$. Consequently $\mathcal{A}_1 \wedge \mathcal{A}_2 \in \mu_c$.

N_3) If $\mathcal{A} \in \mu_c$, then there exists some finite $\mathcal{B} \in \mu$ with $\mathcal{B} < \mathcal{A}$. Since obviously $\text{int}_\mu = \text{int}_{\mu_c}$ and $\{\text{int}_\mu B: B \in \mathcal{B}\} \in \mu$ is a finite refinement of $\{\text{int}_\mu A: A \in \mathcal{A}\}$, $\{\text{int}_{\mu_c} A: A \in \mathcal{A}\} \in \mu_c$.

(C) $\mathcal{A} \in \mu_c$ implies the existence of some finite $\mathcal{B} \in \mu$ with $\mathcal{B} < \mathcal{A}$. Since $\mathcal{B} < \mathcal{B}$, $\mathcal{B} \in \mu_c$.

2) By definition $\mu_c \subset \mu$ and thus $1_X: (X,\mu) \to (X,\mu_c)$ is a uniformly continuous map.

3) If (X',μ') is a contigual nearness space and $f: (X,\mu) \to (X',\mu')$ is a uniformly continuous map, then $f: (X,\mu_c) \to (X',\mu')$ is also a uniformly continuous map; for if $\mathcal{A} \in \mu'$, then there exists some finite $\mathcal{B} \in \mu'$ with $\mathcal{B} < \mathcal{A}$ and $f^{-1}\mathcal{B}$ is a finite element of μ such that $f^{-1}\mathcal{B} < f^{-1}\mathcal{A}$. Consequently, $f^{-1}\mathcal{A} \in \mu_c$.

<u>3.1.3.4 Theorem</u>. If (X,μ) is a topological nearness space, then the following are equivalent:

(a) (X,μ) is contigual.

(b) (X,μ) is compact.

<u>Proof</u>. (a) ⇒ (b). Let $\mathcal{U}$ be an open cover of (X,X_μ) . Since (X,μ) is topological, $\mathcal{U} \in \mu$. By (a) there is some finite $\mathcal{V} \in \mu$ with $\mathcal{V} \subset \mathcal{U}$. Thus (X,X_μ) is compact.

(b) ⇒ (a). Let $\mathcal{A} \in \mu$. Then $\{\text{int}_\mu A: A \in \mathcal{A}\}$ is an open cover of X in (X,X_μ) for which there exists by (b) some finite subcover $\mathcal{B} \subset \{\text{int}_\mu A: A \in \mathcal{A}\}$. Obviously $\mathcal{B} \in \mu$ and $\mathcal{B} < \mathcal{A}$. Consequently, (X,μ) is contigual.

<u>3.1.3.5 Theorem</u>. If (X,μ) is a uniform nearness space, then the following are equivalent:

(a) (X,μ) is contigual.

(b) (X,μ) is totally bounded.

Proof. (a) ⇒ (b). If $V \in \mathcal{W}_\mu$, then there exists some $\mathcal{B} \in \mu$ such that $\bigcup_{B\in\mathcal{B}} B\times B \subset V$. Furthermore, there exists some finite $\mathcal{C} \in \mu$ with $\mathcal{C} < \mathcal{B}$ and for each $C \in \mathcal{C}$,

$$C\times C \subset \bigcup_{B\in\mathcal{B}} B\times B \subset V .$$

Hence $\mathcal{C}$ is a finite cover of X by V-small sets. Consequently, $(X,\mathcal{W}_\mu)$ is totally bounded.

(b) ⇒ (a). Let $\mathcal{A} \in \mu = \mu_{\mathcal{W}_\mu}$. Then there exists some $V \in \mathcal{W}_\mu$ such that $\mathcal{A}_V < \mathcal{A}$. Furthermore there exists some symmetric $V' \in \mathcal{W}_\mu$ with $V'^2 \subset V$. By (b) there is some finite $E \subset X$ such that $V'[E] = \bigcup_{x\in E} V'(x) = X$. Then

$$\mathcal{A}_{V'} < \{V(x)\colon x \in E\} < \mathcal{A}_V < \mathcal{A}$$

($z \in X$ implies the existence of some $x \in E$ with $z \in V'(x)$, hence $V'(z) \subset V'^2(x) \subset V(x)$).
Thus $\{V(x)\colon x \in E\} \in \mu_{\mathcal{W}_\mu} = \mu$ is finite and $\{V(x)\colon x \in E\} < \mathcal{A}$. Consequently, (X,μ) is contigual.

3.1.3.6 Definition. A nearness space (X,μ) is called proximal provided that (X,μ) is uniform and contigual.

3.1.3.7 Theorem. If (X,μ) is a topological nearness space, then the following are equivalent:

(a) (X,μ) is a proximal N_1-space.

(b) (X,μ) is a compact Hausdorff space.

Proof. (a) ⇒ (b). Since (X,μ) is contigual and topological, (X,μ) is compact by 3.1.3.4. A uniform N_1-space is a separated uniform space, hence $(X,\mu_t) = (X,\mu)$ is a Hausdorff space. Thus (X,μ) is a compact Hausdorff space.

(b) ⇒ (a). By 3.1.3.4 (X,μ) is contigual and since $(X,\mu_t) = (X,\mu)$ is a Hausdorff space, (X,μ) is a fortiori an N_1-space. Since (X,μ) is topological, μ consists precisely of all those covers of X which are refined by an open (with respect to X_μ) cover. As well-known a compact Hausdorff space is uniquely uniformizable, i.e. there is a unique uniformity $\mathcal{W}$ on X such that $(\mu_{\mathcal{W}})_t = \mu_t = \mu$; this uniformity consists exactly of all neighbourhoods of the diagonal Δ in $(X,X_\mu) \times (X,X_\mu)$. $\mu_{\mathcal{W}}$ is a U-Near-structure and $\mu_{\mathcal{W}} = \mu$, i.e. (X,μ) is uniform:

α) $A \in \mu_{\mathcal{W}}$ implies the existence of some $W \in \mathcal{W}$ with $A_W < A$. Hence $W(x)$ is a neighbourhood of x with respect to X_μ for each $x \in X$. Thus $\{\text{int}_\mu W(x) : x \in X\} < A$, i.e. $A \in \mu$.

β) If $A \in \mu$, then there exists some open (with respect to X_μ) cover $\mathcal{B}$ of X with $\mathcal{B} < A$. Since (X,X_μ) is paracompact (as a compact Hausdorff space) there exists some open cover $\mathcal{C}$ of X such that $\mathcal{C} \ast\!< \mathcal{B}$. Then $W = \bigcup_{C \in \mathcal{C}} C \times C$ is an open neighbourhood of the diagonal Δ in $(X,X_\mu) \times (X,X_\mu)$, i.e. $W \in \mathcal{W}$. Furthermore, $A_W = \{W(x) : x \in X\} = \{\text{St}(\{x\},\mathcal{C}) : x \in X\} < A$. Thus $A \in \mu_{\mathcal{W}}$.

3.1.3.8 Remarks. ① The full subcategory Pr-Near of Near whose objects are the proximal nearness spaces is isomorphic to the category Tb-Unif of totally bounded uniform spaces (and uniformly continuous maps) [by 3.1.3.5 the isomorphism is obtained by the isomorphism between Unif and U-Near] . It is well-known that the category Prox of proximity spaces and δ-maps is isomorphic to Tb-Unif and Tb-Unif is bireflective in Unif . Thus Pr-Near *is bireflective in* U-Near *and* since U-Near is bireflective in Near , *also bireflective in* Near (this conclusion is admissible because the considered categories are topological). Furthermore, Pr-Near is also bireflective in C-Near , i.e. Prox is bireflective in Cont (Pr-Near is closed under formation of products and subspaces in Near because Pr-Near is bireflective in Near . Products and subspaces in C-Near are formed as in Near , since C-Near is bireflective in Near . Thus Pr-Near is epireflective in C-Near . The indiscrete objects of C-Near are those nearness spaces (X,μ) , for which $\mu' = \{\{X\}\}$ is a base for μ . However, these ones are uniform, i.e. they belong to Pr-Near .

Consequently, Pr-Near is bireflective in C-Near) .

(2) 3.1.3.4 and 3.1.3.5 show that the concept "contigual" may be regarded as a generalization of the concept "compact" for topological spaces as well as a generalization of the concept "totally bounded" for uniform spaces. Generally a nearness space (X,μ) is called totally bounded provided that (X,μ) satisfies the condition (C) (i.e. the concepts "totally bounded" and "contigual" are used synonymously). Then the totally bounded uniform nearness spaces are just identical with the proximal nearness spaces. A nearness space (X,μ) is called compact provided that (X,μ) is topological and contigual.

(3) By (1) Pr-Near is bireflective in Near . *If* $(X,\mu) \in |\underline{\text{Near}}|$, *then* $1_X\colon (X,\mu) \to (X,(\mu_u)_c)$ *is the bireflection of* (X,μ) *with respect to* Pr-Near . For this purpose it suffices to show that *the contigual reflection of a uniform space is uniform* (then the composition of the two reflections $1_X\colon (X,\mu) \to (X,\mu_u)$ and $1_X\colon (X,\mu_u) \to (X,(\mu_u)_c)$ yields the proximal reflection): Let (X,μ) be a uniform space and $\mathcal{A} \in \mu_c$. Then there exists some finite $\mathcal{B} \in \mu$ with $\mathcal{B} < \mathcal{A}$. Furthermore, there exists some $\mathcal{C} \in \mu$ with $\mathcal{C} *< \mathcal{B}$. On $\mathcal{C}$ an equivalence relation R is defined by

$$C_1 \; R \; C_2 \Leftrightarrow \forall B \in \mathcal{B}\,[(C_1 \subset B \Leftrightarrow C_2 \subset B) \text{ and } (St(C_1,\mathcal{C}) \subset B \Leftrightarrow St(C_2,\mathcal{C}) \subset B)]$$

If $\mathcal{B}$ has at most n elements, then there are at most 4^n equivalence classes with respect to R . Put $\tilde{C} = \bigcup\{C' \in \mathcal{C}: C' \; R \; C\}$ for every $C \in \mathcal{C}$. Then $\tilde{\mathcal{C}} = \{\tilde{C}: C \in \mathcal{C}\}$ is finite. Since obviously $\mathcal{C} < \tilde{\mathcal{C}}$, $\tilde{\mathcal{C}} \in \mu$ and thus $\tilde{\mathcal{C}} \in \mu_c$. Furthermore, $\tilde{\mathcal{C}} *< \mathcal{A}$ [For each $C \in \mathcal{C}$, there exists some $B \in \mathcal{B}$ such that $St(C,\mathcal{C}) \subset B$. Additionally $St(\tilde{C},\tilde{\mathcal{C}}) \subset B$ (if $\tilde{C}_1 \in \tilde{\mathcal{C}}$ such that $\tilde{C}_1 \cap \tilde{C} \neq \emptyset$, then there exists some $x \in \tilde{C}_1 \cap \tilde{C}$, i.e. there exist $C_2 \in \mathcal{C}$ and $C_3 \in \mathcal{C}$ with $x \in C_2 \cap C_3$ and $C_2 \; R \; C_1$ as well as $C_3 \; R \; C_1$. Hence $C_3 \subset St(C_2,\mathcal{C}) \subset B$ and thus $C_4 \subset B$ for each C_4 such that $C_4 \; R \; C_3$. Finally $\tilde{C}_1 = \tilde{C}_2 \subset B$ and thus $St(\tilde{C},\tilde{\mathcal{C}}) \subset B$)] .

(4) a) If C: Near → C-Near denotes the contigual bireflector, then $C((X,\mu)) = (X,\mu_c)$ is called *the underlying contigual space of the nearness space* (X,μ) .

b) If Pr: Near → Pr-Near denotes the proximal bireflector, then $Pr((X,\mu)) = (X,(\mu_u)_c)$ is called *the underlying proximity space of the nearness space* (X,μ).

3.1.3.9 Examples for $\mathbb{R}$. In the theory of topological spaces only one topology on $\mathbb{R}$ is of special interest, namely the usual topology. In the theory of nearness spaces there are at least four particularly interesting Near-structures on $\mathbb{R}$, all of them inducing the usual topology:

(1) The T-Near-structure $\mu_{\underline{\mathbb{R}}}$ belonging to the usual topology $\underline{\mathbb{R}}$. The nearness space $\mathbb{R}_t = (\mathbb{R},\mu_{\underline{\mathbb{R}}})$ is topological and uniform[14] but not pseudometrizable (exercise).

(2) The U-Near-structure $\mu_{\mathcal{W}}$ belonging to the "natural" uniformity $\mathcal{W}$ ($\mathcal{W}$ is induced by the Euclidean metric). The nearness space $\mathbb{R}_u = (\mathbb{R},\mu_{\mathcal{W}})$ is pseudometrizable (the pseudometric is even a metric), but not topological.

(3) The finite elements of $\mu_{\mathcal{W}}$ form a base for $\mu_p = (\mu_{\mathcal{W}})_c$ (cf. (2) and 3.1.3.3). $\mathbb{R}_p = (\mathbb{R},\mu_p)$[15] is neither topological nor pseudometrizable, but proximal, i.e. contigual and uniform.

(4) The finite elements of $\mu_{\underline{\mathbb{R}}}$ form a base for $\mu_f = (\mu_{\underline{\mathbb{R}}})_c$. $\mathbb{R}_f = (\mathbb{R},\mu_f)$ is neither topological nor pseudometrizable, but contigual and uniform, i.e. proximal.

3.2 The category P-Near and its subcategories

3.2.1. Prenearness spaces

3.2.1.1 Definition. 1) Let X be a set and let μ be a non-empty set of non-empty covers of X satisfying N_1) (cf. 1.1.6 (5)). Then (X,μ) is called a prenearness space.

2) Let (X,μ) and (X',μ') be prenearness

---

14) $\mu_{\underline{\mathbb{R}}}$ is the "fine uniformity" for $\underline{\mathbb{R}}$ since $(\mathbb{R},\underline{\mathbb{R}})$ is paracompact.

15) μ_p is the usual proximity on $\mathbb{R}$.

spaces and let $f: X \to X'$ be a map. Then $f: (X,\mu) \to (X',\mu')$ is called a uniformly continuous map provided that $f^{-1}\mathcal{A} \in \mu$ for each $\mathcal{A} \in \mu'$, where $f^{-1}\mathcal{A} = \{f^{-1}[A]: A \in \mathcal{A}\}$.

3.2.1.2 Remarks. ① The prenearness spaces together with the uniformly continuous maps form a topological category denoted by P-Near (If X is a set, $((X_i,\mu_i))_{i\in I}$ a family of prenearness spaces and $(f_i: X \to X_i)_{i\in I}$ a family of maps, then $\mu = (\{f_i^{-1}\mathcal{A}_i: \mathcal{A}_i \in \mu_i$ and $i \in I\}) := \{\mathcal{B} \subset P(X):$ there exist $i \in I$ and $\mathcal{A}_i \in \mu_i$ with $f_i^{-1}\mathcal{A}_i < \mathcal{B}\}$ is a P-Near-structure on X which is initial with respect to $(X,f_i,(X_i,\mu_i),I)$.).

② If X is a set, $((X_i,\mu_i))_{i\in I}$ a family of prenearness spaces and $(f_i: X_i \to X)_{i\in I}$ a family of maps, then $\mu = \{\mathcal{A}: \mathcal{A}$ is a cover of X and $f_i^{-1}\mathcal{A} \in \mu_i$ for each $i \in I\}$ is the final P-Near-structure on X with respect to $((X_i,\mu_i),f_i,X,I)$.

3.2.1.3 Theorem. Let (X,μ) be a prenearness space. Then $\mathrm{int}_\mu: P(X) \to P(X)$ defined by $\mathrm{int}_\mu A = \{x \in X: \{A, X\setminus\{x\}\} \in \mu\}$ for each $A \subset X$ satisfies the following conditions:

(1) $\mathrm{int}_\mu X = X$

(2) $\mathrm{int}_\mu A \subset A$ for each $A \subset X$

(3) $A \subset B \subset X$ implies $\mathrm{int}_\mu A \subset \mathrm{int}_\mu B$.

Proof. cf. proof of 3.1.1.3 (2) (b).

3.2.2 Seminearness spaces

3.2.2.1 Theorem. The full subcategory S-Near of P-Near whose objects are the seminearness spaces (cf. 3.1.2.3) is bicoreflective in P-Near. In particular, if (X,μ) is a prenearness space and the set of all $\mathcal{A} \subset P(X)$ for which there exist finitely many elements $\mathcal{A}_1,\ldots,\mathcal{A}_n$ of μ such that $\mathcal{A}_1\wedge\ldots\wedge\mathcal{A}_n < \mathcal{A}$ is denoted by μ_s, then $1_X: (X,\mu_s) \to (X,\mu)$ is the bicoreflection of (X,μ) with respect to S-Near .

Proof. Obviously (X,μ_s) is a seminearness space and by $\mu \subset \mu_s$, $1_X: (X,\mu_s) \to (X,\mu)$ is a uniformly continuous map. Now let (X',μ') be a seminearness space and $f: (X',\mu') \to (X,\mu)$ a uniformly continuous map. Then also $f: (X',\mu') \to (X,\mu_s)$ is a uniformly continuous map (Namely, if $\mathcal{A} \in \mu_s$, then there exist finitely many elements $\mathcal{A}_1,\dots,\mathcal{A}_n$ of μ such that $\mathcal{A}_1 \wedge \dots \wedge \mathcal{A}_n < \mathcal{A}$. Hence $f^{-1}\mathcal{A}_1 \wedge \dots \wedge f^{-1}\mathcal{A}_n < f^{-1}\mathcal{A}$ [If $B \in f^{-1}\mathcal{A}_1 \wedge \dots \wedge f^{-1}\mathcal{A}_n$, then there are $A_i \in \mathcal{A}_i$ $(i \in \{1,\dots,n\})$ with $B = f^{-1}[A_1] \cap \dots \cap f^{-1}[A_n] = f^{-1}[A_1 \cap \dots \cap A_n]$ and there exists some $A \in \mathcal{A}$ such that $A_1 \cap \dots \cap A_n \subset A$. Thus $B \subset f^{-1}[A]]$. Hence $f^{-1}\mathcal{A} \in \mu'$ since $f^{-1}\mathcal{A}_i \in \mu'$ for each $i \in \{1,\dots,n\}$ and (X',μ') satisfies $N_2)$ and $N_1)$.) . Therefore everything has been shown.

3.2.2.2 Remarks. (1) S-Near is a topological category (as a bicoreflective subcategory of a topological category).

(2) If X is a set, $((X_i,\mu_i))_{i\in I}$ a family of seminearness spaces and $(f_i: X \to X_i)_{i\in I}$ a family of maps, then $(\{f_i^{-1}\mathcal{A}_i: \mathcal{A}_i \in \mu_i \text{ and } i \in I\})_s$ is the initial S-Near-structure on X with respect to $(X,f_i,(X_i,\mu_i),I)$ (For the construction of the initial structure in a bicoreflective subcategory of a topological category we refer to the proof of 2.2.12).

(3) If X is a set, $((X_i,\mu_i))_{i\in I}$ a family of seminearness spaces and $(f_i: X_i \to X)_{i\in I}$ a family of maps, then $\{\mathcal{A}: \mathcal{A}$ is a cover of X and $f_i^{-1}\mathcal{A} \in \mu_i$ for each $i \in I\}$ is the final S-Near-structure on X with respect to $((X_i,\mu_i),f_i,X,I)$. (For the construction of the final structure in a bicoreflective subcategory of a topological category we refer to 2.2.13 (1)) .

3.2.2.3 Proposition. Let (X,μ) be a seminearness space. Then $\operatorname{int}_\mu: \mathcal{P}(X) \to \mathcal{P}(X)$ defined by $\operatorname{int}_\mu A = \{x \in X: \{A, X\setminus\{x\}\} \in \mu\}$ satisfies the following conditions:

(1) $\operatorname{int}_\mu X = X$

(2) $\operatorname{int}_\mu A \subset A$ for each $A \subset X$

(3) $A \subset B \subset X$ implies $\text{int}_\mu A \subset \text{int}_\mu B$

(4) $\text{int}_\mu(A \cap B) = \text{int}_\mu A \cap \text{int}_\mu B$ for each $(A,B) \in P(X) \times P(X)$

Proof. cf. proof of 3.1.1.3 (2)(b).

3.2.2.4 Remark. $\text{int}_\mu: P(X) \to P(X)$ is not yet an interior operator. Hence it does not define a topological space, but a closure space in the sense of E. Čech (cf. his book on "Topological spaces").

3.2.2.5 Theorem. The category Near is bireflective in S-Near.

Proof. Let (X,μ_o) be a seminearness space. For each ordinal number α, μ_α is defined by transfinite induction as follows:

(1) Let μ_o be the given S-Near-structure,

(2) $\mu_{\alpha+1} = \{A \in \mu_\alpha: \{\text{int}_{\mu_\alpha} A: A \in A\} \in \mu_\alpha\}$,

(3) $\mu_\alpha = \bigcap_{\beta<\alpha} \mu_\beta$ if α is a limit ordinal.

Then hold:

(a) (μ_α) is a decreasing[16] sequence of S-Near-structures on X .

(b) $\mu = \bigcap_\alpha \mu_\alpha$ is a Near-structure on X .

(a): μ_o is an S-Near-structure by assumption. By transfinite induction one yields just (a) because of 3.2.2.3.
(b): Since the transfinite sequence (μ_α) is decreasing, there exists some α with $\mu_\alpha = \mu_{\alpha+1}$. But then μ_α is a Near-structure and $\mu = \mu_\alpha$. Since $\mu \subset \mu_o$, $1_X: (X,\mu_o) \to (X,\mu)$ is a uniformly continuous map.
If (X',μ') is a nearness space and $f: (X,\mu_o) \to (X',\mu')$ is a uniformly continuous map, then it remains to show that $f: (X,\mu) \to (X',\mu')$ is a uniformly continuous map; for then $1_X: (X,\mu_o) \to (X,\mu)$ is the desired bireflection of (X,μ_o) with respect to Near . It suffices to prove:

---

[16] with respect to the inclusion "$\subset$" .

(*) $\bar{\mu}_f = \{f^{-1}A: A \in \mu'\}$ is a base for a Near-structure μ_f on X. (for: $\bar{\mu}_f \subset \mu_o$ ($f: (X,\mu_o) \to (X',\mu')$ is uniformly continuous!) and hence $\mu_f \subset \mu_o$. Thus, since μ_f is a Near-structure, $\mu_f \subset \mu_\alpha$ for each α by transfinite induction. Consequently, $\bar{\mu}_f \subset \mu_f \subset \mu$, i.e. $f: (X,\mu) \to (X',\mu')$ is uniformly continuous).

Thus let $\mu_f = \{A: A$ is a cover of X and there exists some $\mathcal{B} \in \mu'$ with $f^{-1}\mathcal{B} < A\}$. Then μ_f is a non-empty set of non-empty covers and $N_1)$ and $N_2)$ are trivially satisfied. It remains to show $N_3)$: Let $A \in \mu_f$. Then there exists some $\mathcal{B} \in \mu'$ with $f^{-1}\mathcal{B} < A$. Obviously the proof is finished, if it can be shown that

$$f^{-1}\{\text{int}_{\mu'}B: B \in \mathcal{B}\} < \{\text{int}_{\mu_f}A: A \in A\} .$$

If $B \in \mathcal{B}$, then there exists some $A \in A$ with $f^{-1}[B] \subset A$. Hence $f^{-1}[\text{int}_{\mu'}B] \subset \text{int}_{\mu_f}A$ $(x \in f^{-1}[\text{int}_{\mu'}B]$ implies $f(x) \in \text{int}_{\mu'}B$, i.e. $\{B, X' \setminus \{f(x)\}\} \in \mu'$. Thus $\{f^{-1}[B], f^{-1}[X' \setminus \{f(x)\}]\}$ is an element of μ_f refining $\{A, X \setminus \{x\}\}$, i.e. $x \in \text{int}_{\mu_f}A)$.

3.2.2.6 Remarks. (1) The initial structures in Near are formed as in S-Near (cf. 3.2.2.2 (2)) since Near is bireflective in S-Near.

(2) If the bireflection of $(X,\mu) \in |\text{S-Near}|$ with respect to Near is denoted by $1_X: (X,\mu) \to (X,\mu_N)$ and Y is a set, $((Y_i,\mu_i))_{i\in I}$ is a family of nearness spaces and $(f_i: Y_i \to Y)_{i\in I}$ is a family of maps, then $\{A: A$ is a cover of Y and $f_i^{-1}A \in \mu_i$ for each $i \in I\}_N$ is the final Near-struc-structure on Y with respect to $((Y_i,\mu_i), f_i, Y, I)$.

(3) As seen in (2) the formation of final Near-structures is quite complicated and thus hard to handle. But in the case of coproducts (= sums) the situation is much simpler on account of the special properties of the injections. Namely, if I is a set, $((X_i,\mu_i))_{i\in I}$ a family of nearness spaces and the i-th injection $(i \in I)$ is denoted by $j_i: X_i \to X$, where

$X = \bigcup_{i \in I} X_i \times \{i\}$, then

$\mu = \{A\colon A$ is a cover of X and $j_i^{-1}A \in \mu_i$ for each $i \in I\}$

is already a <u>Near</u>-structure on X , i.e. $\mu_N = \mu$ is the desired final <u>Near</u>-structure with respect to $((X_i,\mu_i),j_i,X,I)$.

(4) 3.2.2.5 implies that <u>Near</u> is closed under formation of subobjects and products (in <u>S-Near</u>).

(3) means that <u>Near</u> is also closed under formation of coproducts (in <u>S-Near</u>) .

(5) A seminearness space is nothing else but a quasi-uniform space in the sense of Isbell. Additionally the category <u>S-Near</u> is isomorphic to the category of merotopic spaces and merotopic maps in the sense of Katětov (1965).

3.2.3 Grill-determined prenearness spaces.

<u>3.2.3.1 Definitions</u>. 1) Let X be a set. A non-empty collection $G \subset P(X)$ is called a <u>grill</u> (on X) provided that the following are satisfied:

G_1) $\emptyset \notin G$,

G_2) $A \cup B \in G$ if and only if $A \in G$ or $B \in G$ for each pair $(A,B) \in P(X) \times P(X)$.

2) If (X,μ) is a prenearness space, then $A \subset P(X)$ is called <u>near</u> (in (X,μ)) provided that $\{X \setminus A\colon A \in A\} \notin \mu$.

3) A prenearness space (X,μ) is called <u>grill-determined</u> provided that each near collection of subsets of X is contained in a near grill.

<u>3.2.3.2 Remarks.</u> (1) Obviously a filter on a set X is an ultrafilter if and only if it is a grill.

(2) *If* (X,μ) *is a prenearness space, then* $A \subset P(X)$ *is near if and only if for each* $B \in \mu$ *there is some* $B \in B$ *such that* $B \cap A \neq \emptyset$ *for each* $A \in A$. (1. "⇐": No ele-

ment of $\mathcal{C} = \{X \setminus A: A \in \mathcal{A}\}$ has a non-empty intersection with each $A \in \mathcal{A}$. Hence, by assumption, $\mathcal{C} \notin \mu$, i.e. $\mathcal{A}$ is near.

2. "$\Rightarrow$" (indirectly): If the assumption is not satisfied, then there is some $\mathcal{B} \in \mu$ such that for each $B \in \mathcal{B}$ there exists some $A \in \mathcal{A}$ with $B \cap A = \emptyset$, i.e. $B \subset X \setminus A$. Hence $\mathcal{B} < \{X \setminus A: A \in \mathcal{A}\}$ and by N_1), $\{X \setminus A: A \in \mathcal{A}\} \in \mu$, i.e. $\mathcal{A}$ is not near.)

(3) If the set of all collections of subsets of X which are near is denoted by ξ, then μ is uniquely determined by ξ and vice versa. Especially N_1) corresponds to N_1') for near collections:

N_1') $\mathcal{A} << \mathcal{B}$ and $\mathcal{B} \in \xi$ imply $\mathcal{A} \in \xi$.[17)]

(4) If (X,d) is a metric space, then d induces a uniformity $\mathcal{W}_d$ whose corresponding nearness structure $\mu_{\mathcal{W}_d}$ has the base $\{\mathcal{A}_{V_\varepsilon} : \varepsilon > 0\}$ (where $V_\varepsilon = \{(x,y) \in X \times X: d(x,y) < \varepsilon\}$ and $\mathcal{A}_{V_\varepsilon} = \{V_\varepsilon(x): x \in X\}$). Let A and B be subsets of X. Then $\{A,B\}$ is near in $(X,\mu_{\mathcal{W}_d})$ if and only if the distance of A and B equals zero (where the distance $d(A,B)$ of A and B is defined by $d(A,B) = \inf\{d(x,y): x \in A, y \in B\}$).

(5) If $(X,\mathcal{X})$ is an R_0-space and $\mu_{\mathcal{X}}$ is the corresponding <u>T-Near</u>-structure, then $\{A,B\} \subset \mathcal{P}(X)$ is near in $(X,\mu_{\mathcal{X}})$ if and only if $\bar{A} \cap \bar{B} \neq \emptyset$ (note that the open covers of X form a base for $\mu_{\mathcal{X}}$).

(6) Let X be a set and for each $x \in X$, let $(\{x\})$ be the ultrafilter with the base $\{x\}$. If the set of those subsets $\mathcal{A}$ of $\mathcal{P}(X)$ for which $\{X \setminus A: A \in \mathcal{A}\}$ is not contained in $(\{x\})$ for each $x \in X$ is denoted by μ, then (X,μ) is a grill-determined prenearness space.

<u>3.2.3.3 Proposition</u>. Let X and Y be sets, $f: X \to Y$ a map and $\mathcal{G}$ a grill on X. Then $(f\mathcal{G}) = \{A \subset Y: \text{there exists some } G \in \mathcal{G} \text{ with } A \supset f[G]\}$ is a grill on Y.

---

17) If X is a set and $\mathcal{A},\mathcal{B} \subset \mathcal{P}(X)$, then we say "$\mathcal{A}$ <u>corefines</u> $\mathcal{B}$" (denoted by $\mathcal{A} << \mathcal{B}$) provided that for each $A \in \mathcal{A}$ there is some $B \in \mathcal{B}$ with $B \subset A$.

Proof. Obviously $\emptyset \notin (f\mathcal{G})$. Let A and B be subsets of Y and suppose that $A \cup B \in (f\mathcal{G})$. Then there is some $G \in \mathcal{G}$ with $A \cup B \supset f[G]$, so that $f^{-1}[A \cup B] = f^{-1}[A] \cup f^{-1}[B] \supset f^{-1}[f[G]] \supset G$ and thus $f^{-1}[A] \cup f^{-1}[B] \in \mathcal{G}$ (Since $\mathcal{G}$ is a grill, each subset of X containing an element of $\mathcal{G}$ belongs to $\mathcal{G}$!). Therefore $f^{-1}[A] = G' \in \mathcal{G}$ or $f^{-1}[B] = G'' \in \mathcal{G}$. Consequently, $A \supset f[G']$ or $B \supset f[G'']$, i.e. $A \in (f\mathcal{G})$ or $B \in (f\mathcal{G})$. Conversely, if $A \in (f\mathcal{G})$ or $B \in (f\mathcal{G})$, then since each subset of Y containing an element of $(f\mathcal{G})$ belongs to $(f\mathcal{G})$, $A \cup B \in (f\mathcal{G})$.

3.2.3.4 Proposition. Let (X,μ), (X',μ') be prenearness spaces and let $f: X \to X'$ be a map. If the set of all near collections in (X,μ) resp. (X',μ') is denoted by ξ resp. ξ' , then the following are equivalent:

(1) $f: (X,\mu) \to (X',\mu')$ is uniformly continuous.

(2) $f\mathcal{A} \in \xi'$ for each $\mathcal{A} \in \xi$.

Proof. (1) ⇒ (2): If $\mathcal{A} \in \xi$, then $\{X \setminus A: A \in \mathcal{A}\} \notin \mu$. We have to show that $f\mathcal{A} \in \xi'$, i.e. $\{X' \setminus f[A]: A \in \mathcal{A}\} \notin \mu'$. Suppose $\{X' \setminus f[A]: A \in \mathcal{A}\} \in \mu'$. Then by (1) $\{f^{-1}[X' \setminus f[A]]: A \in \mathcal{A}\} \in \mu$. Since $f^{-1}[X' \setminus f[A]] = X \setminus f^{-1}[f[A]] \subset X \setminus A$, $\{f^{-1}[X' \setminus f[A]]: A \in \mathcal{A}\} <$ $\{X \setminus A: A \in \mathcal{A}\}$. Thus $\{X \setminus A: A \in \mathcal{A}\} \in \mu$ – a contradiction.

(2) ⇒ (1): If $\mathcal{A} \in \mu'$, then $\{X' \setminus A: A \in \mathcal{A}\} \notin \xi'$. We have to show that $f^{-1}\mathcal{A} \in \mu$. Suppose $f^{-1}\mathcal{A} \notin \mu$. Then $\{X \setminus f^{-1}[A]: A \in \mathcal{A}\} \in \xi$ and hence $\{f[X \setminus f^{-1}[A]]: A \in \mathcal{A}\} \in \xi'$ by (2). Since $f[X \setminus f^{-1}[A]] = f[f^{-1}[X' \setminus A]] \subset X' \setminus A$, $\{X' \setminus A: A \in \mathcal{A}\} << \{f[X \setminus f^{-1}[A]]: A \in \mathcal{A}\}$. Thus $\{X' \setminus A: A \in \mathcal{A}\} \in \xi'$ – a contradiction.

3.2.3.5 Remark. By means of nearness spaces (resp. prenearness spaces) one can explain nearness of arbitrary collections of sets. From this fact results their name. The corresponding morphisms are precisely those maps which preserve nearness as shown in 3.2.3.4.

3.2.3.6 Theorem. The full subcategory Grill of P-Near whose objects are the grill-determined prenearness spaces is bicoreflective in <u>P-Near</u> .

<u>Proof</u>. Let (X,μ) be a prenearness space and $\xi = \{\mathcal{A} \subset P(X): \mathcal{A}$ is near in $(X,\mu)\}$. Put $\xi_G = \{\mathcal{A} \in \xi$: there exists some grill $\mathcal{G} \in \xi$ with $\mathcal{A} \subset \mathcal{G}\}$. Then a <u>Grill</u>-structure μ_G on X is defined by

$$\mathcal{A} \in \mu_G \Leftrightarrow \{X \setminus A: A \in \mathcal{A}\} \notin \xi_G$$

(i.e. ξ_G is the set of all collections which are near in (X,μ_G)) . Obviously μ_G is a non-empty set of non-empty covers of X . For the proof of $N_1)$ it suffices to show the axiom $N_1')$ for ξ_G: Thus let $\mathcal{A} << \mathcal{B}$ and $\mathcal{B} \in \xi_G$. Then $\mathcal{A} \in \xi$ because ξ satisfies $N_1')$ by assumption and $\mathcal{B} \in \xi$. Since $\mathcal{B} \in \xi_G$, there exists some grill $\mathcal{G} \in \xi$ with $\mathcal{B} \subset \mathcal{G}$. For each $A \in \mathcal{A}$, there exists some $B \in \mathcal{B}$ such that $B \subset A$. Since $B \in \mathcal{G}$ and $\mathcal{G}$ is a grill, $A \cup B = A \in \mathcal{G}$. Hence $\mathcal{A} \subset \mathcal{G}$. Thus $\mathcal{A} \in \xi_G$. Consequently, (X,μ_G) is a prenearness space. Since each grill belonging to ξ also belongs to ξ_G , (X,μ_G) is grill-determined.

Obviously $\xi_G \subset \xi$ and thus $\mu \subset \mu_G$. Therefore $1_X: (X,\mu_G) \to (X,\mu)$ is a uniformly continuous map. Furthermore, if (X',μ') is a grill-determined prenearness space and $f: (X',\mu') \to (X,\mu)$ a uniformly continuous map, then $f: (X',\mu') \to (X,\mu_G)$ is also a uniformly continuous map: Namely, if $\mathcal{A}$ is near in (X',μ') , then there exists a grill $\mathcal{G}$ which is near in (X',μ') such that $\mathcal{A} \subset \mathcal{G}$. Hence $f\mathcal{A} \in \xi$ with $f\mathcal{A} \subset f\mathcal{G} \in \xi$. Furthermore, $(f\mathcal{G})$ is a grill (cf. 3.2.3.3) containing $f\mathcal{A}$ and belonging to ξ by $(f\mathcal{G}) << f\mathcal{G}$. Thus $f\mathcal{A} \in \xi_G$. Hence everything is proved by means of 3.2.3.4.

<u>3.2.3.7 Remarks</u>. (1) <u>Grill</u> is a topological category as a bicoreflective subcategory of a topological category.

(2) <u>Grill</u> is closed under formation of coproducts and quotient objects in <u>P-Near</u> .

(3) Final structures in <u>Grill</u> are formed as in <u>P-Near</u> .

(4) Initial structures in <u>Grill</u> are obtained

by forming them first in P-Near and then applying the bicoreflecto G: P-Near $\to$ Grill . In particular, a subobject (in P-Near) of a grill-determined prenearness space is again a grill-determined prenearness space.

⑤ Grill *is also bicoreflective in* S-Near ; *for a grill-determined prenearness space is a seminearness space* (Obviously, there is an equivalent formulation of N_2) for near collections of sets, namely

N_2') $\mathcal{A} \vee \mathcal{B} \in \xi$ implies $\mathcal{A} \in \xi$ or $\mathcal{B} \in \xi$,

where $\mathcal{A} \vee \mathcal{B} = \{A \cup B\colon A \in \mathcal{A}$ and $B \in \mathcal{B}\}$. Now let (X,μ) be a grill-determined prenearness space. Further let $\mathcal{A} \vee \mathcal{B} \in \xi$. Then there is some grill $\mathcal{G} \in \xi$ with $\mathcal{A} \vee \mathcal{B} \subset \mathcal{G}$. Suppose now that $\mathcal{A} \notin \xi$ and $\mathcal{B} \notin \xi$. Then by N_1') , $\mathcal{A} \not\subset \mathcal{G}$ and $\mathcal{B} \not\subset \mathcal{G}$. Thus there exist $A \in \mathcal{A}$ with $A \notin \mathcal{G}$ and $B \in \mathcal{B}$ with $B \notin \mathcal{G}$. This is a contradiction because $\mathcal{G}$ is a grill and $A \cup B \in \mathcal{G}$. Consequently, (X,μ) is a seminearness space.). Evidently $1_X\colon (X,\mu_G) \to (X,\mu)$ is the bicoreflection of $(X,\mu) \in |$S-Near$|$ with respect to Grill .

3.2.3.8 Definitions. Let (X,μ) be a prenearness space.

1) $\mathcal{A} \subset P(X)$ is called a Cauchy system[18] (in (X,μ)) provided that for each $\mathcal{U} \in \mu$, there exist $A \in \mathcal{A}$ and $U \in \mathcal{U}$ with $A \subset U$. The set of all Cauchy systems in (X,μ) is denoted by γ_μ or briefly by γ .

2) A filter on X is called a Cauchy filter on (X,μ) provided that it belongs to γ_μ , i.e. it is a Cauchy system.

3.2.3.9 Remarks. ① For uniform spaces the above definition of a Cauchy filter coincides with the usual one . (Let (X,μ) be a uniform [nearness] space and $\mathcal{F}$ be a filter on X:

1. Let $\mathcal{F}$ be a Cauchy filter and $W \in \mathcal{W}_\mu$. Then there exists some $\mathcal{U} \in \mu$ such that $B_\mu = \bigcup_{U\in\mathcal{U}} U \times U \subset W$. Further there are $U \in \mathcal{U}$ and $F \in \mathcal{F}$ with $F \subset U$. Thus $F \times F \subset U \times U \subset B_\mu \subset W$. Consequently, $\mathcal{F}$ is a Cauchy filter in the usual sense.

2. Let $\mathcal{F}$ be a Cauchy filter in the usual sense and $\mathcal{U} \in \mu$.

18) or occasionally a collection with arbitrary small members.

Then there exists some $\mathcal{W} \in \mu$ with $\mathcal{W} *< \mathcal{U}$. Further there is some $F \in \mathcal{F}$ with $F \times F \subset W$, where $W = \bigcup_{V \in \mathcal{W}} V \times V$. Since $F \neq \emptyset$, there exists some $x \in F$. Furthermore, there exists some $U \in \mathcal{U}$ such that $St(\{x\},\mathcal{W}) \subset U$. Thus $F \subset U$ [$y \in F$ implies $(x,y) \in F \times F \subset W$ so that $y \in W(x) = St(\{x\},\mathcal{W}) \subset U$].)

(2) If (X,d) is a metric space, then d induces a uniformity $\mathcal{W}_d$ for which the collection $\{\mathcal{A}_{V_\varepsilon} : \varepsilon > 0\}$ is a base for the corresponding nearness structure $\mu_{\mathcal{W}_d}$ ($V_\varepsilon = \{(x,y): d(x,y) < \varepsilon\}$ and $\mathcal{A}_{V_\varepsilon} = \{V_\varepsilon(x): x \in X\}$). $\mathcal{A} \subset \mathcal{P}(X)$ is a Cauchy system in $(X,\mu_{\mathcal{W}_d})$ if and only if for each positive real number ε, there is some $A \in \mathcal{A}$ such that $d(A) = \sup\{d(x,y): x,y \in A\} < \varepsilon$, i.e. $\inf\{d(A): A \in \mathcal{A}\} = 0$.

(3) If $(X,\mathcal{X})$ is an R_0-space and $\mu_{\mathcal{X}}$ is the corresponding <u>T-Near</u>-structure, then $\mathcal{A} \subset \mathcal{P}(X)$ is a Cauchy system in $(X,\mu_{\mathcal{X}})$ if and only if $\mathcal{A}$ <u>converges</u> in $(X,\mathcal{X})$, i.e. if there is some $x \in X$ such that the neighbourhood filter of x corefines $\mathcal{A}$ (If $\mathcal{A}$ is a filter, then this definition of convergence coincides with the usual definition of convergence in a topological space!).

(4) μ is uniquely determined by γ; for *if* (X,μ) *is a prenearness space and* $\mathcal{U} \subset \mathcal{P}(X)$, *then* $\mathcal{U} \in \mu$ *if and only if for each* $\mathcal{A} \in \gamma$, *there exist* $A \in \mathcal{A}$ *and* $U \in \mathcal{U}$ *with* $A \subset U$.

(1."⇒": The proof is trivial.

2."⇐" (indirectly): If $\mathcal{U} \notin \mu$, then there does not exist any $\mathcal{V} \in \mu$ with $\mathcal{V} < \mathcal{U}$. Hence for each $\mathcal{V} \in \mu$, we can choose some $A_{\mathcal{V}} \in \mathcal{V}$ such that $A_{\mathcal{V}} \not\subset U$ for each $U \in \mathcal{U}$. Then $\mathcal{A} = \{A_{\mathcal{V}}: \mathcal{V} \in \mu\} \in \gamma$ does not satisfy the desired condition.) On the other hand μ is uniquely determined by the set ξ of all collections of sets which are near. Therefore γ and ξ are uniquely determined by each other. A simple description of this relation is given by the operator sec defined below. Additionally sec clarifies the relationship between grills and filters.

3.2.3.10 Definition. Let X be a set and $\mathcal{A} \subset P(X)$. Then $\sec \mathcal{A} = \{B \subset X : A \cap B \neq \emptyset$ for each $A \in \mathcal{A}\}$.

3.2.3.11 Proposition. Let (X,μ) be a prenearness space and $\mathcal{A} \subset P(X)$. Then the following are satisfied:

(1) $\mathcal{A}$ is a Cauchy system if and only if $\sec \mathcal{A}$ is near.

(2) $\mathcal{A}$ is near if and only if $\sec \mathcal{A}$ is a Cauchy system.

Proof. (1) a) "$\Rightarrow$": If $\mathcal{U} \in \mu$, then there exist $A \in \mathcal{A}$ and $U \in \mathcal{U}$ with $A \subset U$. Each $B \in \sec \mathcal{A}$ meets A and hence U. Thus $\sec \mathcal{A}$ is near by 3.2.3.2 (2).

b) "$\Leftarrow$": If $\mathcal{U} \in \mu$, then there exists some $U \in \mathcal{U}$ with $U \cap B \neq \emptyset$ for each $B \in \sec \mathcal{A}$ (cf. 3.2.3.2 (2)). Thus $X \setminus U \notin \sec \mathcal{A}$. Consequently, there exists some $A \in \mathcal{A}$ such that $A \cap (X \setminus U) = \emptyset$, i.e. $A \subset U$.

(2) a) "$\Rightarrow$": If $\mathcal{U} \in \mu$, then there exists some $U \in \mathcal{U}$ such that $U \cap A \neq \emptyset$ for each $A \in \mathcal{A}$. Hence $U \in \sec \mathcal{A}$.

b) "$\Leftarrow$": If $\mathcal{U} \in \mu$, then there are $U \in \mathcal{U}$ and $B \in \sec \mathcal{A}$ with $B \subset U$. Thus $U \in \sec \mathcal{A}$.

3.2.3.12 Proposition. Let X be a set and $\mathcal{A}$ a non-empty collection of subsets of X. Then the following are satisfied:

(1) If $\mathcal{A}$ is a filter on X, then $\sec \mathcal{A}$ is a grill on X.

(2) If $\mathcal{A}$ is a grill on X, then $\sec \mathcal{A}$ is a filter on X.

Proof. (1) a) $\emptyset \notin \sec \mathcal{A}$ (trivial!).

b) Let $G \in \sec \mathcal{A}$ and $G' \in P(X)$ with $G' \supset G$. Then since $G \cap A \subset G' \cap A$ for each $A \in \mathcal{A}$, $G' \in \sec \mathcal{A}$.

c) (indirectly) Let $G_1, G_2 \in P(X)$ with $G_1 \notin \sec \mathcal{A}$ and $G_2 \notin \sec \mathcal{A}$. Then there exist $A_1, A_2 \in \mathcal{A}$ such that $G_1 \cap A_1 = \emptyset$ and $G_2 \cap A_2 = \emptyset$. Hence $X \setminus G_1 \supset A_1$ and $X \setminus G_2 \supset A_2$. Since $\mathcal{A}$ is a filter, $(X \setminus G_1) \cap (X \setminus G_2) = X \setminus (G_1 \cup G_2) \in \mathcal{A}$. Consequently, $G_1 \cup G_2 \notin \sec \mathcal{A}$.

(2) a) $\emptyset \notin \sec \mathcal{A}$ (trivial!).

b) The fact that $\sec \mathcal{A}$ is closed under formation of supersets is proved analogously to (1) b).

c) (indirectly). Let $F_1, F_2 \in \mathcal{P}(X)$. If $F_1 \cap F_2 \notin \sec \mathcal{A}$, then there exists some $A \in \mathcal{A}$ with $(F_1 \cap F_2) \cap A = \emptyset$, i.e. $X \setminus (F_1 \cap F_2) = (X \setminus F_1) \cup (X \setminus F_2) \supset A$. Hence $(X \setminus F_1) \cup (X \setminus F_2) \in \mathcal{A}$ and further $X \setminus F_1 \in \mathcal{A}$ or $X \setminus F_2 \in \mathcal{A}$ since $\mathcal{A}$ is a grill. Consequently, $F_1 \notin \sec \mathcal{A}$ or $F_2 \notin \sec \mathcal{A}$.

3.2.3.13 Theorem. A prenearness space (X,μ) is grill-determined if and only if each Cauchy system in (X,μ) is corefined by some Cauchy filter.

Proof. 1) "⇒": Let $\mathcal{A}$ be a Cauchy system in (X,μ) . Then $\sec \mathcal{A}$ is near by 3.2.3.11 (1). By assumption there exists some near grill $\mathcal{G}$ with $\sec \mathcal{A} \subset \mathcal{G}$. By 3.2.3.11 and 3.2.3.12 $\mathcal{F} = \sec \mathcal{G}$ is a Cauchy filter satisfying $\mathcal{F} << \mathcal{A}$; for otherwise there is some $F \in \mathcal{F}$ such that $A \not\subset F$ for each $A \in \mathcal{A}$, i.e. for each $A \in \mathcal{A}$ there is some $x_A \in A$ belonging to CF . Therefore $CF \cap A \neq \emptyset$ for each $A \in \mathcal{A}$, i.e. $CF \in \sec \mathcal{A} \subset \mathcal{G}$. On the other hand $F \in \sec \mathcal{G}$, hence $CF \cap F \neq \emptyset$ which is impossible.

2) "⇐": Let $\mathcal{A}$ be near in (X,μ) . Then $\sec \mathcal{A}$ is a Cauchy system (cf. 3.2.3.11 (2)). By assumption, there exists some Cauchy filter $\mathcal{F}$ with $\mathcal{F} << \sec \mathcal{A}$. By 3.2.3.11 and 3.2.3.12 $\mathcal{G} = \sec \mathcal{F}$ is a near grill. Furthermore, $\mathcal{A} \subset \mathcal{G}$; for otherwise there is some $A \in \mathcal{A}$ with $A \notin \mathcal{G} = \sec \mathcal{F}$, hence there exists some $F \in \mathcal{F}$ with $A \cap F = \emptyset$, i.e. $CA \supset F$. Since $\mathcal{F}$ is a filter, $CA \in \mathcal{F}$ and because of $\mathcal{F} << \sec \mathcal{A}$ there exists some $B \in \sec \mathcal{A}$ with $B \subset CA$. Thus since $B \cap A \neq \emptyset$, $A \cap CA \neq \emptyset$ which is impossible.

3.2.3.14 Remark. By the above theorem there is no distinction between grill-determined prenearness spaces and filter merotopic spaces in the sense of Katětov [53].

3.2.3.15 Theorem. A nearness space (X,μ) can be embedded in a topological nearness space if and only if (X,μ) is grill-determined.

Proof. 1) "⇒": Without loss of generality let (X,μ) be a subspace of a topological nearness space (X',μ') and let $i: X \to X'$ be the inclusion map. If $\mathcal{A}$ is a collection of sets which is near in (X,μ), then $i\mathcal{A} = \mathcal{A}$ is near in (X',μ') and thus $\bigcap_{A\in\mathcal{A}} \bar{A}^{X_{\mu'}} \neq \emptyset$ (otherwise for each $x' \in X'$, there is some $A \in \mathcal{A}$ with $x' \notin \bar{A}$ and thus $\{X' \setminus \bar{A}: A \in \mathcal{A}\}$ is an open cover of X' containing no element that meets all $A \in \mathcal{A}$, i.e. $\mathcal{A}$ is not near in (X',μ')). Let $y \in \bigcap_{A\in\mathcal{A}} \bar{A}^{X_{\mu'}}$. Then $\mathcal{A} \subset \mathcal{B} = \{B \subset X: y \in \bar{B}^{X_{\mu'}}\}$. Obviously $\mathcal{B}$ is near in (X,μ) [a) $\mathcal{B} \subset P(X)$ is near in (X,μ) if and only if $\mathcal{B}$ is near in (X',μ').

b) Since $y \in \bigcap_{B\in\mathcal{B}} \bar{B}^{X_{\mu'}}$ it follows that $\bigcap_{B\in\mathcal{B}} \bar{B}^{X_{\mu'}} \neq \emptyset$. Thus $\{\bar{B}^{X_{\mu'}}: B \in \mathcal{B}\}$ is near[19] in (X',μ'). Hence $\mathcal{B}$ is near[20].]. Further $\mathcal{B}$ is a grill:

1. $\emptyset \notin \mathcal{B}$ (trivial).
2. $B \in \mathcal{B}$ and $B' \supset B$ imply $B' \in \mathcal{B}$ since $\bar{B} \subset \overline{B'}$.
3. (indirectly) If $B_1 \notin \mathcal{B}$ and $B_2 \notin \mathcal{B}$, then $y \notin \overline{B_1}$ and $y \notin \overline{B_2}$. Hence $y \notin \overline{B_1} \cup \overline{B_2} = \overline{B_1 \cup B_2}$. Consequently, $B_1 \cup B_2 \notin \mathcal{B}$.

2) "⇐": Let (X,μ) be a grill-determined nearness space and (X,X_μ) be the underlying topological space. For each nonconvergent Cauchy filter $\mathcal{A}$ one adds a point $y_{\mathcal{A}}$ to X. Thereby a set X' is obtained which contains X. If the neighbourhood filter of $A \in \mathcal{A}$ in (X,X_μ) is denoted by $\mathcal{U}_{X_\mu}(A)$, then a filter on X' is defined by

$$\mathcal{U}(y_{\mathcal{A}}) = \{U \subset X': y_{\mathcal{A}} \in U \text{ and there exist } A \in \mathcal{A} \text{ and } V_A \in \mathcal{U}_{X_\mu}(A) \text{ with } V_A \subset U\}$$

---

19) otherwise $\{C\bar{B}: B \in \mathcal{B}\} \in \mu'$ and thus $\bigcup_{B\in\mathcal{B}} C\bar{B} = X'$, i.e. $\bigcap_{B\in\mathcal{B}} \bar{B} = \emptyset$.

20) otherwise $\{CB: B \in \mathcal{B}\} \in \mu'$ and thus $\{(CB)^\circ: B \in \mathcal{B}\} = \{C\bar{B}: B \in \mathcal{B}\} \in \mu'$, hence $\{\bar{B}: B \in \mathcal{B}\}$ is not near.

(note that $\mathcal{A}$ is a filter). If the neighbourhood filter of $x \in X$ is denoted by $\mathcal{U}_{X_\mu}(x)$, then a filter on X' is defined by

$$\mathcal{U}(x) = (\mathcal{U}_{X_\mu}(x)) := \{U \subset X' : \text{there is some } V \in \mathcal{U}_{X_\mu}(x) \text{ with } V \subset U\}$$

Obviously $\mathcal{U}: X' \to P(P(X'))$ is a complete neighbourhood system of X'. The given topological space (X, X_μ) is a subspace of the topological space $(X', X_\mathcal{U})$[21] obtained above. Moreover $(X', X_\mathcal{U})$ is an R_0-space. Let (X', μ') be the corresponding topological nearness space. Let γ (resp. γ') be the set of all Cauchy systems belonging to μ (resp. μ'). In order to show that (X,μ) is a subspace of (X',μ') it suffices to prove that $\gamma = \{\mathcal{B} \subset P(X) : \mathcal{B} \in \gamma'\}$:

1. Let $\mathcal{B} \in \gamma$. Then $\mathcal{B} \subset P(X)$ and there exists some Cauchy filter $\mathcal{A}$ on X with $\mathcal{A} << \mathcal{B}$ (cf. 3.2.3.13):

a) If $\mathcal{A}$ converges in (X,μ) (i.e. in (X, X_μ)), then $(i\mathcal{A}) = i(\mathcal{A})$ converges in (X',μ') (i.e. in $(X', X_\mathcal{U})$) ($i: X \to X'$ denotes the inclusion map). Thus $i(\mathcal{A})$ is a Cauchy system in (X',μ') (cf. 3.2.3.9 ③). But obviously $i(\mathcal{A}) << \mathcal{A}$. Consequently, $\mathcal{B} \in \gamma'$[22].

b) If $\mathcal{A}$ does not converge in (X,μ), then $i(\mathcal{A})$ converges in (X',μ') (obviously $\mathcal{U}(y_\mathcal{A}) << i(\mathcal{A})$). By the above arguments (cf. a)) one concludes $\mathcal{B} \in \gamma'$.

2. Let $\mathcal{B} \subset P(X)$ and $\mathcal{B} \in \gamma'$. Then $\mathcal{B}$ converges in (X',μ'), i.e. there exists some $x' \in X'$ such that $\mathcal{U}(x') << \mathcal{B}$:

a) $x' \in X$: Obviously $\mathcal{U}_{X_\mu}(x') << \mathcal{B}$, i.e. $\mathcal{B}$ converges in (X,μ). Then $\mathcal{B} \in \gamma$ (exercise).

b) $x' = y_\mathcal{A}$: Since $\mathcal{B} \subset P(X)$ and $\mathcal{U}(y_\mathcal{A}) << \mathcal{B}$,

$$\mathcal{U}(\mathcal{A}) = \{V_A \in \mathcal{U}_{X_\mu}(A) : A \in \mathcal{A}\} << \mathcal{B} .$$

---

[21] $X_\mathcal{U}$ denotes the topology induced by the complete neighbourhood system $\mathcal{U}$.

[22] If γ is the set of all Cauchy systems in a prenearness space (X,μ), then the axiom

N_1'') $\mathcal{A} << \mathcal{B}$ and $\mathcal{A} \in \gamma$ imply $\mathcal{B} \in \gamma$

corresponds to N_1).

Obviously $\mathcal{U}(\mathcal{A}) \in \gamma$ since $\mathcal{A} \in \gamma$ and the axiom N_3) holds. Then $\mathcal{B}$ belongs also to γ.

3.2.3.16 Remark. According to H.L. Bentley a nearness space (X,μ) is called subtopological provided that (X,μ) can be embedded in a topological nearness space. As well-known a subspace of a topological nearness space is generally not topological (note that a subspace of a paracompact topological space is generally not paracompact). By the above theorem the grill-determined nearness spaces are identical with the subtopological ones. It can be shown that the full subcategory Sub Top of Near consisting of all subtopological nearness spaces is bireflective in Grill (the bireflector is obtained by the restriction of the bireflector from S-Near into Near to Grill) and bicoreflective in Near (the bicoreflector is obtained by the restriction of the bicoreflector from S-Near into Grill to Near). Furthermore, one can show that T-Near ($\cong$ R_0-Top) is bicoreflective in Sub Top (the bicoreflector is obtained by the restriction of the bicoreflector from Near into T-Near to Sub Top) and that C-Near ($\cong$ Cont) is bireflective in Sub Top (the bireflector is obtained by the restriction of the bireflector from Near into C-Near to Sub Top) . [It remains to show that C-Near is a subcategory of Sub Top , i.e. that *every contigual nearness space* (X,μ) *is grill-determined* which can be shown as follows: If $X \neq \emptyset$ (the case that $X = \emptyset$ is trivial) and $\mathcal{A} \subset P(X)$ is near, then $\mathcal{B} = \mathcal{A} \cup \{X\}$ is a non-empty collection of sets which is near. The set $M = \{\mathcal{C} \subset P(X): \mathcal{C}$ is near in (X,μ) and $\mathcal{C} \supset \mathcal{B}\}$ ordered by inclusion is inductively ordered; for if $\mathcal{V} \subset M$ is totally ordered, then $\mathcal{D} = \bigcup_{\mathcal{C}\in\mathcal{V}} \mathcal{C}$ is an upper bound (obviously $\mathcal{B} \subset \mathcal{C} \subset \mathcal{D}$ for each $\mathcal{C} \in \mathcal{V}$ and $\mathcal{D}$ is near; for if $\mathcal{D} \notin \xi$, then $\{X \setminus D: D \in \mathcal{D}\} \in \mu$ and since (X,μ) is contigual, there exist finitely many sets $D_1,\ldots,D_n \in \mathcal{D}$ such that $\{X \setminus D_i: i \in \{1,\ldots,n\}\} \in \mu$, i.e. $\{D_1,\ldots,D_n\} \notin \xi$ in contradiction to the fact that for each D_i, there exists some $\mathcal{C}_i \in \mathcal{V}$ with $D_i \in \mathcal{C}_i$ for each $i \in \{1,\ldots,n\}$ and since $\mathcal{V}$ is totally ordered, there exists some

$k \in \{1,\dots,n\}$ with $\{D_1,\dots,D_n\} \subset C_k \in \xi\}$. Hence M has maximal elements by Zorn's Lemma. Let G be a maximal element of M. Then $A \subset G$ and G is a grill: Since $B \subset G$, it follows that $G \neq \emptyset$, and $\emptyset \notin G$ because G is near. If $G \in G$ and $G \subset G'$, then since G is maximal, $G' \in G$. It remains to show that $G_1 \notin G$ and $G_2 \notin G$ imply $G_1 \cup G_2 \notin G$. At first $G_i \notin G$ implies $G \cup \{G_i\} \notin \xi$ for $i \in \{1,2\}$, i.e. $U_i = \{X \setminus G\colon G \in G\} \cup \{X \setminus G_i\} \in \mu$. Since $U_1 \wedge U_2 < U = \{X \setminus G\colon G \in G\} \cup \{X \setminus (G_1 \cup G_2)\}$, it follows that $U \in \mu$. Thus $G \cup \{G_1 \cup G_2\} \notin \xi$ and consequently, $G_1 \cup G_2 \notin G$.]

<u>3.2.3.17 Proposition</u>. Let $(X_i)_{i\in I}$ be a non-empty family of sets, $\left(\prod_{i\in I} X_i,(p_i)\right)$ the product of this family in the category of sets and let G_i be a grill on X_i for each $i \in I$. Then

$$\otimes G_i = \Big\{G \subset \prod_{i\in I} X_i\colon \text{for each finite cover } B \text{ of } G\text{, there exists some } B \in B \text{ such that } p_i[B] \in G_i \text{ for each } i \in I\Big\}$$

is a grill on $\prod_{i\in I} X_i$ such that $p_j(\otimes G_i) \subset G_j$ for each $j \in I$. The <u>proof</u> is obvious.

<u>3.2.3.18 Proposition</u>. Let $(f_i\colon X_i \to Y_i)_{i\in I}$ be a non-empty family of maps and let

$$\begin{array}{ccc} \prod X_i & \xrightarrow{f=\prod f_i} & \prod Y_i \\ {\scriptstyle p_i}\downarrow & & \downarrow{\scriptstyle q_i} \\ X_i & \xrightarrow{f_i} & Y_i \end{array}$$

be the corresponding product diagram in the category of sets. Furthermore, let G_i be a grill on X_i for each $i \in I$. Then the following are satisfied:

(a) $\otimes (f_i G_i) = (f \otimes G_i)$

(b) If $K = \{i \in I\colon f_i[X_i] \neq Y_i\}$ is finite and H_i is a grill on Y_i such that $f_i^{-1} \mathcal{H}_i \subset \mathcal{G}_i$ for each $i \in I$, then $f^{-1}(\otimes \mathcal{H}_i) \subset \otimes \mathcal{G}_i$.

Proof. (a) is easy to check.

(b) (indirectly). If $f^{-1}(\otimes \mathcal{H}_i) \not\subset \otimes \mathcal{G}_i$, then there exists some $H \in \otimes \mathcal{H}_i$ with $f^{-1}[H] \notin \otimes \mathcal{G}_i$. By 3.2.3.17 there is some finite cover $\mathcal{B}$ of $f^{-1}[H]$ such that for each $B \in \mathcal{B}$, there exists some $j \in I$ with $p_j[B] = A_j \notin \mathcal{G}_j$. Hence there exist some finite subset J of I and $A_j \notin \mathcal{G}_j$ for each $j \in J$ such that $f^{-1}[H] \subset \bigcup_{j\in J} p_j^{-1}[A_j]$. If B_j is the largest subset of A_j with $f_j^{-1}[f_j[B_j]] = B_j$ (B_j may be empty), then $B_j \notin \mathcal{G}_j$ (otherwise $A_j \in \mathcal{G}_j$) and $f^{-1}[H] \subset \bigcup_{j\in J} p_j^{-1}[B_j]$ [23]. Then

$f[f^{-1}[H]] \subset \bigcup_{j\in J} f[p_j^{-1}[B_j]]$ and

$H \subset \bigcup_{j\in J} f[p_j^{-1}[B_j]] \cup \bigcup_{k\in K} q_k^{-1}[Y_k \setminus f_k[X_k]]$. In this finite cover of H there is some element whose i-th projection belongs to $\mathcal{H}_i$ for each $i \in I$ because $H \in \otimes \mathcal{H}_i$. If this element equals $q_k^{-1}[Y_k \setminus f_k[X_k]]$ for some $k \in K$, then $q_k[q_k^{-1}[Y_k \setminus f_k[X_k]]] =$ $= Y_k \setminus f_k[X_k] \in \mathcal{H}_k$ and by assumption, $f_k^{-1}[Y_k \setminus f_k[X_k]] = \emptyset \in \mathcal{G}_k$ which is impossible. If this element equals $f[p_j^{-1}[B_j]]$ for some $j \in J$, then $q_j[f[p_j^{-1}[B_j]]] = f_j[p_j[p_j^{-1}[B_j]]] =$ $f_j[B_j] \in \mathcal{H}_j$ and by assumption, $f_j^{-1}[f_j[B_j]] = B_j \in \mathcal{G}_j$ which is impossible (cf. above).

3.2.3.19 Remarks. (1) If (X,μ) (resp. (X',μ')) is a prenearness space, $f\colon (X,\mu) \to (X',\mu')$ a uniformly continuous map and ξ (resp. ξ') the corresponding set of near collections of sets, then f is a quotient map in P-Near if and only if f is surjective and $\xi' = \{\mathcal{A} \subset \mathcal{P}(X') : f^{-1}\mathcal{A} \in \xi\}$. Since Grill is bicoreflective in P-Near, the quotient maps are formed in Grill as in P-Near.

---

[23] $x \in f^{-1}[H]$ implies the existence of some $j \in J$ with $p_j(x) = x_j \in A_j$. Since $f_j^{-1}[f_j[B_j \cup f_j^{-1}[\{f_j(x_j)\}]]] = B_j \cup f_j^{-1}[\{f_j(x_j)\}]$ contains x_j, it follows that $x_j \in B_j$ by means of the maximality of B_j, i.e. $x \in p_j^{-1}[B_j]$.

② Let $((X_i,\mu_i))_{i\in I}$ be a family of prenearness spaces and let ξ_i be the corresponding set of near collections of sets for each $i \in I$. If $((\prod X_i,\mu),(p_i))$ is the product of this family formed in <u>P-Near</u>, then any collection $\mathcal{A} \subset \mathcal{P}(\prod X_i)$ is near if and only if $p_i\mathcal{A} \in \xi_i$ for each $i \in I$. The set of all near collections of sets obtained in this way is denoted by ξ. If all (X_i,μ_i) are grill-determined, then a collection $\mathcal{A} \subset \mathcal{P}(\prod X_i)$ is near in the product of the family $((X_i,\mu_i))_{i\in I}$ formed in <u>Grill</u> provided that there is some grill $\mathcal{G} \in \xi$ with $\mathcal{A} \subset \mathcal{G}$.

<u>3.2.3.20 Theorem</u>. In <u>Grill</u> every product of quotient maps is a quotient map.

<u>Proof</u>. Let $(f_i\colon (X_i,\mu_i) \to (X_i',\mu_i'))_{i\in I}$ be any non-empty family of quotient maps in <u>Grill</u> and let

$$\begin{array}{ccc} (P,\mu) & \xrightarrow{f=\prod f_i} & (P',\mu') \\ {\scriptstyle p_i}\downarrow & & \downarrow{\scriptstyle p_i'} \\ (X_i,\mu_i) & \xrightarrow{f_i} & (X_i',\mu_i') \end{array}$$

be the corresponding product diagram in <u>Grill</u>. Since all f_i are surjective, f is surjective. If $\mathcal{A}$ is near in (P',μ'), then there exists some near grill $\mathcal{H}$ with $\mathcal{A} \subset \mathcal{H}$. Then $p_i'\mathcal{H}$ is near in (X_i',μ_i') and $f_i^{-1}p_i'\mathcal{H}$ is near in (X_i,μ_i) for each $i \in I$ because each f_i is a quotient map. Thus for each $i \in I$, there is some grill $\mathcal{G}_i$ on X_i which is near in (X_i,μ_i) such that $f_i^{-1}p_i'\mathcal{H} \subset \mathcal{G}_i$. Then $\otimes\mathcal{G}_i$ is near in (P,μ) (since $p_j \otimes \mathcal{G}_i \subset \mathcal{G}_j$ for each $j \in I$, $p_j \otimes \mathcal{G}_i << \mathcal{G}_j$ and hence $p_j \otimes \mathcal{G}_i$ is near in (X_j,μ_j) for each $j \in I$) and by 3.2.3.18 (b),

$$f^{-1}(\otimes(p_i'\mathcal{H})) \subset \otimes\, \mathcal{G}_i\ .$$

On the other hand $f^{-1}\mathcal{A} \subset f^{-1}\mathcal{H} \subset f^{-1}(\otimes(p_i'\mathcal{H})) \subset \otimes\,\mathcal{G}_i$ (note: $\mathcal{H} \subset \otimes(p_i'\mathcal{H})$). Therefore $f^{-1}\mathcal{A}$ is near in (P,μ).
Conversely, if $\mathcal{A} \subset \mathcal{P}(P')$ such that $f^{-1}\mathcal{A}$ is near in (P,μ), then since f is a surjective uniformly continuous map,

$ff^{-1}A = A$ is near in (P',μ'). Now the proof is finished (cf. 3.2.3.19 ①).

3.2.3.21 Remark. If $((X_i,\mu_i))_{i\in I}$ is a family of prenearness spaces and the set of all collections which are near in (X_i,μ_i) is denoted by ξ_i, then for the set ξ of all collections which are near in the coproduct $((X,\mu),(j_i))$ of this family holds:

$$\xi = \{A \subset P(X): j_i^{-1}A \in \xi_i \text{ for some } i \in I\}$$

(The coproduct is formed in P-Near!). A corresponding assertion holds for Grill since the coproducts in Grill are formed as in P-Near.

3.2.3.22 Theorem. In the category Grill the following is satisfied:

$$X \times \coprod_{i\in I} Y_i \cong \coprod_{i\in I} X \times Y_i .$$

Proof. Consider the following commutative diagram in Grill

$$\begin{array}{ccc}
Y_i = (Y_i,\eta_i) & \xrightarrow{j_i} & (Y,\eta) = \coprod_{i\in I} Y_i \\
\uparrow p_i & & \uparrow p_Y \\
X \times Y_i = (X\times Y_i,\eta_i') & \xrightarrow{1_X\times j_i} & (X\times Y,\eta') = X \times \coprod_{i\in I} Y_i \\
\downarrow p_X & & \downarrow p_X \\
X = (X,\xi) & \xrightarrow{1_X} & (X,\xi) = X
\end{array},$$

in which the top row represents a non-empty coproduct and the columns are products in Grill and in which all Grill-structures are described by collections of sets which are near. It must be shown that the middle row represents a coproduct in Grill. Obviously $(1_X\times j_i: X\times Y_i \to X\times Y)$ is a coproduct in the category of sets [24]. Thus it remains to show that for any $A \in \eta'$, $(1_X \times j_k)^{-1} A \in \eta'_k$ for some $k \in I$ (note 3.2.3.21). At first A is contained in some grill $G \in \eta'$. $p_X G \in \xi$ and $p_Y G \in \eta$ are grills since p_X and p_Y are surjective (thus $p_X G = (p_X G)$ and $p_Y G = (p_Y G)$ which is easily verified). Since η is the final Grill-structure with respect to (j_i), $j_k^{-1} p_Y G \in \eta_k$ for some $k \in I$; hence $j_k^{-1} p_Y G \subset H$ for some grill $H \in \eta_k$. By 3.2.3.18 (b),

$$(*) \qquad (1_X \times j_k)^{-1} (p_X G \otimes p_Y G) \subset p_X G \otimes H .$$

Obviously $p_X G \otimes H \in \eta'_k$ (note that $p_X G \otimes H$ is a grill for which $p_X (p_X G \otimes H) \subset p_X G \in \xi$, hence $p_X (p_X G \otimes H) \in \xi$ and $p_k (p_X G \otimes H) \subset H \in \eta_k$, hence $p_k (p_X G \otimes H) \in \eta_k$). Since on the other hand $A \subset G \subset p_X G \otimes p_Y G$ (which is easily checked), we obtain in connection with (*)

$$(1_X \times j_k)^{-1} A \in \eta'_k .$$

---

24) The category of sets is cartesian closed (cf. the next chapter).

CHAPTER IV

CARTESIAN CLOSED TOPOLOGICAL CATEGORIES

The category Top of topological spaces and continuous maps fails to have some desirable properties, e.g. the product of two quotient maps need not be a quotient map and there is in general no natural function space topology, i.e. Top is not cartesian closed. Because of this fact, which is inconvenient for investigations in algebraic topology (homotopy theory), functional analysis (duality theory) or topological algebra (quotients), Top has been substituted either by well-behaved subcategories or by more convenient supercategories. Unfortunately most of these categories suffer from other deficiencies. Some of them are too small, e.g. the coreflective hull of all (compact) metrizable spaces in Top [whose objects are called sequential spaces] or too big, e.g. the category of quasi-topological spaces introduced by E. Spanier (the quasitopologies on a fixed set in general form a proper class!). Another well-behaved candidate namely the coreflective hull of all compact Hausdorff spaces in Top [whose objects are called compactly generated spaces] has not been described by suitable axioms.

The category Grill introduced in the preceding chapter is free from the above mentioned deficiencies. In this chapter the relations of Grill to other nice cartesian closed topological categories (e.g. to the categories Conv of convergence spaces, Lim of limit spaces and PsTop of pseudotopological spaces, each of which contains Top) are investigated. For topological categories several equivalent characterizations of cartesian closedness are given. In the next chapter further results on cartesian closedness will be obtained even for non-topological categories.

4.1 Definition and equivalent characterizations

4.1.1 Definition. A category C is called cartesian closed provided that the following conditions are satisfied:

(1) For each pair (A,B) of C-objects, there exists a product $A \times B$ in C.

(2) For each C-object A holds: For each C-object B, there exists some C-object B^A and some C-morphism $e_{A,B} : A \times B^A \longrightarrow B$ such that for each C-object C and each C-morphism $f : A \times C \longrightarrow B$, there exists a unique C-morphism $\bar{f} : C \longrightarrow B^A$ such that the diagram

$$\begin{array}{ccc} A \times B^A & \xrightarrow{e_{A,B}} & B \\ & {}_{1_A \times \bar{f}}\nwarrow \quad \nearrow_{f} & \\ & A \times C & \end{array}$$

commutes.

4.1.2 Remark. Each C-object A defines a functor $F_A : C \longrightarrow C$ in the following way:

$F_A(B) = A \times B$ for each $B \in |C|$

$F_A(f) = 1_A \times f$ for each $f \in \text{Mor}\, C$.

Instead of F_A one writes also $A \times -$. Then the assertion (2) in 4.1.1 means that the functor $A \times - : C \longrightarrow C$ has a right adjoint for each $A \in |C|$.

4.1.3 Definition. 1) Let C be a category. A class-indexed family $(f_i : B_i \longrightarrow B)_{i \in I}$ of C-morphisms is called an epi-sink provided that for any pair (α,β) of C-morphisms with domain B such that $\alpha \circ f_i = \beta \circ f_i$ for each $i \in I$, it follows that $\alpha = \beta$.

2) Let C be a topological category. An epi-sink $(f_i : B_i \longrightarrow B)_{i \in I}$ is called final provided that the C-structure of B is final with respect to the family $(f_i)_{i \in I}$.

4.1.4 Theorem. Let C be a topological category. Then the following are equivalent:

(1) C is cartesian closed.

(2) For any $A \in |\mathcal{C}|$ and any set-indexed family $(B_i)_{i \in I}$ of $\mathcal{C}$-objects the following are satisfied:

(a) $A \times \coprod_{i \in I} B_i \cong \coprod_{i \in I} (A \times B_i)$ (more exactly: $A \times -$ preserves coproducts.)

(b) If f is a quotient map, then so is $1_A \times f$, i.e. $A \times -$ preserves quotient maps.

(3) (a) For any $A \in |\mathcal{C}|$ and any set-indexed family $(B_i)_{i \in I}$ of $\mathcal{C}$-objects the following is satisfied:

$A \times \coprod_{i \in I} B_i \cong \coprod_{i \in I} (A \times B_i)$ (more exactly: $A \times -$ preserves coproducts.)

(b) In $\mathcal{C}$ the product $f \times g$ of any two quotient maps f and g is a quotient map.

(4) For each $\mathcal{C}$-object A holds: For any final epi-sink $(f_i : B_i \longrightarrow B)_{i \in I}$ in $\mathcal{C}$, $(1_A \times f_i : A \times B_i \longrightarrow A \times B)_{i \in I}$ is a final epi-sink, i.e. $A \times -$ preserves final epi-sinks.

(5) For each pair $(A,B) \in |\mathcal{C}| \times |\mathcal{C}|$, the set $[A,B]_{\mathcal{C}}$ can be endowed with the structure of a $\mathcal{C}$-object denoted by B^A such that the following are satisfied:

(a) The *evaluation map* $e_{A,B} : A \times B^A \longrightarrow B$ defined by $e_{A,B}(a,g) = g(a)$ for each $(a,g) \in A \times B^A$ is a $\mathcal{C}$-morphism.

(b) For each $\mathcal{C}$-object C, the map $\psi : (B^A)^C \longrightarrow B^{A \times C}$ defined by $\psi(f) = e_{A,B} \circ (1_A \times f)$ for each $f \in [C, B^A]_{\mathcal{C}}$ is surjective.

<u>Proof.</u> (5) $\Rightarrow$ (1): trivial.

(2) $\Rightarrow$ (3): Let $f : A \longrightarrow B$ and $g : C \longrightarrow D$ be quotient maps in $\mathcal{C}$. Then $f \times g = (1_B \times g) \circ (f \times 1_C)$ is a quotient map as a composite of two quotient maps (note that if $1_C \times f$ is a quotient map, then so is $f \times 1_C$; for the following diagram

$$\begin{array}{ccc} C \times A & \xrightarrow{\;1_C \times f\;} & C \times B \\ {\scriptstyle i_1}\downarrow & & \downarrow{\scriptstyle i_2} \\ A \times C & \xrightarrow{\;f \times 1_C\;} & B \times C \end{array}$$

in which i_1, i_2 are the canonical isomorphisms [i.e. $i_1((c,a)) = (a,c)$ for each $(c,a) \in C \times A$ and $i_2((c,b)) = (b,c)$ for each $(c,b) \in C \times B$] is commutative, hence $f \times 1_C = i_2 \circ (1_C \times f) \circ i_1^{-1}$).

(3) $\Rightarrow$ (4): Let $A \in |\mathcal{C}|$ and let $(f_i : B_i \longrightarrow B)_{i \in I}$ be a final epi-sink in $\mathcal{C}$:

(a) Let I be a set: If $(\coprod_{i\in I} B_i, (j_i)_{i\in I})$ is the coproduct of the family $(B_i)_{i\in I}$ in C, then the C-morphism $f : \coprod_{i\in I} B_i \longrightarrow B$ uniquely determined by $f\circ j_i = f_i$ for each $i \in I$ is a quotient map. (Obviously f is an epimorphism and if $h : B \longrightarrow C$ is a map for which $h\circ f$ is a C-morphism, then $(h\circ f)\circ j_i = h\circ f_i$ is also a C-morphism for each $i \in I$. Thus by assumption, h is a C-morphism.) Then applying (3)(b), $1_A \times f : A \times \coprod_{i\in I} B_i \longrightarrow A \times B$ is a quotient map. Furthermore, $(A \times \coprod_{i\in I} B_i, (1_A \times j_i)_{i\in I})$ is the coproduct of the family $(A \times B_i)_{i\in I}$. Since $A \times -$ is a functor and $f\circ j_i = f_i$ for each $i \in I$, the diagram

$$\begin{array}{ccc} A\times\coprod B_i & \xrightarrow{\;1_A\times f\;} & A\times B \\ {\scriptstyle 1_A\times j_i}\nwarrow & & \nearrow{\scriptstyle 1_A\times f_i} \\ & A\times B_i & \end{array}$$

is commutative for each $i \in I$. Since additionally the C-structure of $A \times \coprod_{i\in I} B_i$ is final with respect to $(1_A \times j_i)_{i\in I}$, it results that the C-structure of $A \times B$ is final with respect to $(1_A \times f_i)_{i\in I}$. Since $(f_i : B_i \longrightarrow B)_{i\in I}$ is an epi-sink in C, $\bigcup_{i\in I} f_i[B_i] = B$ (1. If $|B| < 2$, then the assertion is trivial. 2. Let $|B| \geq 2$. If $B \neq \bigcup_{i\in I} f_i[B_i]$, then there would be $b_o \in \bigcup_{i\in I} f_i[B_i]$ and $b_1 \in B\setminus\bigcup_{i\in I} f_i[B_i]$. If $\{b_o, b_1\}$ is endowed with the indiscrete C-structure, then one obtains a C-object Z. Hence $\alpha : B \longrightarrow Z$ defined by $\alpha(b) = b_o$ for each $b \in B$ and $\beta : B \longrightarrow Z$ defined by $\beta(b) = \begin{cases} b_o & \text{for } b \in \bigcup_{i\in I} f_i[B_i] \\ b_1 & \text{otherwise} \end{cases}$ are C-morphisms such that $\alpha\circ f_i = \beta\circ f_i$ for each $i \in I$. Obviously $\alpha \neq \beta$ in contradiction to the fact that $(f_i)_{i\in I}$ is an epi-sink.). Then it is easily verified that $(1_A \times f_i : A \times B_i \longrightarrow A \times B)_{i\in I}$ is an epi-sink (for if $\alpha, \beta : A \times B \longrightarrow D$ are C-morphisms such that $\alpha\circ(1_A \times f_i) = \beta\circ(1_A \times f_i)$ for each $i \in I$

and if $(a,b) \in A \times B$, then since $B = \bigcup_{i \in I} f_i[B_i]$, there is some $i \in I$ and some $b_i \in B_i$ with $f_i(b_i) = b$, hence $\alpha((a,b)) = \alpha((a, f_i(b_i))) = \alpha((1_A \times f_i)((a,b_i))) = \beta((1_A \times f_i)((a,b_i))) = \beta((a, f_i(b_i))) = \beta((a,b))$; consequently $\alpha = \beta$.).

(b) Let I be a proper class. Then by Cat top$_2$), there is a set $K \subset I$ such that the C-structure of B is final with respect to $(f_i)_{i \in K}$ (note that the final structure of B with respect to (f_i) is the supremum [with respect to $\leq$] of the final structures on the underlying set of B generated by each f_i). As under (a) one concludes that the C-structure of $A \times B$ is final with respect to $(1_A \times f_i)_{i \in K}$ and therefore it is final with respect to $(1_A \times f_i)_{i \in I}$. If $(1_A \times f_i)_{i \in K}$ is an epi-sink, then so is $(1_A \times f_i)_{i \in I}$.

(4) $\Rightarrow$ (5): Let $(A,B) \in |C| \times |C|$. Then $[A,B]_C$ is endowed with the final C-structure with respect to the class of all maps $f_i : C_i \longrightarrow [A,B]_C$ from C-objects C_i into $[A,B]_C$ for which the map $\psi(f_i) : A \times C_i \longrightarrow B$ is a C-morphism. The resulting object is denoted by B^A. The resulting sink is a (final) epi-sink: Let $\alpha \circ f_i = \beta \circ f_i$ for each i, where α and β are C-morphisms with domain B^A. If $e \in [A,B]_C$, then $P = \{e\}$ is endowed with the uniquely determined C-structure. A map $f : P \longrightarrow [A,B]_C$ is defined by $f(e) = e$. Then $\psi(f) = e_{A,B} \circ (1_A \times f) = e \circ p_A$, where $p_A : A \times P \longrightarrow A$ denotes the projection, is a C-morphism. Thus $P = C_i$ and $f = f_i$ for a suitable i; especially $\alpha(e) = (\alpha \circ f_i)(e) = (\beta \circ f_i)(e) = \beta(e)$. Hence $\alpha = \beta$. By assumption the C-structure of $A \times B^A$ is final with respect to the family of all maps $1_A \times f : A \times C \longrightarrow A \times B^A$ for which $e_{A,B} \circ (1_A \times f)$ is a C-morphism. Therefore $e_{A,B} : A \times B^A \longrightarrow B$ is a C-morphism. In order to prove that ψ is surjective suppose that $g : A \times C \longrightarrow B$ is a C-morphism. For each $c \in C$, the map $g_c : A \longrightarrow B$ defined by $g_c(a) = g((a,c))$ for each $a \in A$ is a C-morphism (note that a constant map between C-objects is a C-morphism [cf. 1.2.2.3]). For the map $\bar{g} : C \longrightarrow [A,B]_C$ defined by $\bar{g}(c) = g_c$ for each $c \in C$, $\psi(\bar{g}) = e_{A,B} \circ (1_A \times \bar{g}) = g$ is a C-morphism, hence $\bar{g} : C \longrightarrow B^A$ is a C-morphism with $\psi(\bar{g}) = g$.

(1) $\Rightarrow$ (2): It is a well-known theorem in category theory [cf. appendix] that a left adjoint (here: the functor $A \times -$) preserves coproducts and coequalizers (more generally: colimits). In a topological category C every quotient map (= extremal epimorphism) is a coequalizer of two C-morphisms: If $f : (X,\xi) \longrightarrow (X',\xi')$

is a quotient map in C, then f coincides with the natural map $\omega : (X,\xi) \longrightarrow (X/\pi_f,\eta)$ (up to an isomorphism), where $\pi_f \subset X \times X$ consists precisely of those pairs of points (x,y) for which $f(x) = f(y)$ and η is the final C-structure with respect to ω. Let $X \times X$ be endowed with the initial C-structure with respect to the projections $p_i : X \times X \longrightarrow X$ $(i = 1,2)$ and π_f with the initial C-structure with respect to the inclusion map $i : \pi_f \longrightarrow X \times X$. Then $\alpha = p_1|_{\pi_f}$ and $\beta = p_2|_{\pi_f}$ are C-morphisms with codomain X and π_f is the finest equivalence relation on X for which $\alpha((x,y))$ and $\beta((x,y))$ are equivalent for each $(x,y) \in \pi_f$. Then by 1.2.1.8 b), ω is the coequalizer of α and β.

4.1.5 Corollary ("rules"). Let C be a cartesian closed topological category. Then the following are satisfied:

(1) First exponential law: $A^{B\times C} \cong (A^B)^C$

(2) Second exponential law: $(\prod_{i\in I} A_i)^B \cong \prod_{i\in I} A_i^B$

(3) Third exponential law: $A^{\coprod_{i\in I} B_i} \cong \prod_{i\in I} A^{B_i}$

(4) Distributive law: $A \times \coprod_{i\in I} B_i \cong \coprod_{i\in I} A \times B_i$

Proof. (4) has been proved (cf. 4.1.4 (2)(a)).

(2) follows from the fact that the right adjoint of $B \times -$ denoted by $\cdot^B$ preserves products (more generally: limits) (note that every right adjoint has this property [cf. appendix]).

(1) If (F_1,G_1) and (F_2,G_2) are pairs of adjoint functors and if the composite $F_2 \circ F_1$ is defined, then $(F_2\circ F_1, G_1\circ G_2)$ is again a pair of adjoint functors (exercise!). Especially $(B\times -, \cdot^B)$ as well as $(C\times -, \cdot^C)$ are pairs of adjoint functors and consequently $(B\times - \circ C\times -, \cdot^C \circ \cdot^B)$ is a pair of adjoint functors. Since $B\times - \circ C\times - \approx (B\times C)\times -$ and $\cdot^{B\times C}$ is the corresponding right adjoint, the functors $\cdot^C \circ \cdot^B$ and $\cdot^{B\times C}$ are naturally equivalent so that $\cdot^{B\times C}(A) = A^{B\times C} \cong (\cdot^C \circ \cdot^B)(A) = (A^B)^C$.

(3) As well-known (cf. the appendix) the contravariant hom-functor $H_A : C \longrightarrow$ Set assigning to each $B \in |C|$ the set $[B,A]_C$ converts coproducts into products, hence:

$$\prod_{k\in I} H_A(B_k) = \prod_{k\in I} [B_k,A]_C \cong [\coprod_{i\in I} B_i, A]_C = H_A(\coprod_{i\in I} B_i)$$

The one-to-one correspondence $(r_k)_{k\in I} \longleftrightarrow [r_i]_{i\in I}$ describes this isomorphism (i.e. this bijective map), where the following diagram

$$\begin{array}{ccc} \coprod B_i & \xrightarrow{[r_i]} & A \\ & {\scriptstyle e_j}\nwarrow \quad \nearrow{\scriptstyle r_j} & \\ & B_j & \end{array} \qquad (e_j\text{: natural injection})$$

commutes for each $j \in I$. Now the above bijection is "lifted" to an isomorphism in $\mathcal{C}$. Obviously

$$m : A^{\coprod_{i\in I} B_i} \longrightarrow \prod_{k\in I} A^{B_k}$$

defined by $m(f) = (f\circ e_k)_{k\in I}$ is a $\mathcal{C}$-morphism with $m([r_i]) = (r_k)$ (note: $p_i \circ m = \mathrm{Hom}(e_i, 1_A)$ for each $i \in I$ is a $\mathcal{C}$-morphism [cf. 4.1.6], where $p_i : \prod_{k\in I} A^{B_k} \longrightarrow A^{B_i}$ denotes the i-th projection). Since $\mathcal{C}$ is cartesian closed, for any $g : \coprod_{i\in I} B_i \times \prod_{k\in I} A^{B_k} \longrightarrow A$, there is precisely one $\mathcal{C}$-morphism $g^* : \prod_{k\in I} A^{B_k} \longrightarrow A^{\coprod_{i\in I} B_i}$ such that the diagram

$$\begin{array}{ccc} \coprod_{i\in I} B_i \times A^{\coprod_{i\in I} B_i} & \xrightarrow{e_{\coprod B_i, A}} & A \\ {\scriptstyle 1_{\coprod B_i}\times g^*}\nwarrow & & \nearrow{\scriptstyle g} \\ & \coprod_{i\in I} B_i \times \prod_{k\in I} A^{B_k} & \end{array}$$

commutes. In order to get an inverse of m we need g such that

(*) $\quad g((b,(r_k))) = r_j(b_j)$ with $b = (b_j, j)$

(note: $\coprod_{i\in I} B_i = \bigcup_{i\in I} B_i \times \{i\}$). Then $g^*((r_k)) = [r_i]$ (namely if $b \in \coprod_{i\in I} B_i$, then $b = (b_j, j) = e_j(b_j)$ and $g^*((r_k))(b) = e_{\coprod B_i, A}(b, g^*((r_k))) =$
$= (e_{\coprod B_i, A} \circ 1_{\coprod B_i} \times g^*)((b,(r_k))) = g((b,(r_k))) = r_j(b_j) = [r_i]\circ e_j(b_j) =$
$= [r_i](b))$, i.e. g^* is the inverse of m, that means m is an isomorphism. Thus it remains to construct $g: \coprod_{i\in I} B_i \times \prod_{k\in I} A^{B_k} \to A$ such that (*) is satisfied. For each $C \in |\mathcal{C}|$, $C \times - \approx - \times C$, hence $- \times C$ also preserves coproducts. Thus for $C = \prod_{k\in I} A^{B_k}$, there exists an isomorphism $h: \coprod_{i\in I} B_i \times \prod_{k\in I} A^{B_k} \longrightarrow \coprod_{i\in I} (B_i \times \prod_{k\in I} A^{B_k})$ such that for each $i \in I$, the diagram

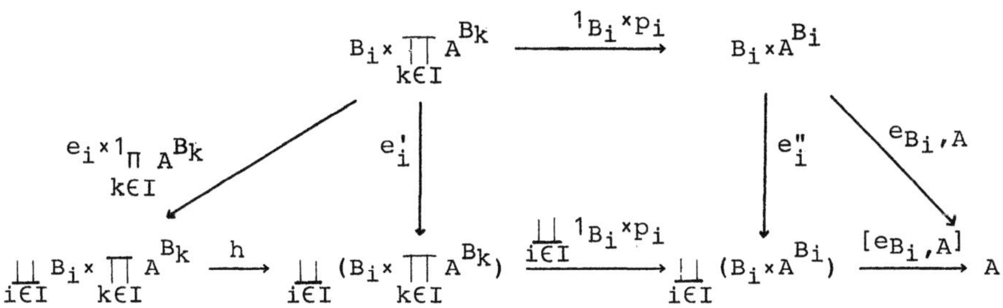

commutes, where e_i, e_i' and e_i'' are the natural injections and p_i the projections. If $b = (b_j, j) \in \coprod_{i \in I} B_i$, then $h((b,(r_k))) = e_j'((b_j,(r_k))) = ((b_j,(r_k)),j)$.

Then $g = [e_{B_i},A] \circ \coprod_{i \in I} 1_{B_i} \times p_i \circ h$ is a $\mathcal{C}$-morphism such that $g((b,(r_k))) =$
$= [e_{B_i},A](\coprod_{i \in I} 1_{B_i} \times p_i(e_j'((b_j,(r_k))))) = [e_{B_i},A](e_j'' \circ 1_{B_j} \times p_j((b_j,(r_k)))) =$
$= [e_{B_i},A](e_j''((b_j,r_j))) = e_{B_j,A}((b_j,r_j)) = r_j(b_j)$, i.e. (*) is fulfilled.

<u>4.1.6 Remark.</u> For every cartesian closed topological category $\mathcal{C}$, there exists an *internal Hom-functor* $\mathrm{Hom} : \mathcal{C}^* \times \mathcal{C} \to \mathcal{C}$ which is defined as follows:

1) $\mathrm{Hom}(A,B) = B^A$ for each $(A,B) \in |\mathcal{C}^* \times \mathcal{C}| = |\mathcal{C}^*| \times |\mathcal{C}|$,
2) $\mathrm{Hom}(f,g) : \mathrm{Hom}(A',B) \to \mathrm{Hom}(A,B')$ for each $(f,g) \in [A,A']_{\mathcal{C}} \times [B,B']_{\mathcal{C}}$ is defined by

$$\mathrm{Hom}(f,g)(u) = g \circ u \circ f \text{ for each } u \in [A',B]_{\mathcal{C}}.$$

(It has been shown that $\mathrm{Hom}(A,B) = B^A$ is a $\mathcal{C}$-object. Hence it remains to prove that $\mathrm{Hom}(f,g)$ is a $\mathcal{C}$-morphism:

a) Put $\hat{g} = e_{A',B} \circ (e_{A,A'} \times 1_{B^{A'}})$. Then there is a unique $\mathcal{C}$-morphism $\hat{g}^* : (A')^A \times B^{A'} \to B^A$ such that the diagram

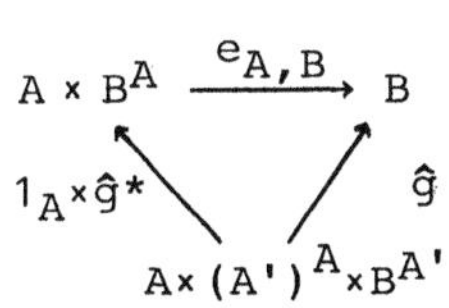

commutes. Define $-\circ f : B^{A'} \to B^A$ by $-\circ f = \hat{g}^* \circ \alpha$, where $\alpha : B^{A'} \to (A')^A \times B^{A'}$ is given by $\alpha(h) = (f,h)$. Then $-\circ f$ is a $\mathcal{C}$-morphism such that

$(-\circ f)(u) = u \circ f$ for each $u \in B^{A'}$ which can be easily checked.

b) Put $\tilde{g} = e_{B,B'} \circ (e_{A,B} \times 1_{(B')^B})$. Then there is a unique C-morphism $\tilde{g}^* : B^A \times (B')^B \longrightarrow (B')^A$ such that the diagram

$$\begin{array}{ccc} A \times (B')^A & \xrightarrow{e_{A,B'}} & B' \\ {\scriptstyle 1_A \times \tilde{g}^*} \nwarrow & & \nearrow {\scriptstyle \tilde{g}} \\ & A \times B^A \times (B')^B & \end{array}$$

commutes. Define $g\circ- : B^A \to (B')^A$ by $g\circ- = \tilde{g}^* \circ \beta$, where $\beta : B^A \longrightarrow B^A \times (B')$ is given by $\beta(k) = (k,g)$. Then $g\circ-$ is a C-morphism such that $(g\circ-)(u) = g\circ u$ for each $u \in B^A$ which can be easily checked.

Thus $\mathrm{Hom}(f,g) = (g\circ-)\circ(-\circ f)$ is a C-morphism.)

4.2 Examples

4.2.1. The category Grill of grill-determined prenearness spaces (and uniformly continuous maps) is cartesian closed (cf. 3.2.3.20, 3.2.3.22 and 4.1.4(3)).

4.2.2. The category Set is a cartesian closed (topological) category. It suffices to prove that for any epi-sink $(f_i : X_i \longrightarrow X)_{i\in I}$ in Set and for any set A, $(1_A \times f_i : A \times X_i \longrightarrow A \times X)_{i\in I}$ is also an epi-sink. But this is easy to check [cf. the last part under "(3) ⇒ (4)" a) in the proof of 4.1.4].

4.2.3. Let X be a set and $F(X)$ the set of all filters on X. If $q \subset F(X) \times X$ satisfies the conditions Lim_1) and Lim_2) (cf. 1.1.6 ④), then (X,q) is called

a) a *convergence space* provided that the following condition is satisfied:

(C) $(F,x) \in q$ implies $(F \cap \dot{x}, x) \in q$,

b) a *limit space* provided that Lim_3) is satisfied (cf. 1.1.6 ④),

c) a *pseudotopological space* provided that the following condition is satisfied:

(PsT) $(\mathcal{F},x) \in q$ whenever $(\mathcal{G},x) \in q$ for each ultrafilter $\mathcal{G} \supset \mathcal{F}$.

Instead of $(\mathcal{F},x) \in q$ one usually writes $\mathcal{F} \to x$ and one says: $\mathcal{F}$ converges to x. The spaces defined above form the objects of the categories Conv, Lim and PsTop; in each case the morphisms are all continuous maps f, i.e. those carrying filters convergent to x to filters convergent to f(x). Each of these categories is a topological category (Let C be one of them. If X is a set, $((Y_i,q_i))_{i \in I}$ is any family of C-objects and $(f_i : X \to X_i)_{i \in I}$ is any family of maps, then $q = \{(\mathcal{F},x) \in \mathcal{F}(X) \times X : (f_i(\mathcal{F}),f_i(x)) \in q_i$ for each $i \in I\}$ is a C-structure on X which is initial with respect to $(X,f_i,(Y_i,q_i),I)$.).

4.2.3.1. Conv is cartesian closed.

(1. An epi-sink $(f_i : (X_i,q_i) \to (X,q))_{i \in I}$ in Conv is final if and only if for each $y \in X$, $(\mathcal{F},y) \in q$ implies that there exist f_i and $\mathcal{E}$ such that $\mathcal{F} \supset f_i(\mathcal{E})$, where $(\mathcal{E},x) \in q_i$ and $f_i(x) = y$.

2. Let $(f_i : (X_i,q_i) \to (Y,q))_{i \in I}$ be a final epi-sink in Conv and $(Z,r) \in |\text{Conv}|$. Then $(1_Z \times f_i : (Z \times X_i, r \times q_i) \to (Z \times Y, r \times q))_{i \in I}$ is an epi-sink in Conv. It remains to show that $(1_Z \times f_i)_{i \in I}$ is final: Let $(z,y) \in Z \times Y$ and $(\mathcal{F},(z,y)) \in r \times q$. Then by the definition of the product structure, $(p_Y(\mathcal{F}),y) \in q$ and $(p_Z(\mathcal{F}),z) \in r$, where p_Y resp. p_Z denotes the projection. By assumption, $p_Y(\mathcal{F}) \supset f_i(\mathcal{E})$ because of 1., where $(\mathcal{E},x) \in q_i$ and $f_i(x) = y$. If $p_Z(\mathcal{F}) \times \mathcal{E}$ denotes the product filter, then $(p_Z(\mathcal{F}) \times \mathcal{E},(z,x)) \in r \times q_i$ with $(1_Z \times f_i)((z,x)) = (z,y)$ and $\mathcal{F} \supset (1_Z \times f_i)(p_Z(\mathcal{F}) \times \mathcal{E})$.)

4.2.3.2. Lim is cartesian closed.

(1. An epi-sink $(f_i : (X_i,q_i) \to (X,q))_{i \in I}$ in Lim is final if and only if for each $x \in X$, it follows from $(\mathcal{F},x) \in q$ that

$$\mathcal{F} \supset f_{i_1}(\mathcal{E}_{i_1}) \cap f_{i_2}(\mathcal{E}_{i_2}) \cap \ldots \cap f_{i_n}(\mathcal{E}_{i_n})$$

for finitely many filters $\mathcal{E}_{i_1},\ldots,\mathcal{E}_{i_n}$ with $(\mathcal{E}_{i_k},x_{i_k}) \in q_{i_k}$ and $f_{i_k}(x_{i_k}) = x$ for $k \in \{1,\ldots,n\}$, where $\{i_1,\ldots,i_n\} \subset I$.

2. Corresponding to 4.2.3.1.2. it is shown that 4.1.4(4) is valid for $\mathcal{C}$ = Lim.)

4.2.3.3. PsTop is cartesian closed.

(1. An epi-sink $(f_i : (X_i, q_i) \longrightarrow (X,q))_{i \in I}$ in PsTop is final if and only if for each $y \in X$ and each ultrafilter $\mathcal{F}$ on X, $(\mathcal{F},y) \in q$ implies $\mathcal{F} = f_i(\mathcal{E})$ for some ultrafilter $\mathcal{E}$ such that $(\mathcal{E},x) \in q_i$ and $f_i(x) = y$.

2. The validity of 4.1.4 (4) for $\mathcal{C}$ = PsTop is shown analogously to 4.2.3.1.2. However, $p_Z(\mathcal{F}) \times \mathcal{E}$ has to be replaced by an ultrafilter containing this filter.)

4.2.3.4 Remarks. (1) If $(X,\mathcal{X})$ is a topological space, then a limit structure $q_{\mathcal{X}}$ on X is defined by: $(\mathcal{H},x) \in q_{\mathcal{X}}$ if and only if $\mathcal{H}$ is finer than the neighbourhood filter of x. $(X,q_{\mathcal{X}})$ is even a pseudotopological space (note that each filter on X is the intersection of all ultrafilters containing it). A pseudotopological space (X,q) is called topologizable provided that there is a topology $\mathcal{X}$ on X with $q_{\mathcal{X}} = q$. If the full subcategory of PsTop whose object class is the class of all topologizable pseudotopological spaces is denoted by T-PsTop, then T-PsTop $\cong$ Top (note: For each pseudotopological space (X,q), a topology $\mathcal{X}_q$ on X is defined by

$$O \in \mathcal{X}_q \iff \text{For each } x \in O \text{ and each filter } \mathcal{F} \text{ on } X \text{ with } (\mathcal{F},x) \in q,\ O \in \mathcal{F}$$

Then we obtain: 1) $\mathcal{X}_{q_{\mathcal{X}}} = \mathcal{X}$ for each topology $\mathcal{X}$

2) $q_{\mathcal{X}_q} = q$ for each topologizable pseudotopological structure q.).

Obviously every limit space is a convergence space. Furthermore, every pseudotopological space is a limit space (namely, if Lim_3) was not satisfied for a pseudotopological space (X,q), then there would be filters $\mathcal{F},\mathcal{G}$ on X and some $x \in X$ such that $(\mathcal{F},x) \in q$ and $(\mathcal{G},x) \in q$ but $(\mathcal{F} \cap \mathcal{G},x) \notin q$. Thus there would exist some ultrafilter $\mathcal{U}$ containing $\mathcal{F} \cap \mathcal{G}$ such that $(\mathcal{U},x) \notin q$. Especially, $\mathcal{U} \not\supset \mathcal{F}$ and $\mathcal{U} \not\supset \mathcal{G}$, i.e. there would exist $F \in \mathcal{F}$ and $G \in \mathcal{G}$ with $F \notin \mathcal{U}$ and $G \notin \mathcal{U}$. On the other hand we would have $F \cup G \in \mathcal{U}$. So it would follow that $F \in \mathcal{U}$ or

$G \in \mathcal{U}$ since $\mathcal{U}$ is an ultrafilter. This is a contradiction.). Thus it follows that

$$\underline{\text{Conv}} \supset \underline{\text{Lim}} \supset \underline{\text{PsTop}} \supset \underline{\text{Top}}$$

In this list each category is a bireflective subcategory of the preceding ones. This is easily checked by applying the usual criteria.

(2) A convergence space (X,q) is called <u>symmetric</u> provided that the following is satisfied:

(S) $(\mathcal{F},y) \in q$ and $x \in \bigcap_{F \in \mathcal{F}} F$ imply $(\mathcal{F},x) \in q$.

The full subcategory of <u>Conv</u> whose object class is the class of symmetric convergence spaces is denoted by <u>S-Conv</u>.
If (X,μ) is a grill-determined prenearness space and ξ (resp. γ) is the set of all near collections in (X,μ) (resp. all Cauchy-systems in (X,μ)), then the following are equivalent:

(1) For each $\mathcal{A} \in \xi$ which is non-empty, there exists some $x \in X$ with $\mathcal{A} \cup \{\{x\}\} \in \xi$.
(2) For each near grill $\mathcal{G}$, there is some $x \in X$ with $\mathcal{G} \cup \{\{x\}\} \in \xi$.
(3) For each $\mathcal{A} \in \gamma$, there exists some $x \in X$ with $\{A \cup \{x\} : A \in \mathcal{A}\} \in \gamma$.[25)]
(4) For each Cauchy-filter $\mathcal{F}$ there is some $x \in X$ with $\mathcal{F} \cap \dot{x} = \{F \cup \{x\} : F \in \mathcal{F}\} \in \gamma$.

The full subcategory of <u>Grill</u> whose objects are those grill-determined prenearness spaces which satisfy one (and thus each) of the four equivalent conditions mentioned above is denoted by <u>C-Grill</u>. According to Katětov the objects of <u>C-Grill</u> are called *localized filtermerotopic spaces*. It can be shown that <u>S-Conv</u> $\cong$ <u>C-Grill</u> (note: If γ is the set of all Cauchy-systems in $(X,\mu) \in |\underline{\text{C-Grill}}|$, then an <u>S-Conv</u>-structure q_γ on X is defined by

$$(\mathcal{F},x) \in q_\gamma \quad \text{iff} \quad \mathcal{F} \cap \dot{x} \in \gamma.$$

Conversely, if $(X,q) \in |\underline{\text{S-Conv}}|$, then the set of all Cauchy systems

---

25) The restriction to a non-empty $\mathcal{A}$ is superfluous since $\{\{x\}\} \in \gamma$ for each $x \in X$.

in $(X,\mu_q) \in |\underline{\text{C-Grill}}|$ is defined by

$$\gamma_q = \{A \subset P(X) : \text{there exists some filter } F \text{ convergent in } (X,q) \text{ with } F << A\}.$$

We obtain: $\gamma = \gamma_{q_\gamma}$ and $q = q_{\gamma_q}$.). Moreover, C-Grill is bicoreflective in Grill. (Let $(X,\xi) \in |\underline{\text{Grill}}|$ and put

$$\xi_C = \{A \subset P(X) : \text{there exists some } x \in X \text{ with } A \cup \{\{x\}\} \in \xi\} \cup \{\emptyset\}.$$

Then $(X,\xi_C) \in |\underline{\text{C-Grill}}|$ and $1_X : (X,\xi_C) \longrightarrow (X,\xi)$ is the C-Grill-coreflection of (X,ξ).)

Thus final structures in C-Grill are formed as in Grill (cf.2.2. 13①). Since C-Grill is closed under formation of products in Grill (which is easily checked), it follows (by means of 4.1.4(4)) from the cartesian closedness of Grill that C-Grill (and thus S-Conv) is also cartesian closed.

4.2.4. A preordered set is a pair $(X,\leq)$, where X is a set and $\leq$ is a reflexive and transitive relation on X. The full subcategory of Rere[26] whose objects are the preordered sets is denoted by PrOrd. It is a topological category (initial structures are formed as in Rere).

PrOrd is cartesian closed.

(1. An epi-sink $(f_i : B_i \longrightarrow C)_{i \in I}$ is final in PrOrd if and only if $c \leq c'$ is valid in C if and only if there exists an (f_i)-chain from c to c', i.e. a finite chain $c = w_o \leq w_1 \leq \ldots \leq w_n = c'$, such that for $k = 0,1,\ldots,n-1$, there is a pair $b_k \leq b_k'$ in some B_{i_k} with $f_{i_k}(b_k) = w_k$ and $f_{i_k}(b_k') = w_{k+1}$.

2. Let A be an object and $(f_i : B_i \longrightarrow C)_{i \in I}$ an epi-sink in PrOrd. In order to prove that $(1_A \times f_i : A \times B_i \longrightarrow A \times C)_{i \in I}$ is a final episink suppose $(a,c) \leq (a',c')$ holds in $A \times C$. Then we have $a \leq a'$ in A and $c \leq c'$ in C. Thus there is an (f_i)-chain from c to c', say $w_o = c, w_1, \ldots, w_n = c'$. Then $(a,w_o),(a,w_1),\ldots,(a',w_n)$ is a $(1_A \times f_i)$-chain from (a,c) to (a',c'). On the other hand

26) cf. 1.1.6⑨

if $z_0,\dots,z_n$ is a $(1_A \times f_i)$-chain from (a,c) to (a',c'), then $z_0 \le z_1 \le \dots \le z_n$ and $(a,c) \le (a',c')$. By means of 1. it follows that $(1_A \times f_i)_{i\in I}$ is a final epi-sink.)

4.2.5. The categories Born and Simp (cf. 1.1.6 (8) and (10)) are cartesian closed.

(1. An epi-sink $(f_i : (X_i,\mathcal{B}_i) \longrightarrow (X,\mathcal{B}))_{i\in I}$ in Born is final if and only if each $B \in \mathcal{B}$ is contained in some finite union of sets $f_i[M_i]$ with $M_i \in \mathcal{B}_i$.

2. Let $(Z,\mathcal{C}) \in |\text{Born}|$ and let $(f_i : (X_i,\mathcal{B}_i) \longrightarrow (X,\mathcal{B}))_{i\in I}$ be a final epi-sink in Born. If $A \subset Z \times X$ is a bounded set with respect to the product structure $\mathcal{C} \times \mathcal{B}$, then

$$p_X[A] \subset f_{i_1}[M_1] \cup \dots \cup f_{i_n}[M_n]$$

where $M_k \in \mathcal{B}_{i_k}$ for each $k \in \{1,\dots,n\}$. Thus $p_Z[A] \times M_k \in \mathcal{C} \times \mathcal{B}_{i_k}$ and $A \subset \bigcup_{k=1}^{n} (1_Z \times f_{i_k})[p_Z[A] \times M_k]$.

3. The proof for Simp is similar.)

4.2.6. The categories S-Near, Near and Unif are not cartesian closed:

4.2.6.1 Example. Let X be the set [0,1] endowed with the topological nearness structure induced by the usual topology on [0,1] (i.e. the topology induced by the usual metric) [additionally this structure is uniform since [0,1] is paracompact]. Further let for each $n \in \mathbb{N}$, Y_n be the uniquely determined nearness space whose underlying set is $\{n\}$. In S-Near each of $X \times \coprod_{n\in\mathbb{N}} Y_n$ and $\coprod_{n\in\mathbb{N}} X \times Y_n$ has the underlying set $[0,1] \times \mathbb{N}$. Put $A = \{0\} \times \mathbb{N}$ and $B = \{(\frac{1}{n},n) : n \in \mathbb{N}\}$. Then $\{A,B\}$ is near in $X \times \coprod_{n\in\mathbb{N}} Y_n$ but not in $\coprod_{n\in\mathbb{N}} X \times Y_n$. Thus $X \times \coprod_{n\in\mathbb{N}} Y_n \not\cong \coprod_{n\in\mathbb{N}} X \times Y_n$ in S-Near. Since Near and Unif are closed under formation of products and coproducts in S-Near (cf. 3.2.2.6 (4) for Near; similar results hold for Unif) none of the three categories S-Near, Near and Unif is cartesian closed.

4.2.7. The category Top is not cartesian closed.

This is an immediate consequence of the following proposition (note that there are topological T_1-spaces which are completely regular but not locally compact, e.g. $\mathbb{Q}$ with the usual topology).

Proposition. Let $(X,\mathcal{X})$ be a completely regular topological T_1-space. If $(X,\mathcal{X}) \times - : \underline{\text{Top}} \longrightarrow \underline{\text{Top}}$ has a right adjoint, then $(X,\mathcal{X})$ is locally compact.

Proof. By assumption, $(X,\mathcal{X}) \times - : \underline{\text{Top}} \longrightarrow \underline{\text{Top}}$ has a right adjoint denoted by $\cdot^{(X,\mathcal{X})}$. From the proof of 4.1.4, it follows that for each topological space $(Y,\mathcal{Y})$, the set $C(X,Y) = [(X,\mathcal{X}),(Y,\mathcal{Y})]_{\underline{\text{Top}}}$ can be endowed with a topology $C(\mathcal{X},\mathcal{Y})$ such that the evaluation map $e_{X,Y} : X \times C(X,Y) \longrightarrow Y$ becomes continuous (where $\cdot^{(X,\mathcal{X})}(Y,\mathcal{Y}) = (Y,\mathcal{Y})^{(X,\mathcal{X})} = (C(X,Y),C(\mathcal{X},\mathcal{Y}))$.

In the following we choose $(Y,\mathcal{Y}) = ([0,1]$, usual topology). Further let $g : X \longrightarrow [0,1]$ be defined by $g(x) = 0$ for each $x \in X$ and let $\varepsilon \in \mathbb{R}$ with $0 < \varepsilon < 1$. Now, if $x \in X$, then there are $U \in \overset{\circ}{\mathcal{U}}(x)$ and $V \in \overset{\circ}{\mathcal{U}}(g)$ with $e_{X,Y}[U \times V] \subset U_\varepsilon(0) = [0,\varepsilon)$ since $e_{X,Y}$ is continuous. It remains to prove that $\bar{U}$ is compact. Let $(O_i)_{i \in I}$ be any family of sets open in $(X,\mathcal{X})$ such that $\bigcup_{i \in I} (O_i \cap \bar{U}) = \bar{U}$. Without loss of generality $X \setminus \bar{U}$ belongs to this family. Thus $(O_i)_{i \in I}$ is an open cover of $(X,\mathcal{X})$. Let

$\mathcal{A} = \{A \subset X : A$ is non-empty and closed in $(X,\mathcal{X})$ and there is some $i \in I$ with $A \subset O_i\}$.

Further let

$$(A,O) = \{f \in C(X,[0,1]) : f[A] \subset O\}$$

for each $A \in \mathcal{A}$ and each $O \in \mathcal{Y}$. Then $\{(A,O) : A \in \mathcal{A}, O \in \mathcal{Y}\}$ is a subbase for a topology $\mathcal{Z}$ on $C(X,[0,1])$.

The evaluation map

$$e_{X,Y} : (X,\mathcal{X}) \times (C(X,[0,1]),\mathcal{Z}) \longrightarrow ([0,1], \text{usual topology})$$

is continuous; for if $c \in X$, $f \in C(X,[0,1])$ and $O \in \overset{\circ}{\mathcal{U}}(f(c))$, then there is some $i_c \in I$ with $c \in O_{i_c} \cap f^{-1}[O] \in \overset{\circ}{\mathcal{U}}(c)$ and since $(X,\mathcal{X})$ is regular, there is a closed neighbourhood A of c with $A \subset O_i \cap f^{-1}[O]$ which by the definition of (A,O) implies $e_{X,Y}[A \times (A,O)] \subset O$ ($A \times (A,O)$

is a neighbourhood of (c,f)). In 4.1.1(2) put $A = (X,\mathcal{X})$, $B = (Y,\mathcal{Y})$ and $C = (C(X,[0,1]),\mathcal{Z})$. Then there exists (cf. remark 4.1.2) a unique continuous map $h : (C(X,[0,1]),\mathcal{Z}) \longrightarrow (C(X,[0,1]),C(\mathcal{X},\mathcal{Y}))$ such that $e_{X,Y} \circ (1_X \times h) = e_{X,Y}$. Thus for each $c \in X$ and each $f \in C(X,[0,1])$, it follows that $h(f)(c) = e_{X,Y}((1_X \times h)((c,f))) = e_{X,Y}((c,f)) = f(c)$ so that $h = 1_{C(X,[0,1])}$. Hence, $C(\mathcal{X},\mathcal{Y}) \subset \mathcal{Z}$. Therefore $V \in \mathring{\mathcal{U}}_{\mathcal{Z}}(g)$ for the above $V \in \mathring{\mathcal{U}}_{C(\mathcal{X},\mathcal{Y})}(g)$. Consequently, there are finitely many $(A_1,O_1),\dots,(A_n,O_n)$ with

$$g \in (A_1,O_1) \cap \dots \cap (A_n,O_n) \subset V.$$

Since $g(x) = 0$ for each $x \in X$, it follows that (because of $A_i \neq \emptyset$) $g[A_i] = \{0\}$ for each $i \in \{1,\dots,n\}$ and $\{0\} \subset \bigcap_{i=1}^{n} O_i$. Furthermore, $U \subset A_1 \cup \dots \cup A_n$; for otherwise there would exist some $z \in U \setminus (A_1 \cup \dots \cup A_n)$ and since $(X,\mathcal{X})$ is completely regular, there would be a continuous function $f : X \longrightarrow [0,1]$ such that $f[A_1 \cup \dots \cup A_n] = \{0\}$ and $f(z) = 1$. Hence it would follow that $f[A_k] = \{0\} \subset \bigcap_{i=1}^{n} O_i$ so that $f \in (A_1,O_1) \cap \dots \cap (A_n,O_n)$. Thus we would have $f \in V$ and $(z,f) \in U \times V$ and consequently, $e_{X,Y}((z,f)) = f(z) = 1 \in [0,\varepsilon)$ which is a contradiction. From this fact, it follows immediately that

$$\bar{U} \subset A_1 \cup \dots \cup A_n.$$

Since $A_k \in \mathcal{A}$, there is some $i_k \in I$ with $A_k \subset O_{i_k}$. Thus $\bar{U} \subset O_{i_1} \cup \dots \cup O_{i_n}$.

4.2.8. The category UConv of uniform convergence spaces and uniformly continuous maps is a topological category which is cartesian closed.

[Definition. A *uniform convergence space* is a pair $(X,\mathcal{J}_X)$, where X is a set and $\mathcal{J}_X$ is a set of filters on $X \times X$ such that the following conditions are satisfied:

UC_1) $\dot{x} \times \dot{x} \in \mathcal{J}_X$ for each $x \in X$ where $\dot{x} = \{A \subset X : x \in A\}$,

UC_2) $\mathcal{F} \in \mathcal{J}_X$ and $\mathcal{G} \supset \mathcal{F}$ imply $\mathcal{G} \in \mathcal{J}_X$,

UC_3) $\mathcal{F} \in \mathcal{J}_X$ and $\mathcal{G} \in \mathcal{J}_X$ imply $\mathcal{F} \cap \mathcal{G} \in \mathcal{J}_X$,

UC_4) $\mathcal{F} \in \mathcal{J}_X$ implies $\mathcal{F}^{-1} \in \mathcal{J}_X$ where $\mathcal{F}^{-1} = \{F^{-1} : F \in \mathcal{F}\}$ ($F^{-1} = \{(a,b) : (b,a) \in F\}$),

UC5) $\mathcal{F} \in \mathcal{J}_X$ and $\mathcal{G} \in \mathcal{J}_X$ imply $\mathcal{F} \circ \mathcal{G} \in \mathcal{J}_X$ where $\mathcal{F} \circ \mathcal{G}$ is the filter generated by the filter base $\{F \circ G : F \in \mathcal{F}, G \in \mathcal{G}\}$ ($F \circ G = \{(a,c) : \exists b \in X$ with $(a,b) \in F$ and $(b,c) \in G\}$).

A *uniformly continuous map* $f : (X,\mathcal{J}_X) \longrightarrow (Y,\mathcal{J}_Y)$ between uniform convergence spaces is a map $f : X \longrightarrow Y$ with $(f \times f)(\mathcal{J}_X) \subset \mathcal{J}_Y$.]

Let X be any set, $((X_i,\mathcal{J}_{X_i}))_{i\in I}$ any family of uniform convergence spaces and $(f_i : X \longrightarrow X_i)_{i\in I}$ any family of maps. Then $\mathcal{J}_X = \{\mathcal{F} \subset \mathcal{P}(X \times X) : \mathcal{F}$ filter and $(f_i \times f_i)(\mathcal{F}) \in \mathcal{J}_{X_i}$ for each $i \in I\}$ is the initial UConv-structure on X with respect to $(X, f_i, (X_i,\mathcal{J}_{X_i}), I)$.

Put $[(X,\mathcal{J}_X),(Y,\mathcal{J}_Y)]_{\underline{UConv}} = \mathcal{U}(X,Y)$ and let $\mathcal{W} = \{\Phi : \Phi$ is a filter on $\mathcal{U}(X,Y) \times \mathcal{U}(X,Y)$ with $\Phi(\mathcal{F}) \in \mathcal{J}_Y$ for each $\mathcal{F} \in \mathcal{J}_X\}$ where $\Phi(\mathcal{F})$ is the filter generated by the filter base $\{A(F) : A \in \Phi, F \in \mathcal{F}\}$ with $A(F) = \{(f(a),g(b)) : (f,g) \in A, (a,b) \in F\}$. Then $\mathcal{W}$ is a UConv-structure on $\mathcal{U}(X,Y)$ by means of which one can prove that UConv is cartesian closed. Especially, the evaluation map, then, becomes uniformly continuous.

4.2.8.1 Remark. A uniform convergence space $(X,\mathcal{J}_X)$ is called a *principal uniform convergence space* provided that there is a filter $\mathcal{F}$ on $X \times X$ such that $\mathcal{J}_X = [\mathcal{F}]$, where $[\mathcal{F}] = \{\mathcal{G} : \mathcal{G}$ filter on $X \times X$ with $\mathcal{G} \supset \mathcal{F}\}$. If the full subcategory of UConv whose objects are the principal uniform convergence spaces is denoted by PrUConv, then Unif $\cong$ PrUConv (note that a filter $\mathcal{F}$ on $X \times X$ is a uniform structure of X if and only if $[\mathcal{F}]$ is a UConv-structure). Moreover PrUConv is bireflective in UConv; namely if $(X,\mathcal{J}_X) \in |\underline{UConv}|$ and the finest uniformity of X which is coarser than each $\mathcal{F} \in \mathcal{J}_X$ is denoted by $\mathcal{W}$, then $1_X : (X,\mathcal{J}_X) \longrightarrow (X,[\mathcal{W}])$ is the desired bireflection of $(X,\mathcal{J}_X)$ with respect to PrUConv (If $(X',[\mathcal{W}']) \in |\underline{PrUConv}|$ and $f : (X,\mathcal{J}_X) \longrightarrow (X',[\mathcal{W}'])$ is uniformly continuous, then $(f \times f)(\mathcal{J}_X) \subset [\mathcal{W}']$. Hence $(f \times f)(\mathcal{F}) \supset \mathcal{W}'$ for each $\mathcal{F} \in \mathcal{J}_X$, i.e. for each $W' \in \mathcal{W}'$, we have $(f \times f)^{-1}[W'] \in \mathcal{F}$ for each $\mathcal{F} \in \mathcal{J}_X$. Thus the initial uniformity of X with respect to f is contained in each $\mathcal{F} \in \mathcal{J}_X$, therefore it is contained in the finest uniformity of this kind, namely in $\mathcal{W}$. This means that $f : (X,[\mathcal{W}]) \longrightarrow (X',[\mathcal{W}'])$ is uniformly continuous.).

4.2.9. The monocoreflective (= bicoreflective) hull of $\underline{\text{CompT}_2}$ in Top (whose objects are called *compactly generated topological spaces*) denoted by CGTop is a cartesian closed topological category . (If X and Y are compactly generated topological spaces and C(X,Y) denotes the set of all continuous maps from X to Y, then a topology $\mathcal{C}(X,Y)$ on C(X,Y) is defined, the so-called *compact open topology*, for which $\{(C,O) : C \subset X$ compact and $O \subset Y$ open$\}$ is a subbase, where $(C,O) = \{f \in C(X,Y) : f[C] \subset O\}$. Let $K : \text{Top} \to \text{CGTop}$ be the bicoreflector. Then $K((C(X,Y),\mathcal{C}(X,Y)))$ is the desired function space Y^X which is needed for proving that CGTop is cartesian closed. Especially, the evaluation map $e_{X,Y} : X \circledast Y^X \to Y$, then, becomes continuous, where $\circledast$ stands for the formation of products in CGTop [products in CGTop are formed by forming them first in Top and then applying the bicoreflector K].)

CHAPTER V

TOPOLOGICAL FUNCTORS

A careful analysis of the similarities between various types of topological categories reveals that they are due to some common properties of the corresponding forgetful functors into the category Set of sets and maps. Therefore these functors might be called "topological". But in this case other interesting functors like the forgetful functor from the category Haus of Hausdorff spaces into Set or the forgetful functor from the category Unif into the category Top (assigning to each uniform space its underlying topological space) would be excluded. The basic new idea is that the concept of a topological functor is not an absolute one depending on a functor $T\colon A \to C$ alone but a relative one depending on the functor $T\colon A \to C$ and an additional factorization structure on C. Thus this chapter starts with the study of factorization structures where it is essential that factorizations of sources (source means class-indexed family of C-morphisms starting from any fixed C-object) are studied instead of factorizations of single morphisms. By the way the construction of the E-reflective hull of a (full and isomorphism-closed) subcategory of a category C supplied with a suitable factorization structure is obtained without any smallness- or completeness-restrictions! After some general discussions concerning topological functors an important special case of a topological functor $T\colon A \to$ Set is considered where the existence of initial A-structures is required only for so-called mono-sources. Adding some other "technical" conditions corresponding to those ones for topological categories we get the concept of an initially structured category A with a forgetful functor $T\colon A \to$ Set . Then every topological category is an initially structured category and even Haus is no longer excluded. In the last part of this chapter the basic properties of these categories are proved and cartesian closedness is studied in the realm of initially structured categories. Especially there exist natural function space objects in a cartesian closed initially structured category A (e.g. the set $[A,B]_A$ of all A-morphisms from A to B may be endowed with a suitable A-structure such that the evaluation map becomes an A-morphism). It turns out that even the category Ord of ordered sets which fails to be topological is an initially structured cartesian closed category. The same is true for Hausdorff limit spaces and separated uniform convergence spaces. Finally, an initially structured category is characterized as an extremal epireflective subcategory of a topological category, i.e. its object class is a relative disconnectedness (with respect to some topological category).

5.1 Factorization structures

5.1.1 Definitions. Let C be a category.

(1) A source in C is a pair $(X,(f_i)_{i\in I})$, where $X \in |C|$ and $(f_i)_{i\in I}$ is a family of C-morphisms $f_i: X \to X_i$ indexed by some class I[27]. Sometimes one writes $(f_i: X \to X_i)_{i\in I}$ instead of $(X,(f_i)_{i\in I})$.

(2) A source $(X,(f_i)_{i\in I})$ in C is called a mono-source in C provided that for any pair $Y \overset{\alpha}{\underset{\beta}{\rightrightarrows}} X$ of C-morphisms such that $f_i\circ\alpha = f_i\circ\beta$ for each $i \in I$, it follows that $\alpha = \beta$, i.e. provided that $(f_i^*: X_i \to X)_{i\in I}$ is an epi-sink in C^* .

(3) A source $(X,(f_i)_{i\in I})$ in C is called extremal provided that for each source $(Y,(g_i)_{i\in I})$ in C and each C-epimorphism $e: X \to Y$ such that for each $i \in I$, the diagram

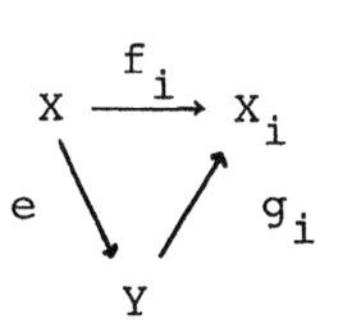

commutes, e must be an isomorphism.

(4) If E is a class of C-morphisms which is closed under com-

27) I may be a proper class, a set or the empty class.

position with isomorphisms and M is a conglomerate of sources in C which is closed under composition with isomorphisms, then the pair (E,M) is called a factorization structure on C provided that the following are satisfied:

(a) For each source $(X,(f_i)_{i\in I})$ in C, there exist $e: X \to Y$ in E and $(Y,(m_i)_{i\in I})$ in M such that $f_i = m_i \circ e$ for each $i \in I$; briefly, each source has an (E,M)-*factorization*.

(b) For any two C-morphisms f and e and any two sources $(Y,(m_i)_{i\in I})$ and $(Z,(f_i)_{i\in I})$ in C such that $e \in E$, $(Y,(m_i)_{i\in I}) \in M$ and $f_i \circ e = m_i \circ f$ for each $i \in I$, there exists a unique C-morphism $g: Z \to Y$ such that the diagram

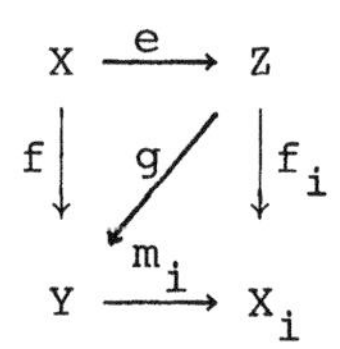

commutes for each $i \in I$; briefly: C satisfies the (E,M)-*diagonalization property*.

(5) C is called an (E,M)-category provided that (E,M) is a factorization structure on C.

5.1.2 Remarks. (1) Sometimes a single C-morphism is considered to be a source indexed by a one element class and vice versa.

(2) Products (more generally: limits) are extremal mono-sources (Let $(X,(p_i)_{i\in I})$ be the product of the family $(X_i)_{i\in I}$ in C, then $(X,(p_i)_{i\in I})$ is a mono-source. In order to show that this source is extremal let

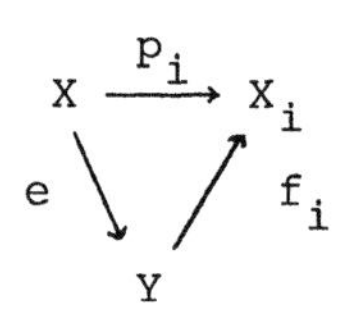

be a factorization where e is an epimorphism. By the definition of product there is a unique C-morphism $g: Y \to X$ such that

$p_i \circ g = f_i$ for each $i \in I$. Hence

$$p_i \circ g \circ e = f_i \circ e = p_i = p_i \circ 1_X$$

so that $g \circ e = 1_X$ since $(X,(p_i))$ is a mono-source. Since e is an epimorphism, it follows immediately that e is an isomorphism.).

(3) a) (X,f) is a mono-source if and only if f is a monomorphism.

b) (X,f) is an extremal mono-source if and only if f is an extremal monomorphism.

5.1.3 Examples (of factorization structures).

(1) Let $\mathcal{C}$ be an arbitrary category. $\mathcal{E}$ consists of all $\mathcal{C}$-isomorphisms and $\mathcal{M}$ consists of all sources in $\mathcal{C}$. Then $(\mathcal{E},\mathcal{M})$ is a factorization structure on $\mathcal{C}$.

(2) Let $\mathcal{C}$ be a topological category or $\mathcal{C}$ = Haus. $\mathcal{E}$ consists of all extremal epimorphisms and $\mathcal{M}$ consists of all mono-sources in $\mathcal{C}$. Then $(\mathcal{E},\mathcal{M})$ is a factorization structure on $\mathcal{C}$.

A) If $(X,(f_i)_{i\in I})$ is a source in $\mathcal{C}$ and an equivalence relation R on X is defined by

$$x\,R\,y \quad \text{iff} \quad f_i(x) = f_i(y) \quad \text{for each } i \in I ,$$

then $\omega: X \to X/_R$ becomes a quotient map in the usual way. If $g_i: X/_R \to X_i$ is defined by $g_i \circ \omega = f_i$ for each $i \in I$, then g_i is a well-defined $\mathcal{C}$-morphism and $(X/_R,(g_i)_{i\in I})$ is a mono-source; for if $\alpha,\beta: D \to X/_R$ are $\mathcal{C}$-morphisms such that $g_i \circ \alpha = g_i \circ \beta$ for each $i \in I$, then $\alpha = \beta$ [otherwise there would exist some $d \in D$ with $\alpha(d) \neq \beta(d)$. Hence there would be $x,x' \in X$ such that $\alpha(d) = \omega(x)$, $\beta(d) = \omega(x')$ and $(x,x') \notin R$, i.e. there would exist some $j \in I$ with $f_j(x) \neq f_j(x')$ so that $g_j(\omega(x)) = g_j(\alpha(d)) \neq g_j(\omega(x')) = g_j(\beta(d))$. Thus $g_j \circ \alpha \neq g_j \circ \beta$ in contradiction to the assumption.].

B) If $h_i \circ f = f_i$ is any (extremal epi, mono-sources)-factorization of $(X,(f_i))$, then there exists some isomorphism $j: X/_R \to Z$ such that the diagram

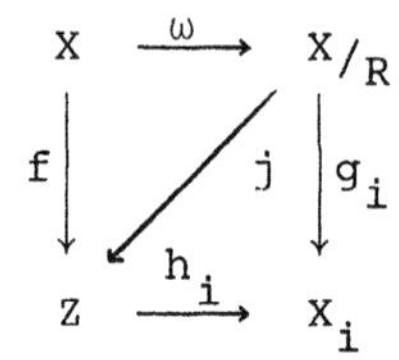

commutes for each $i \in I$ [f can be identified with a quotient map $X \to X/_{R'}$ (up to isomorphism). Since $(Z,(h_i))$ is a mono-source, for $x,y \in X$, the assertion "$f(x) = f(y)$" is equivalent to "$h_i(f(x)) = h_i(f(y))$ for each $i \in I$" so that $x\ R\ y$ is equivalent to $x\ R'\ y$, i.e. $R' = R$ and $f = \omega$. Then $h_i \circ \omega = g_i \circ \omega$ for each $i \in I$ and consequently $h_i = g_i$ for each $i \in I$ since ω is an epimorphism.].

C) For each $i \in I$, let

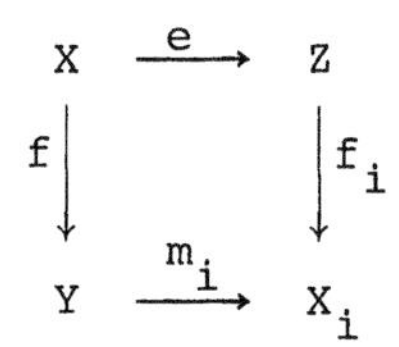

be a commutative diagram in $\mathcal{C}$ with $e \in \mathcal{E}$ and $(Y,(m_i)) \in \mathcal{M}$. Let $f = m' \circ e'$ (resp. $f_i = m_i'' \circ e''$) be some $(\mathcal{E},\mathcal{M})$-factorization of (X,f) (resp. $(Z,(f_i))$) . Since the composition of extremal epimorphisms in $\mathcal{C}$ is again an extremal epimorphism and the composition of a monomorphism and a mono-source in $\mathcal{C}$ is again a mono-source, it follows that $m_i'' \circ (e'' \circ e)$ and $(m_i \circ m') \circ e'$ are $(\mathcal{E},\mathcal{M})$-factorizations of the same source in $\mathcal{C}$. Thus by B), there exists some isomorphism $j: A \to B$ such that the diagram

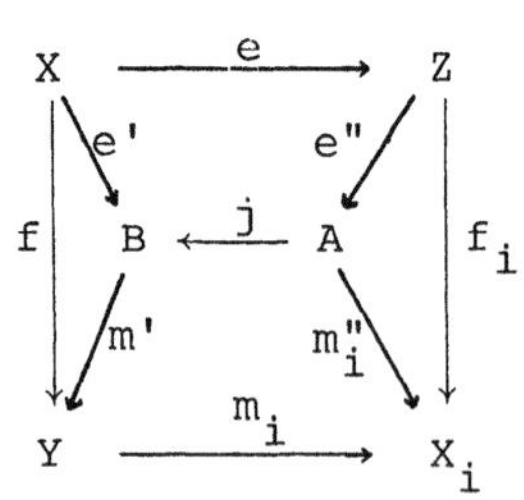

commutes for each $i \in I$. Then $m' \circ j \circ e''$ is the desired diagonal morphism which is uniquely determined since e is an epimorphism.)

5.1.4 Theorem. Let C be a category and (E,M) a factorization structure on C . Then the following are satisfied:

(1) (E,M)-factorizations are uniquely determined (up to isomorphisms).

(2) $E \cap M$ is the class of all C-isomorphisms.

(3) E is a class of C-epimorphisms.

(4) Every extremal source in C belongs to M .

(5) If f,g and h are C-morphisms such that $h = g \circ f$, then the following are satisfied:

(a) If $h \in E$ and f is a C-epimorphism, then $g \in E$.

(b) $f \in E$ and $g \in E$ imply $h \in E$, i.e. E is closed under composition.

(6) If $(X,(f_i)_{i\in I})$ is a source in C and $((X,(g_j)_{j\in J}), (Z_j,(k_{j_i})_{i\in I_j})_{j\in J})$ is a factorization of $(X,(f_i)_{i\in I})$, i.e. $\bigcup_{j\in J} I_j = I$ and $f_i = k_{j_i} \circ g_j$ for each $j \in J$ and each $i \in I_j$, then the following hold:

(a) $(X,(f_i)_{i\in I}) \in M$ implies $(X,(g_j)_{j\in J}) \in M$.

(b) $(X,(g_j)_{j\in J}) \in M$ and $(Z_j,(k_{j_i})_{i\in I_j}) \in M$ for each $j \in J$ imply $(X,(f_i)_{i\in I}) \in M$.

(7) If $(X,(f_i)_{i\in I})$ is a source in C and there is some $J \subset I$ such that $(X,(f_j)_{j\in J}) \in M$, then $(X,(f_i)_{i\in I}) \in M$.

(8) E and M determine each other by the diagonalization property.

Proof. (3) Suppose that E would contain a C-morphism $e: X \to A$ which is not an epimorphism. Then there would exist C-morphisms $r,s: A \to A^*$ with $r \neq s$ but $r \circ e = s \circ e = k$. Let $I = \mathrm{Mor}\, C$ and for each $i \in I$, let $A_i = A^*$ and $f_i = k: X \to A^*$. Then the source $(X,(f_i)_{i\in I})$ would have an (E,M)-factorization

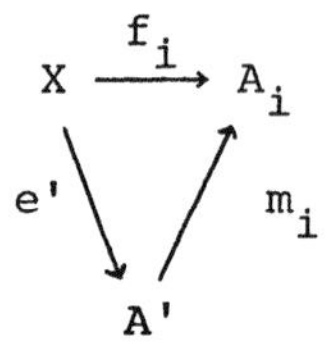

.

For each $f \in I$, we could define

$$g_f = \begin{cases} r & \text{provided } m_f \circ f = s \\ s & \text{otherwise} \end{cases} .$$

Then for each $i \in I$, $g_i \circ e = k = m_i \circ e'$. Hence by the diagonalization property, there would exist a unique $h: A \to A'$ such that for each $i \in I$, the diagram

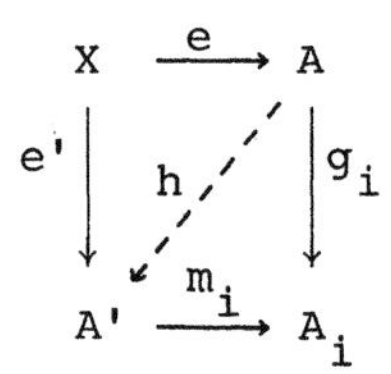

would commute. Especially,

$$m_h \circ h = g_h = \begin{cases} r & \text{if } m_h \circ h = s \\ s & \text{if } m_h \circ h \neq s \end{cases}$$

- a contradiction.

(1) Let $(X, (f_i)_{i \in I})$ be a source in C and let $f_i = m_i \circ e$ as well as $f_i = m_i' \circ e'$ be (E,M)-factorizations of this source. Then there are a unique C-morphism g and a unique C-morphism h such that the diagrams

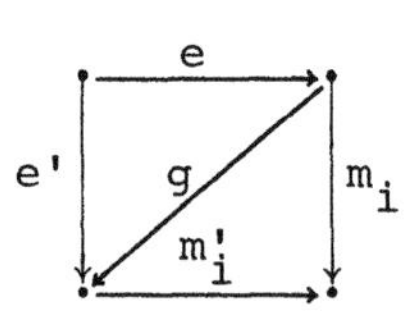

and

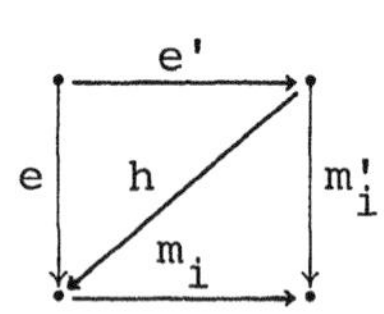

commute for each $i \in I$. Especially, $g \circ e = e'$ and $h \circ e' = e$. Then $g \circ h \circ e' = g \circ e = e' = 1 \circ e'$ and since e' is an epimorphism (note (3)), it follows that $g \circ h = 1$. One concludes analogously that $h \circ g = 1$, i.e. g is an isomorphism.

(2) a) Let $(f: X \to Y) \in \mathcal{E} \cap \mathcal{M}$. Then there is a unique $\mathcal{C}$-morphism $g: Y \to X$ such that the diagram

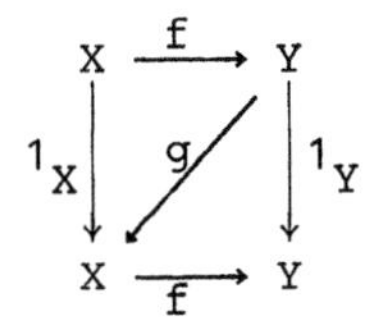

commutes. Thus f is an isomorphism.

b) Let f be a $\mathcal{C}$-isomorphism and $f = m \circ e$ an $(\mathcal{E},\mathcal{M})$-factorization of f. Since f is an extremal monomorphism and e is an epimorphism (cf. (3)), it follows that e is an isomorphism. Thus, $f \in \mathcal{M}$ (because $\mathcal{M}$ is closed under composition with isomorphisms). Furthermore, $m = f \circ e^{-1}$ is an isomorphism as a composite of two isomorphisms. Consequently, $f \in \mathcal{E}$ ($\mathcal{E}$ is closed under composition with isomorphisms!).

(4) Let $(X,(f_i)_{i \in I})$ be an extremal source in $\mathcal{C}$ and $f_i = m_i \circ e$ an $(\mathcal{E},\mathcal{M})$-factorization. Since e is an epimorphism (cf. (3)), e must be an isomorphism by the definition of "extremal". Then $(X,(f_i)_{i \in I})$ belongs to $\mathcal{M}$ since $\mathcal{M}$ is closed under composition with isomorphisms.

(5) (a) Let $g = m \circ e$ be an $(\mathcal{E},\mathcal{M})$-factorization. Then there is a unique $\mathcal{C}$-morphism k such that the diagram

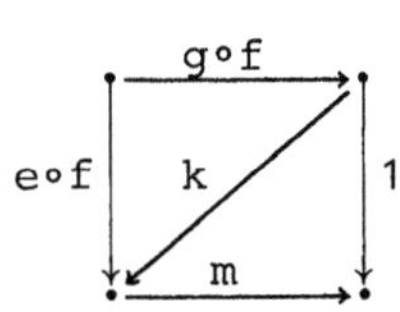

commutes. Hence $k \circ g \circ f = e \circ f$ is an epimorphism as a composite of two epimorphisms. Thus k is an epimorphism which must be an isomorphism because of $m \circ k = 1$. Therefore $m = k^{-1}$ is an isomorphism and consequently, $g \in \mathcal{E}$.

(b) Let $g \circ f = m \circ e$ be an (E,M)-factorization of $g \circ f$. Then there exist C-morphisms k and l such that the diagram

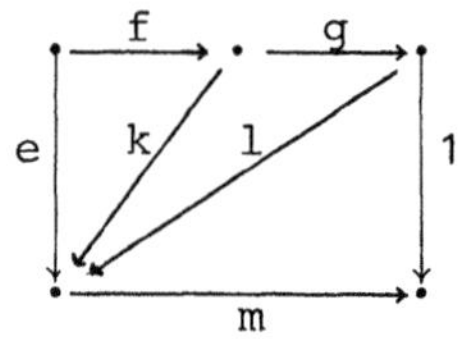

commutes. $l \circ g \circ f = e$ implies that l is an epimorphism which must be an isomorphism because of $m \circ l = 1$. Thus $m = l^{-1}$ is an isomorphism and consequently $g \circ f \in E$.

(6) (a) Let $g_j = m_j \circ e$ be an (E,M)-factorization of $(X,(g_j)_{j \in J})$, i.e. the diagram

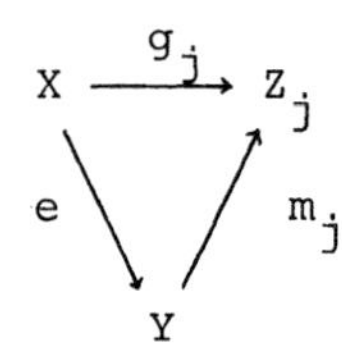

commutes for each $j \in J$. Since $(X,(f_i)_{i \in I}) \in M$, there is a unique C-morphism $h: Y \to X$ such that the diagram

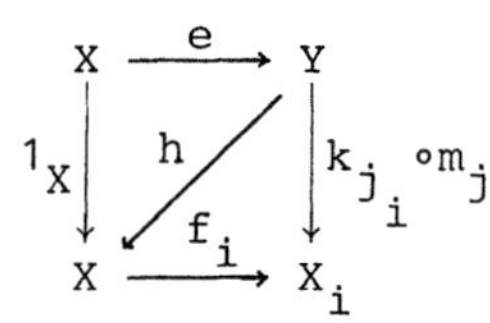

commutes for each $i \in I$. Especially, $h \circ e = 1_X$ so that since e is an epimorphism, it follows that e is an isomorphism. Consequently, $(X,(g_j)_{j \in J}) \in M$.

(b) Let

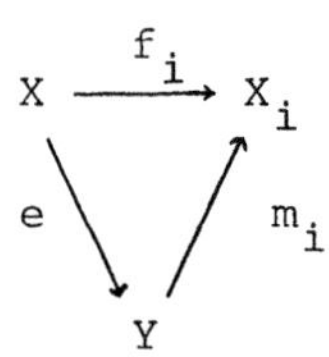

for each $i \in I$ be an (E,M)-factorization of $(X,(f_i)_{i\in I})$. Since $(Z_j,(k_{j_i})_{i\in I_j}) \in M$ for each $j \in J$, it follows that for each $j \in J$, there is a C-morphism $h_j: Y \to Z_j$ such that the diagram

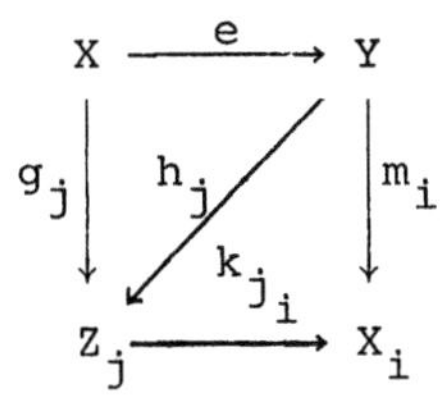

commutes for each $j \in J$ and each $i \in I_j$. Since $(X,(g_j)_{j\in J}) \in M$, there exists a unique C-morphism $k: Y \to X$ such that the diagram

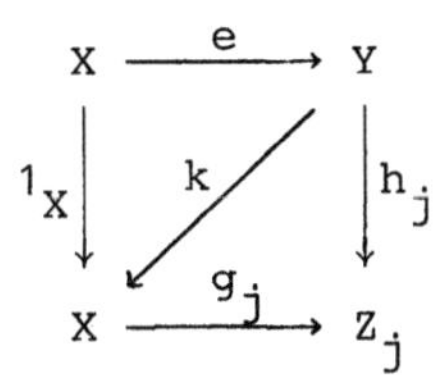

commutes for each $j \in J$. Especially, since e is an epimorphism, it follows from $k \circ e = 1_X$ that e is an isomorphism. Thus, $(X,(f_i)_{i\in I}) \in M$.

(7) Let the source $(X,(f_j)_{j\in J})$ belong to M for $J \subset I$. Further let

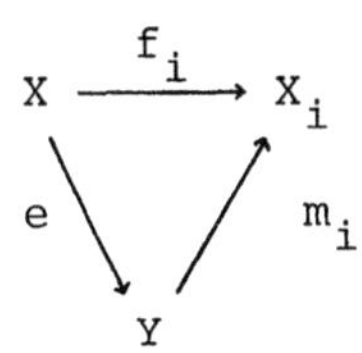

be an (E,M)-factorization of $(X,(f_i)_{i\in I})$. Then there is a unique C-morphism $h: Y \to X$ such that the diagram

$$\begin{array}{ccc} X & \xrightarrow{e} & Y \\ {\scriptstyle 1_X}\downarrow & \overset{h}{\swarrow} & \downarrow{\scriptstyle m_j} \\ X & \xrightarrow{f_j} & X_j \end{array}$$

commutes for each $j \in J$. Since e is an epimorphism, it follows from $h \circ e = 1_X$ that e is an isomorphism. Thus, $(X,(f_i)_{i \in I}) \in M$.

(8) a) E is determined by M via the diagonalization property, i.e. $E = \{e \in \text{Mor } C$: if there exist some class I and C-morphisms f, m_i, f_i such that the diagram

$$\begin{array}{ccc} X & \xrightarrow{e} & Y \\ {\scriptstyle f}\downarrow & & \downarrow{\scriptstyle f_i} \\ Z & \xrightarrow{m_i} & Y_i \end{array}$$

commutes for each $i \in I$ where $(Z,(m_i)_{i \in I}) \in M$, then there is a unique C-morphism $h: Y \to Z$ such that the diagram

$$\begin{array}{ccc} X & \xrightarrow{e} & Y \\ {\scriptstyle f}\downarrow & \overset{h}{\swarrow} & \downarrow{\scriptstyle f_i} \\ Z & \xrightarrow{m_i} & Y_i \end{array}$$

commutes for each $i \in I$ $\}$.

Since obviously each $e \in E$ has the desired property (by the definition of the factorization structure (E,M) on C), it suffices to show that each $e \in \text{Mor } C$ satisfying the property defined above belongs to E. Let $e = m \circ e'$ be the (E,M)-factorization of some $e \in \text{Mor } C$ with this property. Then there is a unique C-morphism $h: Y \to Z$ such that the diagram

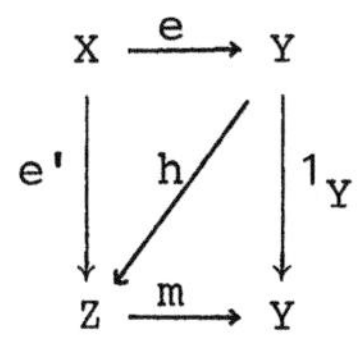

commutes. Since $e' = h \circ e$ is an epimorphism, h is an epimorphism. Thus, since $m \circ h = 1_Y$, it follows that h is an isomorphism. Then $m = h^{-1}$ is an isomorphism and thus $e \in \mathcal{E}$.

b) $\mathcal{M}$ is determined by $\mathcal{E}$ via the diagonalization property, i.e. $\mathcal{M} = \{(Z,(h_i)_{i\in I})$: $(Z,(h_i)_{i\in I})$ is a source in $\mathcal{C}$ and if there exist some source $(Y,(f_i)_{i\in I})$ and $\mathcal{C}$-morphisms f and e with $e \in \mathcal{E}$ such that the diagram

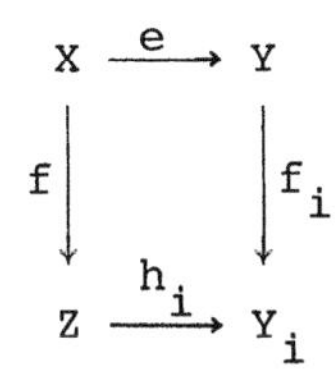

commutes for each $i \in I$, then there exists a unique $\mathcal{C}$-morphism $h: Y \to Z$ such that the diagram

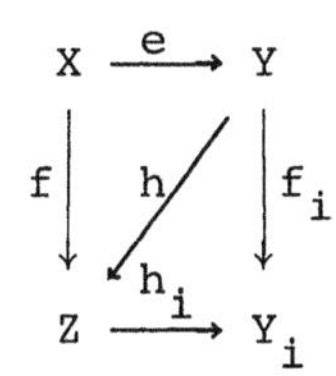

commutes for each $i \in I\}$.

It suffices to show that each source $(Z,(h_i)_{i\in I})$ in $\mathcal{C}$ satisfying the property defined above belongs to $\mathcal{M}$. Let $h_i = m_i \circ e'$ be the $(\mathcal{E},\mathcal{M})$-factorization of such a source. Then there is a unique $\mathcal{C}$-morphism $g: Z' \to Z$ such that the diagram

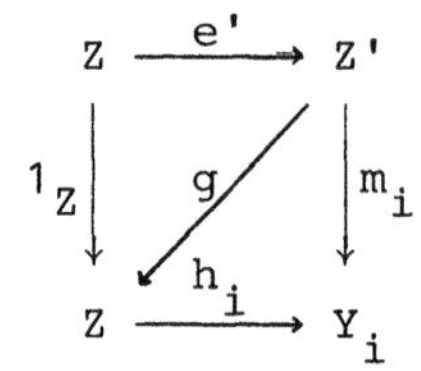

commutes for each $i \in I$. Since e' is an epimorphism, it follows from $g \circ e' = 1_Z$ that e' is an isomorphism; thus $(Z,(h_i)_{i\in I}) \in \mathcal{M}$.

5.1.5 Remark. If $\mathcal{C}$ is a category supplied with a factorization structure $(\mathcal{E},\mathcal{M})$, then every (full and isomorphism-closed) subcategory $\mathcal{A}$ of $\mathcal{C}$ has an *$\mathcal{E}$-reflective hull* $\mathcal{B}$. Especially, $X \in |\mathcal{B}|$ if and only if there exists some source $(X,(m_i: X \to A_i)_{i\in I})$ in $\mathcal{M}$ with $A_i \in |\mathcal{A}|$ for each $i \in I$. Note that *unnatural completeness- and smallness-restrictions do not appear!* (If one defines a full and isomorphism-closed subcategory $\mathcal{B}$ of $\mathcal{C}$ by $|\mathcal{B}| = \{X \in |\mathcal{C}|$: there exists $(X,(m_i: X \to A_i)_{i\in I}) \in \mathcal{M}$ with $A_i \in |\mathcal{A}|$ for each $i \in I\}$, then $\mathcal{A} \subset \mathcal{B}$ [$1_A: A \to A$ belongs to $\mathcal{M}$ for each $A \in |\mathcal{A}|$ since 1_A is an isomorphism] and $\mathcal{B}$ is $\mathcal{E}$-reflective in $\mathcal{C}$; for if $Y \in |\mathcal{C}|$, $(Y,(f_i)_{i\in I})$ is the source of all $\mathcal{C}$-morphisms whose codomains belong to $|\mathcal{A}|$ and

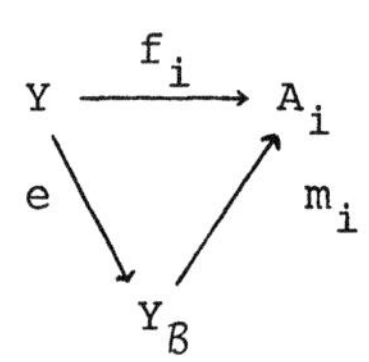

is the $(\mathcal{E},\mathcal{M})$-factorization of $(Y,(f_i)_{i\in I})$, then e is the $\mathcal{E}$-reflection of Y with respect to $\mathcal{B}$: Namely, if $Z \in |\mathcal{B}|$ and $f: Y \to Z$ is a $\mathcal{C}$-morphism, then there exists some source $(Z,(g_j)) \in \mathcal{M}$ where $g_j: Z \to A'_j$ with $A'_j \in |\mathcal{A}|$ for each j . For each j , there is some $i_j \in I$ such that $f_{i_j} = g_j \circ f$ and $A'_j = A_{i_j}$. Hence $g_j \circ f = m_{i_j} \circ e$. By the $(\mathcal{E},\mathcal{M})$-diagonalization property, there exists a unique $\bar{f}: Y_\mathcal{B} \to Z$ such that the

diagram

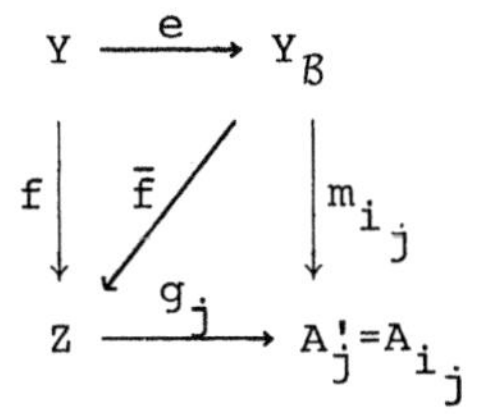

commutes.

If $\mathcal{D}$ is an $\mathcal{E}$-reflective [full and isomorphism-closed] subcategory of $\mathcal{C}$ containing $\mathcal{A}$, then $\mathcal{B} \subset \mathcal{D}$; for if $X \in |\mathcal{B}|$, then there is some source $(X,(m_i)_{i\in I}) \in \mathcal{M}$ such that $m_i\colon X \to A_i$ and $A_i \in |\mathcal{A}| \subset |\mathcal{D}|$ for each $i \in I$. If $e_X\colon X \to X_{\mathcal{D}}$ is the $\mathcal{E}$-reflection of X with respect to $\mathcal{D}$, then for each $i \in I$, there is a unique $\mathcal{C}$-morphism $\bar{m}_i\colon X_{\mathcal{D}} \to A_i$ such that $\bar{m}_i \circ e_X = m_i$. By the $(\mathcal{E},\mathcal{M})$-diagonalization property, there is a unique $\mathcal{C}$-morphism $h\colon X_{\mathcal{D}} \to X$ such that the diagram

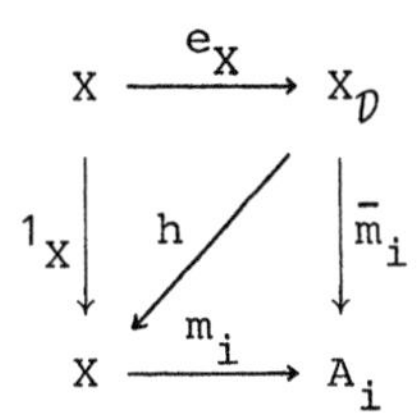

commutes. Thus since e_X is an epimorphism, it follows from $h \circ e_X = 1_X$ that e_X is an isomorphism. Since $\mathcal{D}$ is isomorphism-closed, $X \in |\mathcal{D}|$.)

5.2 Definition and properties of topological functors

5.2.1 Definition. Let $\mathcal{C}$ be a category supplied with a factorization structure $(\mathcal{E},\mathcal{M})$, let $\mathcal{A}$ be any category and let $\mathcal{T}\colon \mathcal{A} \to \mathcal{C}$ be a functor.

(1) A source $(A,(f_i\colon A \to A_i)_{i\in I})$ in $\mathcal{A}$ is called *$\mathcal{T}$-initial* provided that for each source $(B,(g_i\colon B \to A_i)_{i\in I})$ in $\mathcal{A}$ and

each C-morphism $f: T(B) \to T(A)$ such that $T(f_i)\circ f = T(g_i)$ for each $i \in I$, there exists a unique A-morphism $\bar{f}: B \to A$ with $T(\bar{f}) = f$ and $f_i \circ \bar{f} = g_i$ for each $i \in I$.

(2) A source $(A,(f_i: A \to A_i)_{i\in I})$ in A <u>T-lifts</u> a source $(X,(g_i: X \to T(A_i))_{i\in I})$ in C provided that there exists an isomorphism $h: X \to T(A)$ in C with $T(f_i)\circ h = g_i$ for each $i \in I$.

(3) T is called <u>(E,M)-topological</u> provided that for each family $(A_i)_{i\in I}$ of A-objects and each source $(X,(m_i: X \to T(A_i))_{i\in I})$ in M, there exists a T-initial source $(A,(f_i: A \to A_i)_{i\in I})$ in A which T-lifts $(X,(m_i)_{i\in I})$.

(4) T is called <u>absolutely topological</u> provided that T is (E,M)-topological for any factorization structure (E,M) on C.

<u>5.2.2 Examples.</u> (1) The forgetful functor from any topological category into the category C = <u>Set</u> is absolutely topological.

(2) Let F_u: <u>Unif</u> → <u>Top</u> be the forgetful functor. A factorization structure (E,M) on <u>Top</u> is defined by:

E = {surjective continuous maps}

M = {embedding-sources} ,

where a source $((X,\mathcal{X}), (f_i: (X,\mathcal{X}) \to (X_i,\mathcal{X}_i))_{i\in I})$ is called an *embedding-source* provided that it is a mono-source and $\mathcal{X}$ is the initial topology with respect to $(f_i)_{i\in I}$ [If $((X,\mathcal{X}), (f_i: (X,\mathcal{X}) \to (X_i,\mathcal{X}_i))_{i\in I})$ is any source in <u>Top</u>, then the (E,M)-factorization is constructed as follows: An equivalence relation R on X is defined by

$$x \; R \; y \quad \text{iff} \quad f_i(x) = f_i(y) \quad \text{for each} \quad i \in I .$$

Let $g_i: X/_R \to X_i$ be defined by $g_i\circ\omega = f_i$ for each $i \in I$ where $\omega: X \to X/_R$ denotes the natural map. Further let $Y = X/_R$ be endowed with the initial topology $\mathcal{Y}$ with respect to $(g_i)_{i\in I}$. Then $\omega: (X,\mathcal{X}) \to (Y,\mathcal{Y})$ is a surjective continuous map and $((Y,\mathcal{Y}), (g_i: (Y,\mathcal{Y}) \to (X_i,\mathcal{X}_i))_{i\in I})$ is an embedding-source.]. Then F_u *is (E,M)-topological* (note that the initial uniformity induces the initial topology!) *but* <u>not</u> *absolutely topological* (If $(Y,\mathcal{R})$ is any uniform space and $f: (X,\mathcal{X}) \to (Y,\mathcal{Y}_{\mathcal{R}})$ is a continuous map starting from a topological space which is not completely regular, then there is no $\mathcal{F}_u$-initial source in Unif which $\mathcal{F}_u$-lifts $((X,\mathcal{X}),\ f\colon (X,\mathcal{X}) \to (Y,\mathcal{Y}_R))$ [otherwise $(X,\mathcal{X})$ would be a completely regular space].) .

(3) The forgetful functor $T\colon$ Haus $\to$ Set is (extremal epi, mono-source)-topological (If $((X_i,\mathcal{X}_i))_{i\in I}$ is any family of Hausdorff spaces and $(X,\ (m_i\colon X \to X_i)_{i\in I})$ is a mono-source, then $((X,\mathcal{X}),\ (m_i\colon (X,\mathcal{X}) \to (X_i,\mathcal{X}_i))_{i\in I})$ is a T-initial source in Haus [provided $\mathcal{X}$ is the initial topology with respect to $(m_i)_{i\in I}$] which T-lifts $(X,(m_i))$) but not absolutely topological (Given the set $\mathbb{R}$ of real numbers endowed with the usual topology and let $f\colon \{0,1\} \to \mathbb{R}$ be defined by $f(x) = 0$ for each $x \in \{0,1\}$. Then $(\{0,1\},\ f\colon \{0,1\} \to \mathbb{R})$ is a source in Set which is not T-lifted by any T-initial source in Haus ; for there is only one Hausdorff topology on $\{0,1\}$, namely the discrete topology $\mathcal{D}$ and for this topology $\mathcal{D}$, the source $((\{0,1\},\mathcal{D}),\ f\colon (\{0,1\},\mathcal{D}) \to (\mathbb{R}$, usual top.)) is not T-initial [e.g. $h\colon (\mathbb{R}$, usual top.) $\to (\{0,1\},\mathcal{D})$ defined by

$$h(x) = \begin{cases} 0 & \text{for } x < 0 \\ 1 & \text{for } x \geq 0 \end{cases}$$

is not continuous but $f\circ h$ is constant and therefore continuous].).

5.2.3 Remark. Let C be a topological category and let (E,M) be the factorization structure explained in 5.1.3 (2) . Further let A be a (full and isomorphism-closed) extremal epireflective subcategory of C . By 5.1.5 and 2.3.4 (2) holds: Each $X \in |C|$ for which there exists a mono-source $(X,(f_i\colon X \to A_i)_{i\in I})$ with $A_i \in |A|$ for each $i \in I$ belongs to $|A|$. Therefore the forgetful functor $T\colon A \to$ Set is (extremal epi, mono-source)-topological (If $((X_i,\xi_i))_{i\in I}$ is a family of A-objects, $(X,(m_i\colon X \to X_i)_{i\in I})$ is a mono-source and ξ denotes the initial C-structure on X with respect to (m_i) , then $((X,\xi),\ (m_i\colon (X,\xi) \to (X_i,\xi_i))_{i\in I})$ is a T-initial source in A which T-lifts $(X,(m_i))$.).

5.2.4 Theorem. Let C be a category supplied with a factorization structure (E,M) and let $T: A \to C$ be an (E,M)-topological functor. Then T is faithful[28].

Proof. If T were not faithful, then there would be a pair (A,B) of A-objects and morphisms $r: A \to B$ and $s: A \to B$ such that $r \neq s$ but $T(r) = T(s)$. Let I be a proper class which is a disjoint union of K and $\{0\}$. Further let $(A_i)_{i \in I}$ be a family of A-objects defined as follows:

$$A_o = A \quad \text{and} \quad A_k = B \quad \text{for each} \quad k \in K .$$

The source $(T(A), (m_i: T(A) \to T(A_i))_{i \in I})$ defined by $m_o = 1_{T(A)}$ and $m_k = T(r)$ for each $k \in K$ is extremal [from each source factorization $m_i = g_i \circ e$ where e is an epimorphism, it follows that $m_o = 1_{T(A)} = g_o \circ e$, i.e. e is an isomorphism] and so it belongs to M. Then there exists a T-initial source $(D, (f_i: D \to A_i)_{i \in I})$ and an isomorphism $h: T(A) \to T(D)$ with $T(f_i) \circ h = m_i$ for each $i \in I$. In order to prove that $\{f_k: k \in K\}$ is a proper class we choose two distinct elements j and $\tilde{j}$ of K and define a source $(A, (g_i: A \to A_i)_{i \in I})$ by $g_o = 1_A$, $g_j = r$ and $g_k = s$ for each $k \in K \setminus \{j\}$. Because of $m_i = T(g_i)$ we have $T(f_i) \circ h = T(g_i)$ for each $i \in I$. Since $(D, (f_i))$ is T-initial, there exists a morphism $g: A \to D$ such that $f_i \circ g = g_i$ for each $i \in I$. Thus since $g_j \neq g_{\tilde{j}}$, it follows that $f_j \neq f_{\tilde{j}}$. Consequently, $\{f_k: k \in K\} \subset [D,B]_A$ is a proper class in contradiction to the usual definition of a category where the "collection" of all morphisms between two objects has to be a set.

5.2.5 Theorem. Let C be an (E,M)-category and $T: A \to C$ an (E,M)-topological functor. If E_T denotes the class of all morphisms f in A with $T(f) \in E$ and M_T denotes the conglomerate of all T-initial sources $(A, (f_i)_{i \in I})$ in A with $(T(A), (T(f_i))_{i \in I}) \in M$, then A is an (E_T, M_T)-category.

---

28) A functor $F: C \to D$ is called faithful provided that for each pair $(A,B) \in |C| \times |C|$, the map $[A,B]_C \to [F(A),F(B)]_D$ $(f \mapsto F(f))$ is injective.

<u>Proof</u>. 1) Since T is faithful by 5.2.4, E_T is a class of epimorphisms ($\alpha \circ f = \beta \circ f$ with $f \in E_T$ and $\alpha, \beta \in \text{Mor}\, A$ implies $T(\alpha) \circ T(f) = T(\beta) \circ T(f)$ so that since $T(f)$ is an epimorphism, it follows that $T(\alpha) = T(\beta)$. Thus since T is faithful, we have $\alpha = \beta$.).

2) Let $(A, (f_i: A \to A_i)_{i \in I})$ be a source in A and let

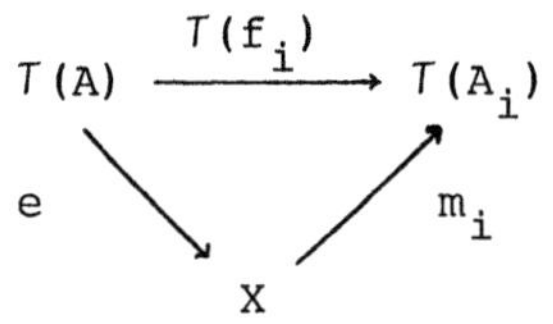

be an (E,M)-factorization of $(T(A), (T(f_i))_{i \in I})$. Then there exists a T-initial source $(B, (g_i: B \to A_i)_{i \in I})$ and an isomorphism $h: X \to T(B)$ with $T(g_i) \circ h = m_i$ for each $i \in I$. Hence, $(B, (g_i)) \in M_T$. Furthermore, let $f = h \circ e: T(A) \to T(B)$ and note that $T(g_i) \circ f = T(g_i) \circ h \circ e = m_i \circ e = T(f_i)$ for each $i \in I$. Then there exists a unique $\bar{f}: A \to B$ such that $T(\bar{f}) = f = h \circ e$ and $g_i \circ \bar{f} = f_i$ for each $i \in I$. Since obviously $T(\bar{f}) = h \circ e \in E$, we have $\bar{f} \in E_T$ so that

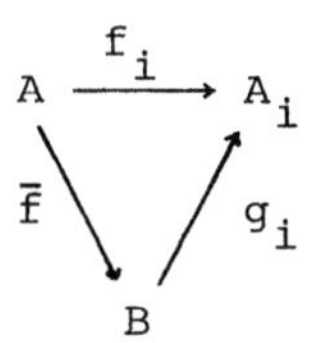

is the desired (E_T, M_T)-factorization of $(A, (f_i))$.

3) Let the diagram

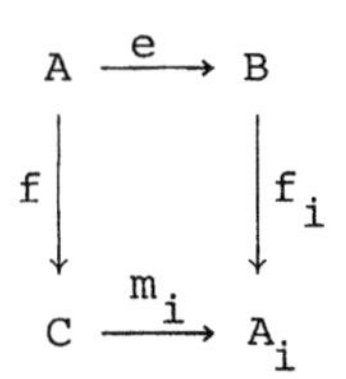

be commutative for each $i \in I$ where $e \in E_T$ and $(C,(m_i)) \in M_T$. Then the diagram

$$\begin{array}{ccc} T(A) & \xrightarrow{T(e)} & T(B) \\ {\scriptstyle T(f)}\downarrow & & \downarrow{\scriptstyle T(f_i)} \\ T(C) & \xrightarrow{T(m_i)} & T(A_i) \end{array}$$

commutes for each $i \in I$ where $T(e) \in E$ and $(T(C),(T(m_i))) \in M$. Hence there exists a unique $g\colon T(B) \to T(C)$ such that the diagram

$$\begin{array}{ccc} T(A) & \xrightarrow{T(e)} & T(B) \\ {\scriptstyle T(f)}\downarrow & \swarrow{\scriptstyle g} & \downarrow{\scriptstyle T(f_i)} \\ T(C) & \xrightarrow[T(m_i)]{} & T(A_i) \end{array}$$

commutes for each $i \in I$. Since $(C,(m_i))$ is T-initial, there exists a unique morphism $\bar{g}\colon B \to C$ such that $T(\bar{g}) = g$ and $m_i \circ \bar{g} = f_i$ for each $i \in I$. Since T is faithful, it follows from $g \circ T(e) = T(\bar{g}) \circ T(e) = T(\bar{g} \circ e) = T(f)$ that $\bar{g} \circ e = f$. Thus the diagram

$$\begin{array}{ccc} A & \xrightarrow{e} & B \\ {\scriptstyle f}\downarrow & {\scriptstyle \bar{g}}\swarrow & \downarrow{\scriptstyle f_i} \\ C & \xrightarrow[m_i]{} & A_i \end{array}$$

commutes for each $i \in I$ (because of 1) $\bar{g}$ is uniquely determined).

<u>5.2.6 Definitions</u>. Let C be an arbitrary category and A a small category (i.e. $|A|$ is a set). A functor $F\colon A \to C$ is called a <u>diagram</u> in C over A (If $|A|$ is finite, then the

diagram is described in the usual way).

1) A <u>lower bound</u> of $\mathcal{F}$ is a pair $(L,(l_A)_{A\in|\mathcal{A}|})$ where L is a $\mathcal{C}$-object and $l_A\colon L \to \mathcal{F}(A)$ is a $\mathcal{C}$-morphism for each $A \in |\mathcal{A}|$ such that for every $\mathcal{A}$-morphism $f\colon A \to A'$, the diagram

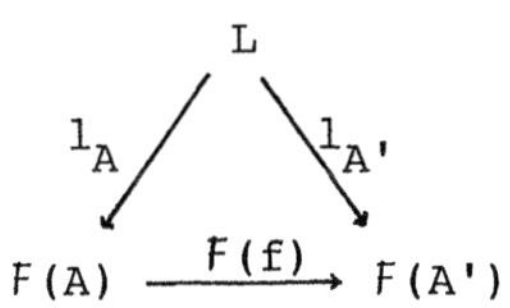

commutes.

1') An <u>upper bound</u> of $\mathcal{F}$ is a pair $(C,(c_A)_{A\in|\mathcal{A}|})$ where C is a $\mathcal{C}$-object and $c_A\colon \mathcal{F}(A) \to C$ is a $\mathcal{C}$-morphism for each $A \in |\mathcal{A}|$ such that for every $\mathcal{A}$-morphism $f\colon A' \to A$, the diagram

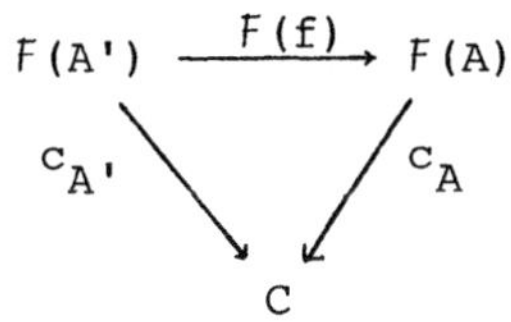

commutes, i.e. such that $(C,(c_A^*)_{A\in|\mathcal{A}|})$ is a lower bound of $\mathcal{F}^*$[29].

2) A lower bound $(L,(l_A)_{A\in|\mathcal{A}|})$ is called a <u>limit of $\mathcal{F}$</u> provided that for every lower bound $(L',(l'_A)_{A\in|\mathcal{A}|})$ of $\mathcal{F}$, there exists a unique morphism $l\colon L' \to L$ such that the diagram

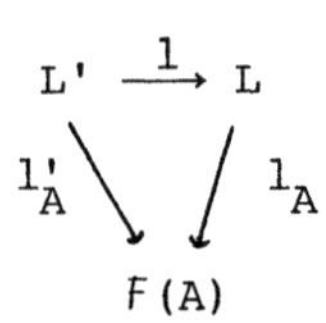

2') An upper bound $(C,(c_A)_{A\in|\mathcal{A}|})$ is called a <u>colimit of $\mathcal{F}$</u> provided that for every upper bound $(C',(c'_A)_{A\in|\mathcal{A}|})$ of $\mathcal{F}$, there exists a unique morphism $c'\colon C \to C'$ such that the diagram

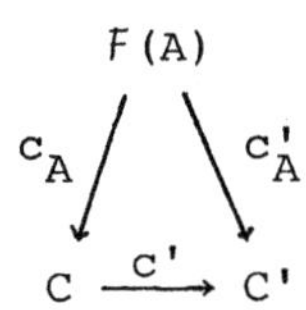

---

29) $\mathcal{F}^*\colon \mathcal{A}^* \to \mathcal{C}^*$ denotes the opposite functor. It is obtained by applying first the dualizing functor $\mathcal{A}^* \to \mathcal{A}^{**} = \mathcal{A}$, then the functor $\mathcal{F}$ and finally the dualizing functor $\mathcal{C} \to \mathcal{C}^*$.

commutes for each $A \in |A|$. commutes for each $A \in |A|$, i.e. provided that $(C,(c_A^*)_{A\in|A|})$ is a limit of F^* .

5.2.7 Remarks. (1) Limits (colimits) are uniquely determined by their defining properties (up to isomorphisms).

(2) Limits are extremal mono-sources (cf. 5.1.2 (2)).

(3) a) Products (coproducts) are limits (colimits) of diagrams over a small discrete category I ("discrete" means that Mor I consists only of identical morphisms); for such a diagram $F: I \to C$ can be considered as a family $(X_i)_{i\in I}$ of C-objects (put $|I| = I$ and $F(i) = X_i$ for each $i \in I$) .

b) Equalizers (coequalizers) are limits (colimits) of diagrams over the category A described by

$$\bullet \rightrightarrows \bullet \ ;$$

for such a diagram can be considered as a pair $X \overset{f}{\underset{g}{\rightrightarrows}} Y$ of C-morphisms.

(4) We say that a category *has limits* (resp. *colimits*) provided that for each small category A , every diagram in C over A has a limit (resp. colimit). A category is complete (resp. cocomplete) if and only if it has limits (resp. colimits) [1. "⇐": Obviously by means of (3) .
2. "⇒": Let A be a small category and let $F: A \to C$ be a diagram. Further let $P = \prod_{A\in|A|} F(A)$ and $Q = \prod_{f\in \mathrm{Mor}\, A} F(C(f))$ where $C(f)$ denotes the codomain of f . For each A-morphism f , there are two morphisms, namely the projection $p_{C(f)} : P \to F(C(f))$ and the morphism $F(f)\circ p_{D(f)}: P \to F(C(f))$ where $D(f)$ denotes the domain of f . Thus there are morphisms $p,q: P \to Q$ defined by $q_f\circ p = p_{C(f)}$ and $q_f\circ q = F(f)\circ p_{D(f)}$ for each $f \in \mathrm{Mor}\, A$ ($q_f: Q \to F(C(f))$ denotes the projection!) . Let (L,h) be the equalizer of p and q and put $p_A\circ h = l_A$ for each $A \in |A|$. Then it is easy to check that $(L,(l_A))$ is a limit of F .]. This characterization is often used for the definition of "complete" (resp. "cocomplete").

<u>5.2.8 Theorem</u>. Let $T: A \to C$ be an (E,M)-topological functor, let $\mathcal{D}: I \to A$ be a diagram and let $(L,(l_i: L \to \mathcal{D}(i))_{i\in|I|})$ be a source in A. Then the following are equivalent:

(1) $(L,(l_i)_{i\in|I|})$ is a limit of $\mathcal{D}$.

(2) $(L,(l_i)_{i\in|I|})$ is T-initial and $(T(L),(T(l_i))_{i\in|I|})$ is a limit of $T\circ\mathcal{D}$.

<u>Proof</u>. (1) ⇒ (2): Let $(L,(l_i))$ be a limit of $\mathcal{D}$. By 5.1.2 ② and 5.1.4 (4), $(L,(l_i))$ belongs to M_T (cf. 5.2.5). Then $(L,(l_i))$ is T-initial and $(T(L),(T(l_i)))$ belongs to M. Obviously $(T(L),(T(l_i)))$ is a lower bound of $T\circ\mathcal{D}$. Now let $(X,(f_i))$ be an arbitrary lower bound of $T\circ\mathcal{D}$ and let

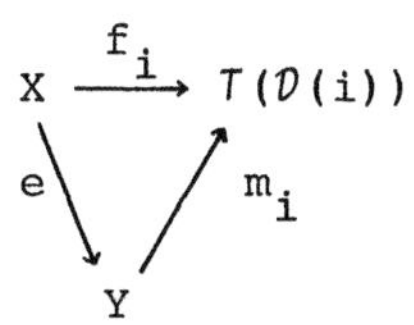

be an (E,M)-factorization of $(X,(f_i))$. Then there exists a T-initial source $(A,(g_i: A \to \mathcal{D}(i))_{i\in|I|})$ and an isomorphism $h: Y \to T(A)$ with $T(g_i)\circ h = m_i$ for each $i \in |I|$. If $f: i \to i'$ is an I-morphism, then in the diagram

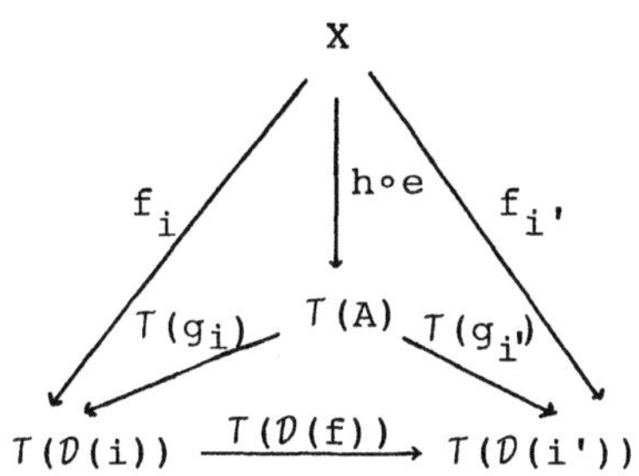

the outer triangle as well as the two upper triangles commute. Since E is isomorphism-closed, $h\circ e \in E$ and thus $h\circ e$ is an epimorphism so that from $T(\mathcal{D}(f))\circ T(g_i)\circ h\circ e = T(\mathcal{D}(f))\circ f_i = f_{i'} = T(g_{i'})\circ h\circ e$, it follows that $T(\mathcal{D}(f))\circ T(g_i) = T(g_{i'})$, i.e. $(T(A),(T(g_i)))$ is a lower bound of $T\circ\mathcal{D}$. But then $(A,(g_i))$

is a lower bound of $\mathcal{D}$ since $\mathcal{T}$ is faithful. Hence there exists a unique morphism $k\colon A \to L$ with $l_i \circ k = g_i$. Then $\mathcal{T}(k) \circ h \circ e\colon X \to \mathcal{T}(L)$ is a morphism such that $\mathcal{T}(l_i) \circ (\mathcal{T}(k) \circ h \circ e) = \mathcal{T}(l_i \circ k) \circ h \circ e = \mathcal{T}(g_i) \circ h \circ e = f_i$ for each $i \in |\mathcal{I}|$. If $s\colon X \to \mathcal{T}(L)$ were a morphism with $\mathcal{T}(l_i) \circ s = f_i$ for each $i \in |\mathcal{I}|$, then the diagram

$$\begin{array}{ccc} X & \xrightarrow{\;h \circ e\;} & \mathcal{T}(A) \\ {\scriptstyle s}\big\downarrow & & \big\downarrow{\scriptstyle \mathcal{T}(l_i \circ k) \,=\, \mathcal{T}(g_i)} \\ \mathcal{T}(L) & \xrightarrow{\;\mathcal{T}(f_i)\;} & \mathcal{T}(\mathcal{D}(i)) \end{array}$$

would commute for each $i \in |\mathcal{I}|$ where $h \circ e \in \mathcal{E}$ and $(\mathcal{T}(L), (\mathcal{T}(l_i))) \in \mathcal{M}$. Thus there would exist a unique morphism $t\colon \mathcal{T}(A) \to \mathcal{T}(L)$ such that the diagram

$$\begin{array}{ccc} X & \xrightarrow{\;h \circ e\;} & \mathcal{T}(A) \\ {\scriptstyle s}\big\downarrow & {\scriptstyle t}\swarrow & \big\downarrow{\scriptstyle \mathcal{T}(g_i)} \\ \mathcal{T}(L) & \xrightarrow{\;\mathcal{T}(l_i)\;} & \mathcal{T}(\mathcal{D}(i)) \end{array}$$

would commute for each $i \in |\mathcal{I}|$. Since $(L, (l_i))$ is $\mathcal{T}$-initial, there would exist a unique morphism $u\colon A \to L$ with $\mathcal{T}(u) = t$ and $l_i \circ u = g_i = l_i \circ k$ for each $i \in |\mathcal{I}|$ (because of $\mathcal{T}(l_i) \circ t = \mathcal{T}(g_i)$ for each $i \in |\mathcal{I}|$). Hence $u = k$ (since $(L, (l_i))$ is a mono-source) and $s = t \circ h \circ e = \mathcal{T}(u) \circ h \circ e = \mathcal{T}(k) \circ h \circ e$. Consequently $(\mathcal{T}(L), (\mathcal{T}(l_i)))$ is a limit of $\mathcal{T} \circ \mathcal{D}$.

(2) $\Rightarrow$ (1): Since $(\mathcal{T}(L), (\mathcal{T}(l_i)))$ is a limit of $\mathcal{T} \circ \mathcal{D}$ and $\mathcal{T}$ is faithful, it follows that $(L, (l_i))$ is a lower bound of $\mathcal{D}$. If $(A, (f_i))$ is an arbitrary lower bound of $\mathcal{D}$, then $(\mathcal{T}(A), (\mathcal{T}(f_i)))$ is a lower bound of $\mathcal{T} \circ \mathcal{D}$. Hence there exists a unique morphism $f\colon \mathcal{T}(A) \to \mathcal{T}(L)$ with $\mathcal{T}(l_i) \circ f = \mathcal{T}(f_i)$ for each $i \in |\mathcal{I}|$. Since $(L, (l_i))$ is $\mathcal{T}$-initial, there exists a unique morphism $g\colon A \to L$ with $\mathcal{T}(g) = f$ and $l_i \circ g = f_i$ for each $i \in |\mathcal{I}|$. The uniqueness of g follows from the uniqueness of f and the faithful-

ness of $\mathcal{T}$. Consequently, $(L,(l_i))$ is a limit of $\mathcal{D}$.

<u>5.2.9 Corollary</u>. If $\mathcal{T}\colon \mathcal{A} \to \mathcal{C}$ is an $(\mathcal{E},\mathcal{M})$-topological functor and $\mathcal{C}$ is complete, then $\mathcal{A}$ is complete.

<u>Proof</u>. Let $\mathcal{D}\colon \mathcal{I} \to \mathcal{A}$ be a diagram in $\mathcal{A}$ over $\mathcal{I}$. Then $\mathcal{T}\circ\mathcal{D}\colon \mathcal{I} \to \mathcal{C}$ is a diagram in $\mathcal{C}$ over $\mathcal{I}$. By assumption (cf. 5.2.7 ④), $\mathcal{T}\circ\mathcal{D}$ has the limit $(L',(l_i'))$ which is an extremal source and therefore belongs to $\mathcal{M}$. Put $\mathcal{D}(i) = A_i$ for each $i \in |\mathcal{I}|$. Then $(A_i)_{i\in|\mathcal{I}|}$ is a family of $\mathcal{A}$-objects and there exists a $\mathcal{T}$-initial source $(A,(f_i\colon A \to A_i)_{i\in|\mathcal{I}|})$ in $\mathcal{A}$ which $\mathcal{T}$-lifts $(L',(l_i'))$. Thus $(A,(f_i))$ is a limit of $\mathcal{D}$ by applying 5.2.8.

<u>5.2.10 Proposition</u>. Let $\mathcal{T}\colon \mathcal{A} \to \mathcal{C}$ be an $(\mathcal{E},\mathcal{M})$-topological functor. Then for each family $(A_i)_{i\in I}$ of $\mathcal{A}$-objects and each sink[30] $((f_i\colon \mathcal{T}(A_i) \to X)_{i\in I}, X)$ in $\mathcal{C}$, there exists a sink $((t_i\colon A_i \to A)_{i\in I}, A)$ in $\mathcal{A}$ and a morphism $e\colon X \to \mathcal{T}(A)$ in $\mathcal{E}$ such that $\mathcal{T}(t_i) = e\circ f_i$ for each $i \in I$ and such that the following condition is satisfied:

(F) For each sink $((g_i\colon A_i \to B), B)$ in $\mathcal{A}$ and each morphism $g\colon X \to \mathcal{T}(B)$ with $\mathcal{T}(g_i) = g\circ f_i$ for each $i \in I$, there exists a morphism $k\colon A \to B$ with $\mathcal{T}(k)\circ e = g$.

<u>Proof</u>. Let $((g_j,((g_{i_j}\colon A_i \to B_j)_{i\in I}, B_j)))_{j\in J}$ be the family of all pairs such that $((g_{i_j}), B_j)$ is a sink and $g_j\colon X \to \mathcal{T}(B_j)$ is a morphism with $g_j\circ f_i = \mathcal{T}(g_{i_j})$ for each $i \in I$. If

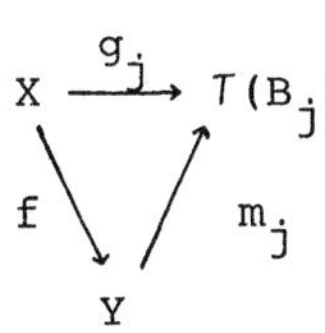

is an $(\mathcal{E},\mathcal{M})$-factorization of $(X,(g_j)_{j\in J})$, then there exists a

30) Dual notion: source (cf. 5.1.1 (1)).

T-initial source $(A,(n_j: A \to B_j)_{j\in J})$ and an isomorphism $h: Y \to T(A)$ with $T(n_j)\circ h = m_j$ for each $j \in J$. Since E is isomorphism-closed, $h\circ f = e$ belongs to E. Furthermore, since $(A,(n_j))$ is T-initial and $T(n_j)\circ(e\circ f_i) = T(g_{i_j})$ for each $i \in I$ and each $j \in J$, it follows that there exists for each $i \in I$ a unique morphism $t_i: A_i \to A$ with $T(t_i) = e\circ f_i$.

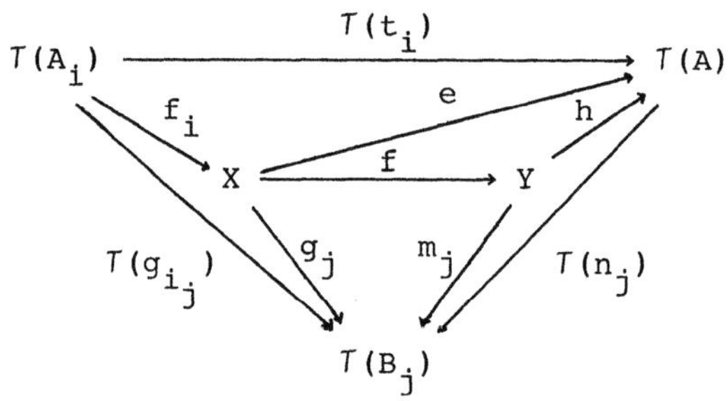

Now the condition (F) is satisfied (consider a suitable $j \in J$).

5.2.11 Theorem. Let $T: A \to C$ be an (E,M)-topological functor and let $D: I \to A$ be a diagram such that $T\circ D$ has a colimit. Then D has a colimit.

Proof. Let $((f_i: T(D(i)) \to X),X)$ be the sink belonging to the colimit $(X,(f_i))$ of $T\circ D$. By 5.2.10 there is a sink $((t_i: D(i) \to A),A)$ in A and a morphism $e: X \to T(A)$ in E such that $T(t_i) = e\circ f_i$ for each $i \in I$ and such that the condition (F) is satisfied.

Then $(A,(t_i))$ is a colimit of D:

a) If $h: i \to i'$ is an I-morphism, then since $(X,(f_i))$ is an upper bound of $T\circ D$, the diagram

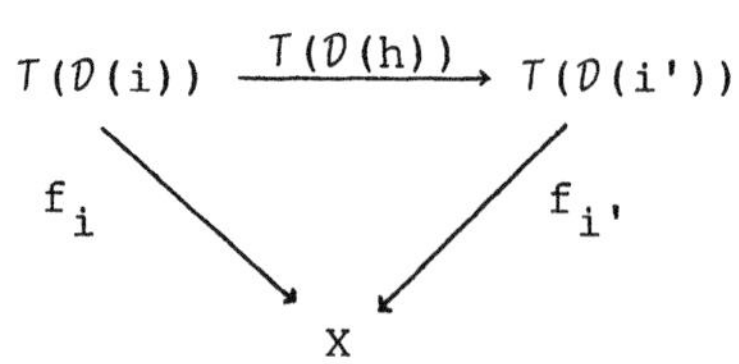

commutes. Hence $\mathcal{T}(t_{i'}) \circ \mathcal{T}(\mathcal{D}(h)) = e \circ f_{i'} \circ \mathcal{T}(\mathcal{D}(h)) = e \circ f_i = \mathcal{T}(t_i)$ from which follows that $t_{i'} \circ \mathcal{D}(h) = t_i$ because $\mathcal{T}$ is a faithful functor. Thus $(A,(t_i))$ is an upper bound of $\mathcal{D}$.

b) Let $(B,(g_i: \mathcal{D}(i) \to B))$ be an upper bound of $\mathcal{D}$. Then $(\mathcal{T}(B),(\mathcal{T}(g_i): \mathcal{T}(\mathcal{D}(i)) \to \mathcal{T}(B)))$ is an upper bound of $\mathcal{T} \circ \mathcal{D}$. Hence there exists a unique morphism $g: X \to \mathcal{T}(B)$ with $g \circ f_i = \mathcal{T}(g_i)$ since $(X,(f_i))$ is a colimit of $\mathcal{T} \circ \mathcal{D}$. Because of (F) there is a morphism $k: A \to B$ with $\mathcal{T}(k) \circ e = g$. In order to show that the diagram

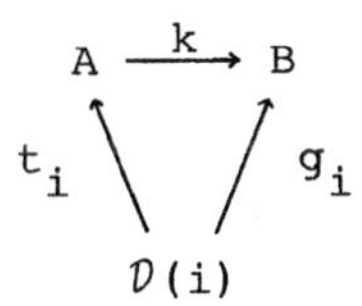

commutes for each $i \in |\mathcal{I}|$ it suffices to prove that $\mathcal{T}(k \circ t_i) = \mathcal{T}(g_i)$ for each $i \in |\mathcal{I}|$ (note: $\mathcal{T}$ is faithful). Since $\mathcal{T}(k) \circ \mathcal{T}(t_i) = \mathcal{T}(k) \circ e \circ f_i = g \circ f_i = \mathcal{T}(g_i)$ for each $i \in |\mathcal{I}|$ and $\mathcal{T}$ is a functor, we have the desired equality. If $k': A \to B$ is a morphism with $k' \circ t_i = g_i$ for each $i \in |\mathcal{I}|$, then $\mathcal{T}(k') \circ \mathcal{T}(t_i) = \mathcal{T}(k') \circ e \circ f_i = \mathcal{T}(g_i) = g \circ f_i$ for each $i \in |\mathcal{I}|$ so that $\mathcal{T}(k') \circ e = g$ since $(X,(f_i))$ is a colimit. Hence $\mathcal{T}(k') \circ e = \mathcal{T}(k) \circ e$ so that $\mathcal{T}(k) = \mathcal{T}(k')$ because e is an epimorphism. Since $\mathcal{T}$ is faithful, it follows that $k = k'$.

5.2.12 Corollary. Let $\mathcal{T}: \mathcal{A} \to \mathcal{C}$ be an $(\mathcal{E},\mathcal{M})$-topological functor. If $\mathcal{C}$ is cocomplete, then $\mathcal{A}$ is cocomplete.

Proof. This is an immediate consequence of 5.2.11.

5.3 Initially structured categories

5.3.1 Definition. A pair $(\mathcal{A},\mathcal{T})$ is called an initially structured category provided that $\mathcal{A}$ is a category and $\mathcal{T}: \mathcal{A} \to \underline{\text{Set}}$

is a functor which is amnestic[31)] and transportable[32)] such that the following hold:

IS_1) $\mathcal{T}$ is (epi, mono-source)-topological.

IS_2) $\mathcal{T}$ has small fibres, i.e. for each $X \in |\underline{Set}|$, $\{A \in |\mathcal{A}| : \mathcal{T}(A) = X\}$ is a set.

IS_3) There is precisely one object P in $\mathcal{A}$ (up to isomorphism) such that $\mathcal{T}(P)$ is a terminal[33)] separator in Set, i.e. $\mathcal{T}(P)$ is a singleton.

Occasionally one writes $\mathcal{A}$ instead of $(\mathcal{A},\mathcal{T})$.

5.3.2 Examples. ① Let $\mathcal{A}$ be a topological category and $\mathcal{T}: \mathcal{A} \to \underline{Set}$ the forgetful functor. Then $(\mathcal{A},\mathcal{T})$ is an initially structured category (cf. 5.2.2 ①).

② Let $\mathcal{A}$ be an extremal epireflective (full and isomorphism-closed) subcategory of a topological category $\mathcal{C}$ and $\mathcal{T}: \mathcal{A} \to \underline{Set}$ the forgetful functor. Then $(\mathcal{A},\mathcal{T})$ is an initially structured category (cf. 5.2.3). Especially, Haus is an initially structured category which is not topological.

5.3.3 Remark. Obviously the condition IS_1) of 5.3.1 can be replaced by the following one:

IS_1') Every source $(X,(f_i: X \to \mathcal{T}(A_i))_{i \in I})$ in Set has an (epi, mono-source)-factorization

$$\begin{array}{ccc} X & \xrightarrow{f_i} & \mathcal{T}(A_i) \\ & {}_{e}\searrow \quad \nearrow_{\mathcal{T}(g_i)} & \\ & \mathcal{T}(B) & \end{array}$$

---

31) A functor $\mathcal{F}: \mathcal{A} \to \mathcal{B}$ is called amnestic provided that any $\mathcal{A}$-isomorphism f is an $\mathcal{A}$-identity iff $\mathcal{F}(f)$ is a $\mathcal{B}$-identity.

32) A functor $\mathcal{F}: \mathcal{A} \to \mathcal{B}$ is called transportable provided that for each $\mathcal{A}$-object A each $\mathcal{B}$-object B and each isomorphism $q: B \to \mathcal{F}(A)$, there exists a unique $\mathcal{A}$-object C and an isomorphism $\bar{q}: C \to A$ with $\mathcal{F}(\bar{q}) = q$.

33) An object X in a category $\mathcal{C}$ is called terminal provided that for every object Y in $\mathcal{C}$, the set of morphisms $[Y,X]_{\mathcal{C}}$ is a singleton.

such that $(B,(g_i: B \to A_i)_{i\in I})$ is a $\mathcal{T}$-initial source in $\mathcal{A}$. (Namely, if IS_1) is satisfied and

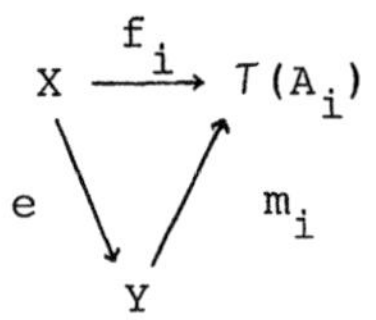

is an (epi, mono-source)-factorization of $(X,(f_i)_{i\in I})$, then there exists a $\mathcal{T}$-initial source $(B,(g_i: B \to A_i)_{i\in I})$ in $\mathcal{A}$ which $\mathcal{T}$-lifts $(Y,(m_i)_{i\in I})$ because $\mathcal{T}$ is (epi,mono-source)-topological. Hence without loss of generality we can put $Y = \mathcal{T}(B)$ and $m_i = \mathcal{T}(g_i)$ for each $i \in I$.
Conversely, if IS_1') is satisfied, then for each family $(A_i)_{i\in I}$ of $\mathcal{A}$-objects and each mono-source $(X,(m_i: X \to \mathcal{T}(A_i))_{i\in I})$ in Set there is an (epi,mono-source)-factorization

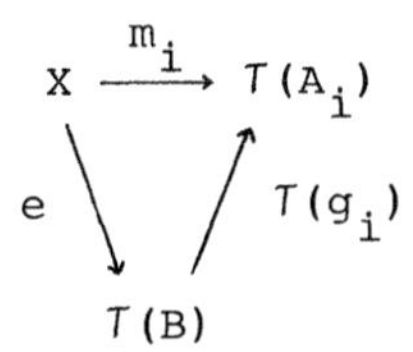

such that $(B,(g_i: B \to A_i)_{i\in I})$ is $\mathcal{T}$-initial. It suffices to show that e is a monomorphism (for then e is an isomorphism and $\mathcal{T}$ is (epi, mono-source)-topological). From $e\circ\alpha = e\circ\beta$, it follows that $m_i\circ\alpha = m_i\circ\beta$ for each $i \in I$, i.e. $\alpha = \beta$ since $(m_i)_{i\in I}$ is a mono-source.)

<u>5.3.4 Theorem</u>. Let $(\mathcal{A},\mathcal{T})$ be an initially structured category. Then the following are satisfied:
(1) $\mathcal{T}$ is faithful.
(2) A source $(A,(f_i: A \to D_i)_{i\in I})$ in $\mathcal{A}$ is a limit of a diagram $\mathcal{D}: \mathcal{I} \to \mathcal{A}$ with $|\mathcal{I}| = I$ if and only if this source is $\mathcal{T}$-initial and $(\mathcal{T}(A),(\mathcal{T}(f_i): \mathcal{T}(A) \to \mathcal{T}(D_i))_{i\in I})$ is a limit of $\mathcal{T}\circ\mathcal{D}$.
(3) For any sink $((f_i: \mathcal{T}(D_i) \to X)_{i\in I},X)$ in Set there exists a sink $((a_i: D_i \to A)_{i\in I},A)$ in $\mathcal{A}$ and an epimorphism $e: X \to \mathcal{T}(A)$

with $e \circ f_i = \mathcal{T}(a_i)$ for each $i \in I$ such that the following condition (F) is satisfied: For each sink $((b_i: D_i \to B)_{i\in I}, B)$ in $\mathcal{A}$ and each morphism $d: X \to \mathcal{T}(B)$ with $d \circ f_i = \mathcal{T}(b_i)$ for each $i \in I$ there exists a (unique) morphism $c: A \to B$ such that $\mathcal{T}(c) \circ e = d$.

(4) $\mathcal{A}$ is complete and cocomplete.

The proof follows immediately from the results on topological functors in 5.2 (cf. 5.2.4, 5.2.8, 5.2.10, 5.2.9 and 5.2.12).

5.3.5 Proposition. Let $(\mathcal{A},\mathcal{T})$ be an initially structured category. Any sink $((f_i: A_i \to C)_{i\in I}, C)$ in $\mathcal{A}$ has

(1) a factorization

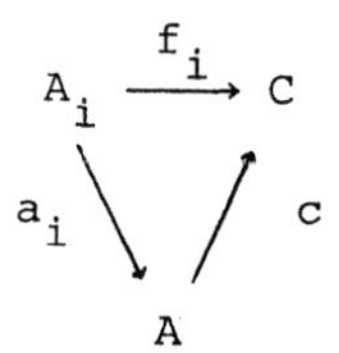

such that $\mathcal{T}(c)$ is an isomorphism and $((a_i: A_i \to A)_{i\in I}, A)$ is $\mathcal{T}$-*final* (i.e. for any sink $((b_i: A_i \to B)_{i\in I}, B)$ and any morphism $f: \mathcal{T}(A) \to \mathcal{T}(B)$ with $f \circ \mathcal{T}(a_i) = \mathcal{T}(b_i)$ for each $i \in I$, there exists a unique morphism $\bar{f}: A \to B$ with $\mathcal{T}(\bar{f}) = f$ and $\bar{f} \circ a_i = b_i$ for each $i \in I$) , and

(2) a factorization $f_i = c \circ a_i$ where c is a monomorphism and $((a_i: A_i \to A)_{i\in I}, A)$ is a $\mathcal{T}$-final epi-sink.

Proof. (1) By 5.3.4 (3) for any $((\mathcal{T}(f_i): \mathcal{T}(A_i) \to \mathcal{T}(C))_{i\in I}, \mathcal{T}(C))$, there exists a sink $((a_i: A_i \to A)_{i\in I}, A)$ and an epimorphism $e: \mathcal{T}(C) \to \mathcal{T}(A)$ with $e \circ \mathcal{T}(f_i) = \mathcal{T}(a_i)$ for each $i \in I$ such that the condition (F) is satisfied. Put $d = 1_{\mathcal{T}(C)}$ in (F). Then there exists a unique $c: A \to C$ with $\mathcal{T}(c) \circ e = 1_{\mathcal{T}(C)}$. Thus e is an isomorphism. Hence $\mathcal{T}(c)$ is an isomorphism. In order to prove that $(a_i)_{i\in I}$ is $\mathcal{T}$-final we choose $d = f \circ e$ in (F) provided that $f: \mathcal{T}(A) \to \mathcal{T}(B)$ is a morphism with $f \circ \mathcal{T}(a_i) = \mathcal{T}(b_i)$ for each $i \in I$ and $((b_i: A_i \to B)_{i\in I}, B)$ is a sink in $\mathcal{A}$. Then there exists a unique $\bar{f}: A \to B$ with $\mathcal{T}(\bar{f}) \circ e = d = f \circ e$ so that

$T(\bar{f}) = f$. Thus since T is faithful, from $T(\bar{f}\circ a_i) = T(\bar{f})\circ T(a_i) =$ $= f\circ T(a_i) = T(b_i)$ for each $i \in I$, it follows that $\bar{f}\circ a_i = b_i$ for each $i \in I$ and from $T(c\circ a_i) = T(c)\circ T(a_i) = T(c)\circ e\circ T(f_i) =$ $= 1\circ T(f_i) = T(f_i)$ for each $i \in I$, it follows that $c\circ a_i = f_i$ for each $i \in I$.

(2) Put $M = \bigcup_{i\in I} T(f_i)[T(A_i)] \subset T(C)$. If for each $i \in I$, $g_i\colon T(A_i) \to M$ is defined by $g_i(z) = T(f_i)(z)$ for each $z \in T(A_i)$, then $((g_i\colon T(A_i) \to M)_{i\in I}, M)$ is a sink in Set and by 5.3.4(3), there exists a sink $((a_i\colon A_i \to A)_{i\in I}, A)$ in A and an epimorphism $e\colon M \to T(A)$ with $e\circ g_i = T(a_i)$ for each $i \in I$ such that the condition (F) is satisfied. If $d\colon M \to T(C)$ denotes the inclusion map, then there exists a unique morphism $c\colon A \to C$ with $T(c)\circ e = d$. Thus e is an isomorphism so that $T(c)$ is a monomorphism. Since T is faithful, c is a monomorphism and we have $c\circ a_i = f_i$ for each $i \in I$ as under (1). The fact that $(a_i)_{i\in I}$ is T-final is shown analogously to (1). Moreover $(a_i)_{i\in I}$ is obviously an epi-sink.

5.3.6 Proposition. Let (A,T) be an initially structured category. Then the following are satisfied:

a) The object P in A given by IS_3) is terminal and a separator in A .
b) If X and Y are A-objects and $g\colon T(X) \to T(Y)$ is a constant morphism, then there exists a unique A-morphism $f\colon X \to Y$ with $T(f) = g$.

Proof. a) Since $T(P)$ is terminal, for each $Z \in |A|$ there exists a unique map $g\colon T(Z) \to T(P)$. By 5.3.4 (3) there is a morphism $f\colon Z \to A$ and an epimorphism $e\colon T(P) \to T(A)$ which is obviously an isomorphism. Hence without loss of generality $P = A$. Moreover f is the unique morphism from Z to P . Thus P is terminal. If $h,k\colon B \to C$ are two distinct A-morphisms, then $T(h)$ and $T(k)$ are distinct. Since $T(P)$ is a separator, there is a morphism $l\colon T(P) \to T(B)$ with $T(h)\circ l \neq T(k)\circ l$. And since every mono-source can be lifted, there exists a morphism $d\colon D \to B$ and an isomorphism $j\colon T(P) \to T(D)$ with $T(d)\circ j = l$. Since P is unique up to isomorphism by $IS_3)$, there exists an isomorphism $i: P \to D$. Then $h \circ c \neq k \circ c$ where $c = d \circ i$; for if $h \circ c = k \circ c$, then $T(h) \circ T(d) = T(k) \circ T(d)$ and thus $T(h) \circ 1 = T(k) \circ 1$ - a contradiction. Hence P is a separator.

b) Let

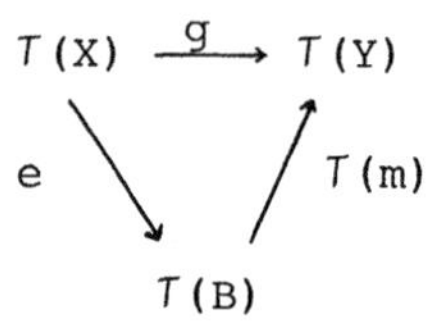

be the (epi,mono-source)-factorization of g existing by $IS_1')$ such that $(B, m: B \to Y)$ is T-initial. Since g is a constant morphism, $e[T(X)] = T(B)$ is a singleton. By $IS_3)$ there exists an isomorphism $j: P \to B$. Since P is terminal by a), there is a unique morphism $1: X \to P$. Put $f = m \circ j \circ 1$. Then $T(f) = T(m) \circ (T(j) \circ T(1)) = T(m) \circ e = g$ ($T(B)$ is terminal!). The uniqueness of f follows from the faithfulness of T .

<u>5.3.7 Definition</u>. Let (A,T) be an initially structured category. An A-morphism $f: A \to B$ is called a

a) T-*embedding* provided that $(A, f: A \to B)$ is a T-initial mono-source.

b) T-*quotient map* provided that $(f: A \to B, B)$ is a T-final epi-sink.

<u>5.3.8 Proposition</u>. Let (A,T) be an initially structured category. Then the following are satisfied:

(1) f is an A-monomorphism if and only if $T(f)$ is a monomorphism in <u>Set</u> .

(2) f is an A-epimorphism whenever $T(f)$ is an epimorphism in <u>Set</u> .

(3) Every extremal monomorphism in A is a T-embedding.

<u>Proof</u>. (1) a) "⇒": Let $T(f)(x) = T(f)(y)$ where $f: X \to Y$ and $x,y \in T(X)$. By 5.3.6 b) there are A-morphisms $\bar{x}, \bar{y}: X \to X$ such that $T(\bar{x})(z) = x$ and $T(\bar{y})(z) = y$ for each $z \in T(X)$.

Thus $\mathcal{T}(f)\circ\mathcal{T}(\bar{x}) = \mathcal{T}(f)\circ\mathcal{T}(\bar{y})$. From this we may conclude that $f\circ\bar{x} = f\circ\bar{y}$, since $\mathcal{T}$ is a faithful functor. Consequently, by assumption, $\bar{x} = \bar{y}$, i.e. $x = y$.

b) "$\Leftarrow$": Let $f\circ\alpha = f\circ\beta$. Then $\mathcal{T}(f)\circ\mathcal{T}(\alpha) = \mathcal{T}(f)\circ\mathcal{T}(\beta)$ so that $\mathcal{T}(\alpha) = \mathcal{T}(\beta)$ and thus $\alpha = \beta$ ($\mathcal{T}$ is faithful).

(2) Analogously to (1) b).

(3) Let $f: X \to Y$ be an extremal monomorphism in $\mathcal{A}$. Further let

$$\begin{array}{ccc} \mathcal{T}(X) & \xrightarrow{\mathcal{T}(f)} & \mathcal{T}(Y) \\ & {}_{e}\searrow \quad \nearrow_{\mathcal{T}(m)} & \\ & \mathcal{T}(Z) & \end{array}$$

be an (epi,mono-source)-factorization existing by IS_1') such that $(Z,m: Z \to Y)$ is a $\mathcal{T}$-initial mono-source. Then by the definition of "$\mathcal{T}$-initial" there is a unique morphism $g: X \to Z$ with $\mathcal{T}(g) = e$ and $m\circ g = f$. By (2) g is an epimorphism and since f is an extremal monomorphism, g is even an isomorphism. Then $(X,f: X \to Y)$ is $\mathcal{T}$-initial and since f is a monomorphism, also a mono-source.

<u>5.3.9 Remark</u>. The converses of 5.3.8 (2) and (3) are not valid which can be seen by considering the category <u>Haus</u> (e.g. the embedding $i: \mathbb{Q} \to \mathbb{R}$ is an epimorphism which is not surjective and a monomorphism which is not extremal).

<u>5.3.10 Proposition</u>. Let $(\mathcal{A},\mathcal{T})$ be an initially structured category. Then the following are satisfied:

a) If $((f_i: A_i \to A)_{i\in I},A)$ is a $\mathcal{T}$-final epi-sink in $\mathcal{A}$, then $((\mathcal{T}(f_i): \mathcal{T}(A_i) \to \mathcal{T}(A))_{i\in I},\mathcal{T}(A))$ is an epi-sink in <u>Set</u> .

b) A sink in $\mathcal{A}$ is an extremal [34] epi-sink if and only if it is a $\mathcal{T}$-final epi-sink.

---

[34] Dual notion: extremal source (cf. 5.1.1(3)).

Proof. a) Obviously we obtain an (epi-sink, extremal-mono)-factorization

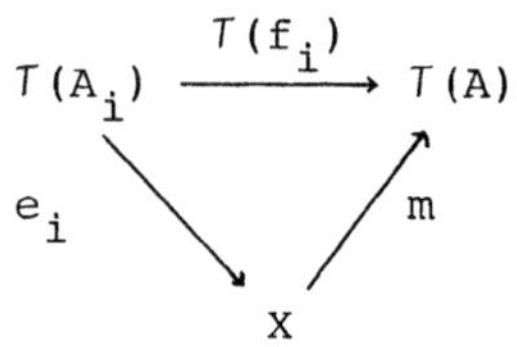

of $((T(f_i): T(A_i) \to T(A))_{i\in I}, T(A))$, provided $X = \bigcup_{i\in I} T(f_i)[T(A_i)]$, $e_i(x_i) = T(f_i)(x_i)$ for each $x_i \in T(A_i)$ and m is the inclusion map. Now we choose some $r: T(A) \to X$ with $r\circ m = 1_X$. Then $(m\circ r)\circ T(f_i) = m \circ r \circ m \circ e_i = m \circ 1_X \circ e_i = m \circ e_i = T(f_i)$ for each $i \in I$. Since $((f_i)_{i\in I}, A)$ is T-final, there exists a unique morphism $d: A \to A$ with $T(d) = m \circ r$ and $d \circ f_i = f_i$ for each $i \in I$. On the other hand $f_i = 1_A \circ f_i$ for each $i \in I$ so that $d = 1_A$ since $((f_i), A)$ is an epi-sink. But then $m \circ r = T(d) = 1_{T(A)}$ so that m is an isomorphism.

b) α) "⇒": Given an extremal epi-sink $((f_i: A_i \to C)_{i\in I}, C)$ in A. Let $f_i = j \circ a_i$ for each $i \in I$ be the factorization existing by 5.3.5(2). Then especially j is a monomorphism and since $((f_i), C)$ is extremal, j is an isomorphism. Thus $((f_i), C)$ is T-final since $((a_i), A)$ is T-final.

β) "⇐": If $((f_i: A_i \to C)_{i\in I}, C)$ is a T-final epi-sink in A and $f_i = m \circ g_i$ for each $i \in I$ is a factorization where m is a monomorphism, then $((T(f_i): T(A_i) \to T(C))_{i\in I}, T(C))$ is an epi-sink in Set by a). Hence $T(m)$ is an epimorphism and thus an isomorphism by 5.3.8(1). Then by the definition of "T-final" there is a morphism k such that $(T(m))^{-1} = T(k)$. Obviously k is the desired inverse of m, i.e. m is an isomorphism.

5.3.11 Proposition. For every A-morphism $f: A \to B$ in an initially structured category (A, T) the following are equivalent:

(1) f is a T-quotient map.

(2) f is a regular epimorphism (i.e. f is a coequalizer of two A-morphisms).

(3) f is an extremal epimorphism.

Proof. (1) $\Rightarrow$ (2): If f is a T-quotient map, then by 5.3.10 a) $T(f): T(A) \to T(B)$ is an epimorphism in Set and thus a coequalizer of $\alpha,\beta: X \to T(A)$ (let $X = \pi_{T(f)} \subset T(A)\times T(A)$ [defined by $(x,y) \in \pi_{T(f)}$ iff $T(f)(x) = T(f)(y)$] and put $\alpha = p_1 \circ i$ and $\beta = p_2 \circ i$ where $i: X \to T(A) \times T(A)$ denotes the inclusion map and p_1 (resp. p_2) is the projection map from $T(A) \times T(A)$ onto the first (resp. second) factor). By IS_1') there are factorizations $\alpha = T(g) \circ e$ and $\beta = T(g') \circ e$ where e is an epimorphism. Thus since $T(f)$ is a coequalizer of α and β, we obtain that $T(f)$ is also a coequalizer of $T(g)$ and $T(g')$. But then f is a coequalizer of g and g' (Since T is faithful, the equation $f \circ g = f \circ g'$ results from the corresponding assertion for $T(f)$. If $h: A \to D$ is a morphism with $h \circ g = h \circ g'$, then there exists a unique morphism $l: T(B) \to T(D)$ with $l \circ T(f) = T(h)$ because $T(f) = CE(T(g),T(g'))$. Since $(f: A \to B, B)$ is T-final, there exists a unique morphism $\bar{h}: B \to D$ with $T(\bar{h}) = h$ and $\bar{h} \circ f = h$. Since f is an epimorphism, $\bar{h}$ is unique with respect to the given data.).

(2) $\Rightarrow$ (3): Let $f = CE(g,g')$.

a) Let $\alpha,\beta: B \to C$ be A-morphisms with $\alpha \circ f = \beta \circ f = h$. Then $h \circ g = \alpha \circ f \circ g = \alpha \circ f \circ g' = h \circ g'$. By assumption there exists a unique morphism $h': B \to \cdot$ with $h' \circ f = h$. Thus $h' = \alpha = \beta$.

b) Let $f = h \circ k$ where h is a monomorphism. Then $h \circ k \circ g = h \circ k \circ g'$ so that $k \circ g = k \circ g'$ because h is a monomorphism. By assumption there exists a unique A-morphism k' such that $k' \circ f = k$. Furthermore, $h \circ k' \circ f = f = 1 \circ f$ so that $h \circ k' = 1$ because f is an epimorphism by a). Thus h is an isomorphism (since 1 is an extremal epimorphism).

"(3) $\Rightarrow$ (1)" has been shown by 5.3.10 b).

5.3.12 Remark. It follows from the proof of 5.3.11 that for any category C, a regular epimorphism is an extremal epimorphism. (Dually: Every regular monomorphism is an extremal monomorphism.)

5.3.13 Proposition. Every initially structured category (A,T) is well-powered.

Proof. Since T is transportable and $IS_2)$ is satisfied, the proof is analogous to the proof of 1.2.2.9.

5.3.14 Remark. Initially structured categories are generally not co-well-powered. It is a non-trivial result of J. Schröder [76] that the category of T_{2a}-spaces (= Urysohn spaces) is not co-well-powered. This is highly remarkable since the category of Hausdorff spaces is co-well-powered (cf. 2.2.5 ③). But it can be shown analogously to the proof for topological categories that every initially structured category is E-co-well-powered where E consists of all A-morphisms f for which $T(f)$ is an epimorphism or of all extremal epimorphisms respectively.

5.3.15 Proposition. Let (A,T) be an initially structured category. Then A is a category with a factorization structure (E,M) where E consists of all extremal epimorphisms and M of all mono-sources.

Proof. By 5.2.5 (A,T) is a category with a factorization structure (E',M') where E' (resp. M') consists of all A-morphisms f for which $T(f)$ is an epimorphism (resp. all T-initial mono-sources) [note that Set is an (epi,mono-source)-category since it is a topological category (cf. 5.1.3 ②)]. If $(f_i: X \to X_i)_{i\in I}$ is a source in A and $f_i = h_i \circ g$ for each $i \in I$ is the corresponding (E',M')-factorization, then $f_i = (h_i \circ c) \circ e$ is the desired (extremal epi, mono-source)-factorization provided $g = c \circ e$ is the (T-final epi-sink,mono)-factorization of g (note also 5.3.11) existing by 5.3.5 (2). If the diagram in A

$$(D) \qquad \begin{array}{ccc} X & \xrightarrow{\ h\ } & Y \\ {\scriptstyle k}\downarrow & & \downarrow{\scriptstyle f_i} \\ Z & \xrightarrow{\ m_i\ } & X_i \end{array}$$

commutes for each $i \in I$ where h is an extremal epimorphism in $\mathcal{A}$ and $(m_i)_{i\in I}$ is a mono-source in $\mathcal{A}$, then applying $\mathcal{T}$ we obtain a commutative diagram in Set where $\mathcal{T}(h)$ is an epimorphism (cf. 5.3.10 a)), and $(\mathcal{T}(m_i))_{i\in I}$ is a mono-source in Set. Since by 5.1.3 (2) Set (as a topological category) is an (epi, mono-source)-category there exists a unique map $1: \mathcal{T}(Y) \to \mathcal{T}(Z)$ completing the obtained diagram in Set commutatively. Then there is a unique $\mathcal{A}$-morphism $\bar{1}: Y \to Z$ with $\mathcal{T}(\bar{1}) = 1$ and $\bar{1}\circ h = k$ because $(h: X \to Y, Y)$ is $\mathcal{T}$-final and $1\circ\mathcal{T}(h) = \mathcal{T}(k)$. Since $\mathcal{T}$ is faithful, it follows from $\mathcal{T}(m_i\circ\bar{1}) = \mathcal{T}(m_i)\circ\mathcal{T}(\bar{1}) = \mathcal{T}(m_i)\circ 1 = \mathcal{T}(f_i)$ for each $i \in I$ that $m_i\circ\bar{1} = f_i$ for each $i \in I$. Moreover $\bar{1}$ is uniquely determined because h is an epimorphism.

5.3.16 Theorem. Every $\mathcal{E}$-reflective (full and isomorphism-closed) subcategory of an initially structured category $(\mathcal{A},\mathcal{T})$ is initially structured provided that $\mathcal{E}$ consists of all $\mathcal{A}$-morphism f for which $\mathcal{T}(f)$ is an epimorphism in Set.

Proof. By 5.2.5 $\mathcal{A}$ is an ($\mathcal{E}$,$\mathcal{T}$-initial mono-source)-category. If $\mathcal{B}$ is an $\mathcal{E}$-reflective (full and isomorphism-closed) subcategory of $\mathcal{A}$, then by 5.1.5, $|\mathcal{B}|$ consists of all $A \in |\mathcal{A}|$ for which there exists a $\mathcal{T}$-initial mono-source $(f_i: A \to B_i)_{i\in I}$ with $B_i \in |\mathcal{B}|$ for each $i \in I$. Therefore since IS_1') is valid for $\mathcal{A}$, we obtain that IS_1') is valid for $\mathcal{B}$. IS_2) is also satisfied for $\mathcal{B}$ because it holds for $\mathcal{A}$. The $\mathcal{A}$-object P existing by IS_3) belongs obviously to $\mathcal{B}$ (for since P is terminal by 5.3.6 a), the class of all morphisms $m: P \to B$ with $B \in |\mathcal{B}|$ is a $\mathcal{T}$-initial mono-source). Consequently, IS_3) holds for $\mathcal{B}$.

5.3.17 Remark. It follows from 5.3.16 that *every extremal epireflective (resp. epireflective) full and isomorphism-closed subcategory of an initially structured (resp. topological) category is initially structured* [cf. 5.3.10 (resp. 1.2.2.4)]. An epireflective (full and isomorphism-closed) subcategory of an initially structured category which is not topological is generally not initially structured, e.g. the category of compact Hausdorff spaces (and continuous maps) is epireflective in Haus but not initially structured (the inclusion map i from the open interval $(0,1)$ into the closed interval $[0,1]$ endowed with the usual topology describes a mono-source in Set which has no initial lift because there is no compact Hausdorff topology on $(0,1)$ for which i is continuous [otherwise $(0,1)$ endowed with the usual topology would be compact]) .

5.3.18 Proposition. For each $\mathcal{T}$-final epi-sink $(f_i\colon A_i \to A)_{i\in I}$ in an initially structured category $(\mathcal{A},\mathcal{T})$, there is a set $J \subset I$ such that $(f_j\colon A_j \to A)_{j\in J}$ is likewise a $\mathcal{T}$-final epi-sink.

Proof. I) For each $X \in |\underline{\text{Set}}|$, $\mathcal{T}_X = \{A \in |\mathcal{A}|\colon \mathcal{T}(A) = X\}$ is a set by IS_2). If we define for $A,B \in \mathcal{T}_X$

$$A \leq B \text{ iff there is an } \mathcal{A}\text{-morphism } f\colon A \to B \text{ with } \mathcal{T}(f) = 1_X ,$$

then $(\mathcal{T}_X , \leq)$ is a pre-ordered set which is even ordered since $\mathcal{T}$ is amnestic.

II) Let $(f_i\colon A_i \to A)_{i\in I}$ be a $\mathcal{T}$-final epi-sink in $\mathcal{A}$. Put $\mathcal{T}(A) = X$. For each fixed $i \in I$, let

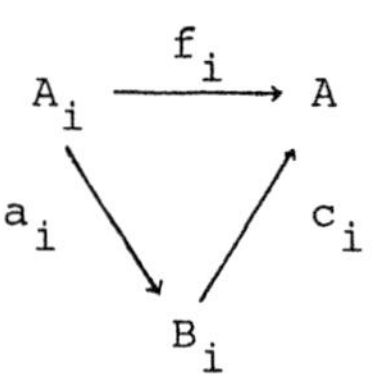

be the factorization existing by 5.3.5(1), i.e. $(a_i\colon A_i \to B_i, B_i)$ is $\mathcal{T}$-final and $\mathcal{T}(c_i)$ is an isomorphism. Since $\mathcal{T}$ is transportable, there exists an $\mathcal{A}$-object C_i and an isomorphism

$q_i\colon C_i \to B_i$ such that $\mathcal{T}(q_i) = (\mathcal{T}(c_i))^{-1}$ and $\mathcal{T}(C_i) = X$. Then we have $A = \sup\{C_i\colon i \in I\}$:

a) $C_i \leq A$ for each $i \in I$ because $c_i \circ q_i\colon C_i \to A$ is an $\mathcal{A}$-morphism such that $\mathcal{T}(c_i \circ q_i) = \mathcal{T}(c_i) \circ \mathcal{T}(q_i) = \mathcal{T}(c_i) \circ (\mathcal{T}(c_i))^{-1} = 1_X$.

b) If $C_i \leq A'$ for each $i \in I$ with $\mathcal{T}(A') = X$, then for each $i \in I$, there exists an $\mathcal{A}$-morphism $h_i\colon C_i \to A'$ with $\mathcal{T}(h_i) = 1_X$ so that $h_i \circ q_i^{-1} \circ a_i\colon A_i \to A'$ is an $\mathcal{A}$-morphism with $\mathcal{T}(h_i) \circ (\mathcal{T}(q_i))^{-1} \circ (\mathcal{T}(c_i))^{-1} \circ \mathcal{T}(f_i) = \mathcal{T}(h_i) \circ (\mathcal{T}(q_i))^{-1} \circ \mathcal{T}(a_i) = \mathcal{T}(h_i \circ q_i^{-1} \circ a_i)$ since $\mathcal{T}(q_i^{-1}) = (\mathcal{T}(q_i))^{-1}$. Hence there exists a unique $\mathcal{A}$-morphism $f\colon A \to A'$ with $\mathcal{T}(f) = \mathcal{T}(h_i) \circ (\mathcal{T}(q_i))^{-1} \circ (\mathcal{T}(c_i))^{-1} = \mathcal{T}(h_i) \circ 1_X = \mathcal{T}(h_i) = 1_X$ and $f \circ f_i = h_i \circ q_i^{-1} \circ a_i$. Thus, $A \leq A'$.

III) If $(g_i\colon X_i \to X)_{i \in I}$ is an epi-sink in <u>Set</u> , then there exists a set $J' \subset I$ such that $(g_{j'}\colon X_{j'} \to X)_{j' \in J'}$ is an epi-sink: We have shown earlier that $X = \bigcup_{i \in I} g_i[X_i]$. For each $x \in X$, we choose $(j', x_{j'})$ such that $x = g_{j'}(x_{j'})$. The set J' of all $j' \in I$ determined in this way satisfies the desired property.

IV) If $(f_i\colon A_i \to A)_{i \in I}$ is a $\mathcal{T}$-final epi-sink in $\mathcal{A}$, then by 5.3.10 a) $(\mathcal{T}(f_i)\colon \mathcal{T}(A_i) \to \mathcal{T}(A))_{i \in I}$ is an epi-sink in <u>Set</u> . Hence by III) there exists a set $J' \subset I$ such that $(\mathcal{T}(f_{j'})\colon \mathcal{T}(A_{j'}) \to \mathcal{T}(A))_{j' \in J'}$ is an epi-sink in <u>Set</u> . Put $\mathcal{T}(A) = X$. Then $\mathcal{T}_X$ is a set by IS_2) and thus according to II), there is a set $J'' \subset I$ such that $\{C_i\colon i \in I\} = \{C_{j''}\colon j'' \in J''\}$. Then $J' \cup J'' = J \subset I$ is a set and $(f_j\colon A_j \to A)_{j \in J}$ is an epi-sink in $\mathcal{A}$ which is $\mathcal{T}$-final: Namely if

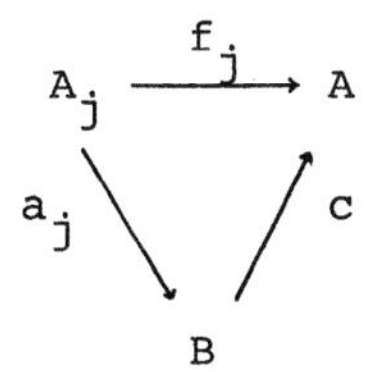

is the factorization existing by 5.3.5 (1) , i.e. $(a_j\colon A_j \to B)_{j \in J}$ is $\mathcal{T}$-final and $\mathcal{T}(c)$ is an isomorphism, then there is an $\mathcal{A}$-object C and an isomorphism $q\colon C \to B$ with $\mathcal{T}(q) = (\mathcal{T}(c))^{-1}$ and $\mathcal{T}(C) = X$ because $\mathcal{T}$ is transportable. Hence $(q^{-1} \circ a_j\colon A_j \to C)_{j \in J}$

is $\mathcal{T}$-final and $\mathcal{T}(q^{-1} \circ a_j) = \mathcal{T}(f_j)$ for each $j \in J$. Thus applying II), $C = \sup \{C_j : j \in J\}$ (since $\mathcal{T}$ is amnestic, "final structures" are unique!). On the other hand $\{C_j : j \in J\} = \{C_{j''} : j'' \in J''\} = \{C_i : i \in I\}$ so that $A = C$. Hence q is the inverse of c , i.e. c is an isomorphism. Consequently, $(f_j)_{j \in J}$ is $\mathcal{T}$-final.

<u>5.3.19 Proposition</u>. Let $(\mathcal{A},\mathcal{T})$ and $(\mathcal{A}',\mathcal{T}')$ be initially structured categories and let $F\colon \mathcal{A} \to \mathcal{A}'$ be a functor preserving colimits. If $(f_i\colon A_i \to A)_{i \in I}$ is a $\mathcal{T}$-final epi-sink in $\mathcal{A}$, then $(F(f_i)\colon F(A_i) \to F(A))_{i \in I}$ is a $\mathcal{T}'$-final epi-sink in $\mathcal{A}'$.

<u>Proof</u>. If $(f_i)_{i \in I}$ is a $\mathcal{T}$-final epi-sink in $\mathcal{A}$, then by 5.3.18 there exists a set $J \subset I$ such that $(f_j)_{j \in J}$ is a $\mathcal{T}$-final epi-sink in $\mathcal{A}$. If $A_j \xrightarrow{k_j} \coprod_{j \in J} A_j \xrightarrow{f} A$ is the obvious factorization of $(f_j)_{j \in J}$ through the coproduct, then f is a $\mathcal{T}$-quotient map and thus a regular epimorphism (cf. 5.3.11). By assumption $\left(F\left(\coprod_{j \in J} A_j\right), (F(k_j))_{j \in J}\right)$ is a coproduct and $F(f)$ is a regular epimorphism. Since colimits are extremal epi-sinks (dualize 5.1.2 (2)) and thus by 5.3.10 b), $\mathcal{T}$-final epi-sinks, it follows that $(F(k_j))_{j \in J}$ and $(F(f))$ are $\mathcal{T}'$-final epi-sinks and hence $(F(f_j) = F(f) \circ F(k_j))_{j \in J}$ is a $\mathcal{T}'$-final epi-sink. But then $(F(f_i))_{i \in I}$ is likewise a $\mathcal{T}'$-final epi-sink.

<u>5.3.20 Theorem</u>. Let $(\mathcal{A},\mathcal{T})$ be an initially structured category. $\mathcal{A}$ is cartesian closed if and only if $A\times-$ preserves $\mathcal{T}$-final epi-sinks for each $A \in |\mathcal{A}|$.

<u>Proof</u>. 1) "⇒": Since $\mathcal{A}$ is cartesian closed, $A\times-$ is a left adjoint and therefore it preserves colimits. Then by 5.3.19 $A\times-$ preserves $\mathcal{T}$-final epi-sinks.

2) "⇐": Let $A,B \in |\mathcal{A}|$. For each $C \in |\mathcal{A}|$ and each $\mathcal{A}$-morphism $f\colon A \times C \to B$, a map $g_{f,C}\colon \mathcal{T}(C) \to [A,B]_{\mathcal{A}}$ is defined by $g_{f,C}(x) = f \circ 1_x$ for each $x \in \mathcal{T}(C)$ where $1_x \in [A,A \times C]_{\mathcal{A}}$

is the uniquely determined A-morphism with $p_A \circ 1_x = 1_A$ and $p_C \circ 1_x = \bar{x}$ (where $\bar{x}: A \to C$ denotes the uniquely determined A-morphism with $T(\bar{x}): T(A) \to T(C)$ defined by $T(\bar{x})(a) = x$ for each $a \in T(A)$ [cf. 5.3.6 b)]) provided that $p_A: A \times C \to A$ and $p_C: A \times C \to C$ denote the projections. The sink $(g_{f,C})$ indexed by f and C is an epi-sink in Set ; namely from $\alpha \circ g_{f,C} = \beta \circ g_{f,C}$ for each pair (f,C) , it follows that $\alpha = \beta$ because for each $m \in [A,B]_A$, we get $m = g_{f,C}(x)$ for suitable f,C and $x \in T(C)$ (Put $C = P \in |A|$ and $T(P) = \{x\}$ and define $f: A \times P \to B$ by $f = m \circ p_A$) . By 5.3.4 (3) there exists a T-final epi-sink $(\bar{g}_{f,C}: C \to B^A)_{(f,C)}$ and an epimorphism $h_{A,B}: [A,B]_A \to T(B^A)$ such that the diagram

$$(D_1) \qquad \begin{array}{ccc} T(C) & \xrightarrow{g_{f,C}} & [A,B]_A \\ & {\scriptstyle T(\bar{g}_{f,C})} \searrow \quad \swarrow {\scriptstyle h_{A,B}} & \\ & T(B^A) & \end{array}$$

commutes for each pair (f,C) (At first since the above diagram is commutative, $(T(\bar{g}_{f,C}))$ is an epi-sink and since T is faithful, then $(\bar{g}_{f,C})$ is also an epi-sink; the fact that the latter one is T-final follows from the condition (F) in 5.3.4 (3) by means of a simple calculation). In order to show that $h_{A,B}$ is additionally a monomorphism we define a mono-source $(m_a: [A,B]_A \to T(B))_{a \in T(A)}$ by $m_a(1) = T(1)(a)$ for each $1 \in [A,B]_A$ and each $a \in T(A)$ such that $m_a \circ g_{f,C} = T(d_{a,f,C})$, where $d_{a,f,C} \in [C,B]_A$ is defined by $d_{a,f,C} = f \circ 1_a$ with the uniquely determined A-morphism $1_a: C \to A \times C$ such that $p_A \circ 1_a = \bar{a}$ and $p_C \circ 1_a = 1_C$ (Obviously, $(m_a)_{a \in T(A)}$ is a mono-source. By 5.3.4 (2), $(T(A \times C), (T(p_A), T(p_C)))$ is the product of $(T(A), T(C))$. Further $T(p_A) \circ T(1_x) = 1_{T(A)}$ and $T(p_C) \circ T(1_x) = T(\bar{x})$. Hence, $T(1_x)(a) = (a,x)$. Analogously we have $T(1_a)(x) = (a,x)$. Thus $m_a(g_{f,C}(x)) = T(g_{f,C}(x))(a) = (T(f) \circ T(1_x))(a) = T(f)(a,x) = T(f)(T(1_a)(x)) = T(d_{a,f,C})(x)$ for each $x \in T(C)$.). If $a \in T(A)$ is fixed, then $(d_{a,f,C})$ is a sink in $\mathcal{A}$ indexed by (f,C). Since 5.3.4 (3) was applied to $(g_{f,C})$, the condition (F) is fulfilled. Thus if we choose $d = m_a\colon [A,B]_{\mathcal{A}} \to T(B)$ with $m_a \circ g_{f,C} = T(d_{a,f,C})$ for each pair (f,C), there exists a unique $c_a\colon B^A \to B$ with $T(c_a)\circ h_{A,B} = m_a$. Then from $h_{A,B}\circ\alpha = h_{A,B}\circ\beta$, it follows that $m_a\circ\alpha = T(c_a)\circ h_{A,B}\circ\alpha = = T(c_a)\circ h_{A,B}\circ\beta = m_a\circ\beta$ for each $a \in T(A)$. Thus $\alpha = \beta$. Consequently, $h_{A,B}$ is a bimorphism in <u>Set</u>, hence an isomorphism. The following diagram

(D_2)

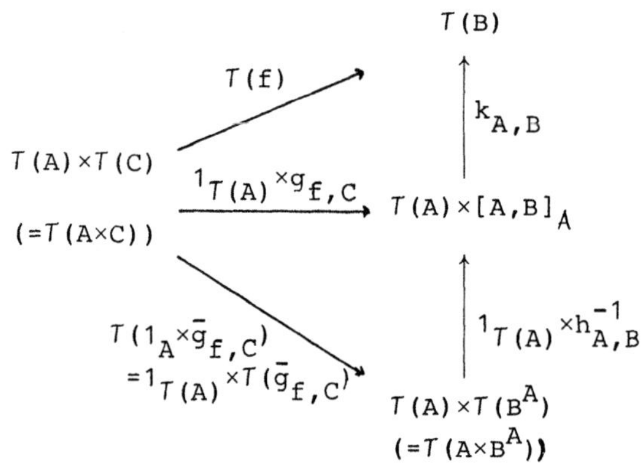

(in which the equalities are valid up to isomorphisms and) in which $k_{A,B}$ is defined by $k_{A,B}((a,r)) = T(r)(a)$ for each $a \in T(A)$ and each $r \in [A,B]_{\mathcal{A}}$ is commutative (obviously the lower triangle of (D_2) is commutative because (D_1) is commutative and $T(A)\times -$ is a functor; further $k_{A,B}\circ(1_{T(A)}\times g_{f,C})((a,c)) = k_{A,B}((a,g_{f,C}(c))) = T(f\circ 1_C)(a) = T(f)((a,c))$). By assumption $(1_A\times\bar{g}_{f,C})_{(f,C)}$ is a T-final epi-sink because $(\bar{g}_{f,C})_{(f,C)}$ is such a one. Thus there exists a unique $\mathcal{A}$-morphism $e_{A,B}\colon A \times B^A \to B$ with $T(e_{A,B}) = k_{A,B}\circ(1_{T(A)}\times h^{-1}_{A,B})$ and $e_{A,B}\circ(1_A\times\bar{g}_{f,C}) = f$ for each

$C \in |A|$ and each A-morphism $f\colon A \times C \to B$. If $e_{A,B}\circ(1_A \times \bar{f}) = f$, then $T(e_{A,B})\circ(1_{T(A)} \times T(\bar{f}))((a,c)) = k_{A,B}((a, h^{-1}_{A,B}(T(\bar{f})(c)))) = T(h^{-1}_{A,B}(T(\bar{f})(c)))(a) = T(h^{-1}_{A,B}(T(\bar{g}_{f,C})(c)))(a)$ for each $a \in T(A)$ and each $c \in T(C)$. Using the faithfulness of T it follows from this $\bar{f} = \bar{g}_{f,C}$.

5.3.21 Remark. It follows from part 2) of the proof of 5.3.20 that *in a cartesian closed initially structured category* (A,T) *the object* B^A *may be interpreted (up to isomorphism) as the set* $[A,B]_A$ *endowed with a suitable* A*-structure, i.e. as a "function space" and that the* A*-morphism* $e_{A,B}$ *is the usual evaluation map (up to isomorphism).* Moreover 4.1.5 (exponential laws and distributive law) and 4.1.6 (internal Hom-functor) hold analogously for initially structured cartesian closed categories.

5.3.22 Theorem. Every extremal epireflective (full and isomorphism-closed) subcategory B of an initially structured cartesian closed category (A,T) is cartesian closed.

Proof. By 5.3.16 B is initially structured. Hence the preceding theorem 5.3.20 may be applied in order to prove the cartesian closedness of B. Now let $B \in |B|$ and let $(f_i\colon B_i \to C)_{i\in I}$ be a $T\circ I$-final epi-sink in B ($I\colon B \to A$ denotes the inclusion functor). Further by 5.3.5 (2) there is a (T-final epi-sink, mono)-factorization $(B_i \xrightarrow{g_i} A \xrightarrow{m} C)_{i\in I}$ of $(f_i)_{i\in I}$ in A. Applying 5.3.15 A is an (extremal epi, mono source)-category and by 5.1.5 it follows that $A \in |B|$. Moreover $(f_i)_{i\in I}$ is an extremal epi-sink in B by 5.3.10 b). Hence m is a B-isomorphism (= A-isomorphism). Thus, $(f_i)_{i\in I}$ is a T-final epi-sink in A. Then by assumption, $(1_B \times f_i)_{i\in I}$ is a T-final epi-sink in A. Since the right adjoint I preserves products, $(1_B \times f_i)_{i\in I}$ is a sink in B. Since the left adjoint $R\colon A \to B$ of I preserves colimits, it follows from 5.3.19 that $(R(1_B \times f_i)\colon R(B\times B_i) \to R(B\times C))_{i\in I}$ is a $T\circ I$-final epi-sink in $\mathcal{B}$ which coincides with $(1_B \times f_i)_{i \in I}$ (up to isomorphism); more exactly the following diagram

$$\begin{array}{ccc}
B\times B_i & \xrightarrow[\cong]{r_{B\times B_i}} & \mathcal{R}(B\times B_i) \\
{\scriptstyle 1_B\times f_i}\downarrow & & \downarrow{\scriptstyle \mathcal{R}(1_B\times f_i)} \\
B\times C & \xrightarrow[\cong]{r_{B\times C}} & \mathcal{R}(B\times C)
\end{array}$$

is commutative for each $i \in I$. Consequently, $(1_B \times f_i)_{i \in I}$ is a $\mathcal{T}\circ\mathcal{I}$-final epi-sink in $\mathcal{B}$.

5.3.23 Examples. (1) The category Ord of ordered sets (and order preserving maps) is a cartesian closed initially structured category which is not topological [for obviously Ord is an extremal epireflective subcategory of PrOrd (it is easy to check that Ord is closed under formation of weak subobjects and products in PrOrd) , hence it is initially structured by 5.3.17 and cartesian closed by 5.3.22 (cf. 4.2.4) , but not topological (e.g. on a two-element set M there is no initial order structure with respect to $f: M \to (\mathbb{R},\leq)$[35] defined by $f(m) = 0$ for each $m \in M$)] .

(2) The categories HConv (Hausdorff convergence spaces), HLim (Hausdorff limit spaces) and HPsTop (Hausdorff pseudotopological spaces) [the Hausdorff property means in each case that limits of filters are unique] are cartesian closed initially structured categories which are not topological; for they are extremal epireflective in Conv, Lim and PsTop respectively which are cartesian closed topological categories (None of the three categories mentioned first is topological because the unique Hausdorff convergence structure

---

35) $\leq$ denotes the usual order on $\mathbb{R}$.

(resp. limit structure or pseudotopological structure) on $\{0,1\}$ is $q = \{(\dot{0},0),(\dot{1},1)\}$ which is not initial with respect to $f\colon \{0,1\} \to (\mathbb{R}, q_{\mathbb{R}})$[36] defined by $f(x) = 0$ for each $x \in \{0,1\}$, e.g. $h\colon (\mathbb{R}, q_{\mathbb{R}}) \to (\{0,1\}, q)$ defined by $h(x) = \begin{cases} 0 & \text{for } x \leq 0 \\ 1 & \text{for } x > 0 \end{cases}$ is not continuous, but $f \circ h$ is continuous.).

③ A uniform convergence space $(X, \mathcal{J}_X)$ is called *separated* provided that the induced limit space $(X, q_{\mathcal{J}_X})$ is a Hausdorff space ($(\mathcal{F},x) \in q_{\mathcal{J}_X}$ iff $\mathcal{F} \times \dot{x} \in \mathcal{J}_X$). The full subcategory <u>SepUConv</u> of <u>UConv</u> whose objects are the separated uniform convergence spaces is a cartesian closed initially structured category since it is obviously extremal epireflective in <u>UConv</u> which is a cartesian closed topological category by 4.2.8 (note that initial uniform convergence structures induce initial limit structures and that <u>HLim</u> is extremal epireflective in <u>Lim</u>). But <u>SepUConv</u> is not topological.

<u>5.3.24.</u> Each of the initially structured categories under 5.3.23 is an extremal epireflective subcategory of a topological category. The question whether every initially structured category is of this kind is answered by the following theorem:

<u>Theorem</u>. For a concrete category C (in the sense of 1.1.1) the following are equivalent:

(1) C is initially structured.

(2) C is an epireflective (full and isomorphism-closed) subcategory of a topological category.

(3) C is an extremal epireflective (full and isomorphism-closed) subcategory of a topological category.

---

36) $q_{\mathbb{R}}$ denotes the usual convergence structure (resp. limit structure or pseudotopological structure) on $\mathbb{R}$.

Proof.

(3) ⇒ (2): trivial

(2) ⇒ (1): cf. 5.3.17.

(1) ⇒ (3): Let $\mathcal{B}$ be the following category: objects are all triples (X,e,A) such that $X \in |\underline{Set}|$, $A \in |\mathcal{C}|$ and $e: X \to T_{\mathcal{C}}(A)$ is an epimorphism where $T_{\mathcal{C}}: \mathcal{C} \to \underline{Set}$ is the corresponding forgetful functor; morphisms from (X,e,A) to (X',e',A') are all pairs (f,g) with $f: X \to X'$, $g: A \to A'$ and $e' \circ f = T_{\mathcal{C}}(g) \circ e$; and the composition of morphisms is defined componentwise. A functor $T_{\mathcal{B}}: \mathcal{B} \to \underline{Set}$ is defined by $T_{\mathcal{B}}((X,e,A)) = X$ and $T_{\mathcal{B}}((f,g)) = f$. Then $T_{\mathcal{B}}$ is absolutely topological: Let $(B_i)_{i\in I} = ((X_i,e_i,A_i))_{i\in I}$ be any family of $\mathcal{B}$-objects and let $(X,(f_i: X \to T_{\mathcal{B}}(B_i))_{i\in I})$ be a source in $\underline{Set}$. Further let

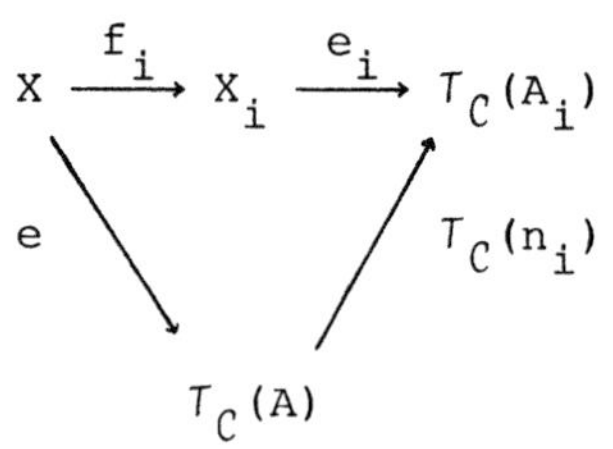

be the (epi, mono-source)-factorization existing by $IS_1')$ such that $(A,(n_i: A \to A_i)_{i\in I})$ is a $T_{\mathcal{C}}$-initial source in $\mathcal{C}$. Then $B = (X,e,A)$ is a $\mathcal{B}$-object and $r_i = (f_i,n_i): B \to B_i$ is a $\mathcal{B}$-morphism for each $i \in I$. The source $(B,(r_i)_{i\in I})$ $T_{\mathcal{B}}$-lifts the source $(X,(f_i)_{i\in I})$. In order to prove that $(B,(r_i)_{i\in I})$ is $T_{\mathcal{B}}$-initial let $(G,(g_i: G \to B_i)_{i\in I})$ be a source and $g: T_{\mathcal{B}}(G) \to T_{\mathcal{B}}(B)$ a morphism with $T_{\mathcal{B}}(r_i) \circ g = T_{\mathcal{B}}(g_i)$ for each $i \in I$. Suppose that $G = (X',e',A')$ and $g_i = (p_i,q_i)$ for each $i \in I$. Then the following diagram

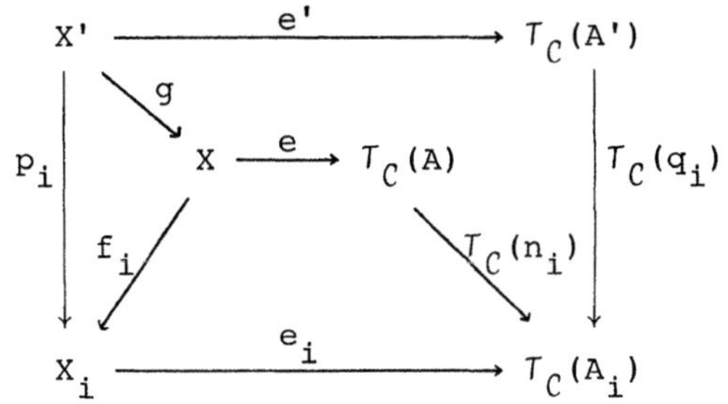

is commutative. Since e' is an epimorphism and $(T_C(A),(T_C(n_i))_{i\in I})$ is a mono-source, there exists a morphism $f: T_C(A') \to T_C(A)$ completing the above diagram commutatively. Since $(A,(n_i)_{i\in I})$ is T_C-initial, there exists a unique morphism $t: A' \to A$ with $T_C(t) = f$ and $n_i \circ t = q_i$ for each $i \in I$. Thus $x = (g,t): (X',e',A') \to (X,e,A)$ is a $\mathcal{B}$-morphism with $T_{\mathcal{B}}(x) = g$ and $r_i \circ x = g_i$ for each $i \in I$. Since $T_{\mathcal{B}}$ is faithful, x is uniquely determined by the properties mentioned above.

Then the full subcategory $\mathcal{B}'$ of $\mathcal{B}$ defined by

$$|\mathcal{B}'| = \{(X,e,A) \in |\mathcal{B}|: T_C(A) \subset X\}$$

obviously satisfies all properties of a topological category with the exception of the uniqueness of initial structures (If $\mathcal{B}$ is not replaced by $\mathcal{B}'$, then $T_C(A)$ is only isomorphic to a subset of X and especially $\mathcal{B}$ does not need to be small-fibred.). Now an equivalence relation $\sim$ on $|\mathcal{B}'|$ is defined by

$$(X,e,A) \sim (X',e',A') \quad \text{iff} \quad \begin{cases} X = X' \text{ and there exists an isomorphism} \\ g: A \to A' \text{ with } T_C(g) \circ e = e' \end{cases}.$$

If $|\mathcal{B}''|$ is a system of representatives of $\sim$ containing $|C''| = \{(T_C(A), 1_{T_C(A)}, A): A \in |C|\}$[37), then the corresponding full subcategory $\mathcal{B}''$ of $\mathcal{B}'$ is a topological category. Moreover the full subcategory C'' of $\mathcal{B}'$ defined by $|C''|$ as above

37) Note that T_C is amnestic.

is an isomorphism-closed extremal epireflective subcategory of $\mathcal{B}''$ and is isomorphic to $\mathcal{C}$ which can be seen as follows:

a) If $(X,e,A) \in |\mathcal{B}''|$ and if there is an isomorphism $(f,g): (T_{\mathcal{C}}(A'), 1_{T_{\mathcal{C}}(A')}, A') \to (X,e,A)$, then f and g are isomorphisms and thus $e = T_{\mathcal{C}}(g) \circ f^{-1}: X \to T_{\mathcal{C}}(A)$ is also an isomorphism. Since $T_{\mathcal{C}}$ is transportable, there exists a unique $\mathcal{C}$-object C and an isomorphism $\bar{e}: C \to A$ with $T_{\mathcal{C}}(\bar{e}) = e$. Then $(T_{\mathcal{C}}(C), 1_{T_{\mathcal{C}}(C)}, C)$ and (X,e,A) are members of the same equivalence class with respect to $\sim$ so that $(X,e,A) = (T_{\mathcal{C}}(C), 1_{T_{\mathcal{C}}(C)}, C)$, i.e. $\mathcal{C}''$ is isomorphism-closed in $\mathcal{B}''$.

b) Obviously $(e,1_A): (X,e,A) \to (T_{\mathcal{C}}(A), 1_{T_{\mathcal{C}}(A)}, A)$ is the extremal epireflection of $(X,e,A) \in |\mathcal{B}''|$ with respect to $\mathcal{C}''$; for if $(f,g): (X,e,A) \to (T_{\mathcal{C}}(A'), 1_{T_{\mathcal{C}}(A')}, A')$ is a $\mathcal{B}''$-morphism, then $(T_{\mathcal{C}}(g), g) \circ (e,1_A) = (f,g)$ and $(e,1_A)$ is an extremal epimorphism, i.e. a quotient map: If $T_{\mathcal{B}''}: \mathcal{B}'' \to \underline{\text{Set}}$ denotes the forgetful functor, then by assumption $T_{\mathcal{B}''}((e,1_A)) = e$ is surjective, where $(e,1_A)$ is a $\mathcal{B}''$-morphism. Given $(X',e',A') \in |\mathcal{B}''|$ and let $f: T_{\mathcal{C}}(A) \to X'$ be a map such that $f \circ e: X \to X'$ is a morphism from (X,e,A) to (X',e',A'), i.e. there is a morphism $h: A \to A'$ with $e' \circ f \circ e = T_{\mathcal{C}}(h) \circ e$. Then $e' \circ f = T_{\mathcal{C}}(h) \circ 1_{T_{\mathcal{C}}(A)}$ since e is an epimorphism. Thus $(f,h): (T_{\mathcal{C}}(A), 1_{T_{\mathcal{C}}(A)}, A) \to (X',e',A')$ is a morphism. Therefore $(1_{T_{\mathcal{C}}(A)}, A)$ is the final $\mathcal{B}''$-structure on $T_{\mathcal{C}}(A)$ with respect to e.

c) A functor $H: \mathcal{C} \to \mathcal{B}''$ is defined by
$H(A) = (T_{\mathcal{C}}(A), 1_{T_{\mathcal{C}}(A)}, A)$ for each $A \in |\mathcal{C}|$ and
$H(f) = (T_{\mathcal{C}}(f), f)$ for each $f \in \text{Mor } \mathcal{C}$.
Obviously H is a full[38] embedding[39]. Then especially $H': \mathcal{C} \to \mathcal{C}''$ defined by $H'(A) = H(A)$ for each $A \in |\mathcal{C}|$ and $H'(f) = H(f)$

---

[38] A functor $H: \mathcal{A} \to \mathcal{B}$ is called <u>full</u> provided that for each pair $(C,D) \in |\mathcal{A}| \times |\mathcal{A}|$, the map $[C,D]_{\mathcal{A}} \to [H(C),H(D)]_{\mathcal{B}}$ $(f \to H(f))$ is surjective.

[39] A functor $H: \mathcal{A} \to \mathcal{B}$ is called an <u>embedding</u> provided that $H_2: \text{Mor } \mathcal{A} \to \text{Mor } \mathcal{B}$ defined by $H_2(f) = H(f)$ for each $f \in \text{Mor } \mathcal{A}$ is injective.

for each $f \in \text{Mor}\, \mathcal{C}$ is an isomorphism which is easy to check.

5.3.25 Remark. By 2.2.5 (2) and the preceding theorem the object class of an initially structured category is always a relative disconnectedness with respect to a suitable topological category.

CHAPTER VI

COMPLETIONS

The most known completion is the construction of the real numbers from the rational numbers. This has been done in two different ways by G. Cantor and R. Dedekind respectively. As well-known Dedekind used cuts while Cantor constructed the reals by means of Cauchy sequences. Cantor's method was generalized by A. Weil in order to construct a completion for each uniform space. At the same time MacNeille used Dedekind's method (via a slight modification of the original definition of "cut") for his embedding of each ordered set in a complete lattice known as MacNeille completion.

In the first part of this chapter we will learn that each ordered set $(X,\leq)$ may be considered to be a concrete category $(\mathcal{A},\mathcal{F})$ over the base category $\mathcal{X}$ consisting of exactly one object and exactly one morphism, where $\mathcal{F}: \mathcal{A} \to \mathcal{X}$ is a faithful, amnestic and transportable functor. $(X,\leq)$ is a complete lattice iff $(\mathcal{A},\mathcal{F})$ is initially complete, i.e. $\mathcal{F}: \mathcal{A} \to \mathcal{X}$ is absolutely topological. Then the MacNeille completion may be interpreted as a "nice" embedding of a concrete category over $\mathcal{X}$ in a concrete initially complete category over $\mathcal{X}$. Thus the question arises whether it is possible to obtain such an embedding for each concrete category $(\mathcal{A},\mathcal{F})$ over any base category, i.e. a MacNeille completion of $(\mathcal{A},\mathcal{F})$. It turns out that such a completion does not always exist. Nevertheless each small concrete category has a MacNeille completion. Necessary and sufficient conditions for the existence of the MacNeille completion are proved. Not even the classical MacNeille completion fulfills the universal property claimed by N. Bourbaki. Therefore a universal initial completion is studied and necessary and sufficient conditions for its existence are proved. It is also shown that this completion differs from the MacNeille completion in general. Dualizing the introduced concepts one obtains the universal final completion while the MacNeille completion is selfdual (as in the classical case).

In the second part of this chapter Weil's construction of a completion of a uniform space is generalized to nearness spaces. But at first a suitable concept of completeness for nearness spaces has to be introduced. Clusters (i.e. non-void maximal near collections) are more appropriate than Cauchy filters in order to explain completeness. Therefore a nearness space will be called complete iff each cluster has an adherence point. Especially, each topological nearness space (= R_0-topological space) is complete. For each nearness space (X,μ) there is a special construction of a complete nearness space (X^*,μ^*) in which (X,μ) is densely embedded. This construction is called the canonical completion of (X,μ). It coincides with Weil's Hausdorff completion (up to isomorphism) for uniform N_1-spaces (= separated uniform spaces). The embedding $j_X: (X,\mu) \to (X^*,\mu^*)$ does not have the universal property in general. Therefore regular nearness spaces are introduced including uniform spaces and regular topological spaces. Then $j_X: (X,\mu) \to (X^*,\mu^*)$ is even an epireflection for each regular N_1-space with respect to the full subcategory of <u>Near</u> consisting of all complete regular N_1-spaces. Last not least well-known extensions and compactifications of topological spaces are obtained via the canonical completion, e.g. the Wallman extension, Hewitt's realcompactification, Alexandroff's one point compactification and the Stone-Čech compactification. Finally it is shown that even every Hausdorff compactification (resp. regular Hausdorff extension) of a topological space may be obtained by means of the canonical completion.

6.1. Initial (and final) completions

6.1.1 Definitions. 1) Let X be a fixed category, called base category. A concrete category over X is a pair (A,F) where A is a category and $F: A \to X$ is a functor which is faithful, amnestic and transportable, called the underlying functor of (A,F).

2) A concrete category (A,F) over X is called initially complete provided that $F: A \to X$ is absolutely topological.

3) If (A,F) and (B,G) are concrete categories over X, then a functor $H: A \to B$ is called

a) concrete provided that $G \circ H = F$,

b) initiality preserving provided that it is concrete and for each F-initial source $(f_i: A \to A_i)_{i \in I}$ in A, the source $(H(f_i): H(A) \to H(A_i))_{i \in I}$ is G-initial in B, and

c) initially dense provided that it is concrete and for each $B \in |B|$, there exists a G-initial source $(B,(g_i: B \to H(A_i))_{i \in I})$.

4) An initial completion of a concrete category (A,F) is an initiality preserving initially dense full[38)] (concrete) embedding[39)] $H: (A,F) \to (B,G)$ from (A,F) into some initially complete category (B,G). Occasionally (B,G) is already called an initial completion of (A,F) provided that $H: (A,F) \to (B,G)$ is an initial completion of (A,F).

6.1.2 Remark. Obviously a concrete category (A,F) is initially complete if and only if for each class-indexed family $(A_i)_{i \in I}$ and each source $(X,(f_i: X \to F(A_i))_{i \in I})$ in the base category X, there exists a (unique) F-initial source $(A,(g_i: A \to A_i)_{i \in I})$ with $F(A) = X$ and $F(g_i) = f_i$ for each $i \in I$. If "source" is replaced by "sink" and "F-initial" by "F-final", then the dual concept <u>finally complete</u> is obtained. Corresponding to the result 1.2.1.1 for X = <u>Set</u> we have the equivalence of initially complete and finally complete for a concrete category (A,F) over an arbitrary base category X. The dual concepts of 6.1.1 3) b), 3) c) and 4) are <u>finality preserving</u>, <u>finally dense</u> and <u>final completion</u>.

<u>6.1.3 Proposition</u>. Let $H: (A,F) \to (B,G)$ be a concrete functor which is full and finally dense. Then H is initiality preserving.

<u>Proof</u>. Let $(A \xrightarrow{f_i} A_i)_{i\in I}$ be an F-initial source in A. In order to show that $(H(A) \xrightarrow{H(f_i)} H(A_i))_{i\in I}$ is G-initial pick some $B \in |B|$ and some X-morphism $f: G(B) \to G(H(A))$ such that for each $i \in I$, there exists a (unique) B-morphism $h_i: B \to H(A_i)$ with $G(h_i) = G(H(f_i))\circ f$. For each B-morphism $a': H(A') \to B$, $h_i\circ a': H(A') \to H(A_i)$ is a B-morphism for each $i \in I$ for which there exists a (unique) A-morphism $h_i^!: A' \to A$ with $H(h_i^!) = h_i\circ a'$ since H is full. Then we have $F(h_i^!) = G(H(h_i^!)) = G(h_i)\circ G(a') = F(f_i)\circ f\circ G(a')$. Thus there is a (unique) A-morphism $h_{a'}: A' \to A$ with $F(h_{a'}) = f\circ G(a')$ since $(A \xrightarrow{f_i} A_i)_{i\in I}$ is F-initial. But then also $G(H(h_{a'})) = F(h_{a'}) = f\circ G(a')$. Hence there exists a (unique) B-morphism $\bar{f}: B \to H(A)$ with $G(\bar{f}) = f$ since G is finally dense. Thereby proving the proposition.

<u>6.1.4 Definition</u>. If $H_1: (A,F) \to (B_1,G_1)$ and $H_2: (A,F) \to (B_2,G_2)$ are initial (resp. final) completions of (A,F), then the following is defined:

1) $H_2 \le H_1$ provided that there exists a (unique) full concrete embedding $H: (B_2,G_2) \to (B_1,G_1)$ with $H_1 = H\circ H_2$.
2) $H_2 \cong H_1$ provided that $H_2 \le H_1$ and $H_1 \le H_2$.

<u>6.1.5 Proposition</u>. 1) The relation $\le$ defined in 6.1.4. 1) is reflexive and transitive.

2) $H_2 \cong H_1$ as defined in 6.1.4. 2) is valid if and only if there is a concrete isomorphism $H\colon (\mathcal{B}_2,G_2) \to (\mathcal{B}_1,G_1)$ with $H_1 = H \circ H_2$.

Proof. 1) is trivial.

2) a) "$\Leftarrow$": Since every isomorphism is a full embedding, the assertion is trivial.

b) "$\Rightarrow$": There exist full concrete embeddings $H\colon (\mathcal{B}_2,G_2) \to (\mathcal{B}_1,G_1)$ and $H'\colon (\mathcal{B}_1,G_1) \to (\mathcal{B}_2,G_2)$ such that the diagram

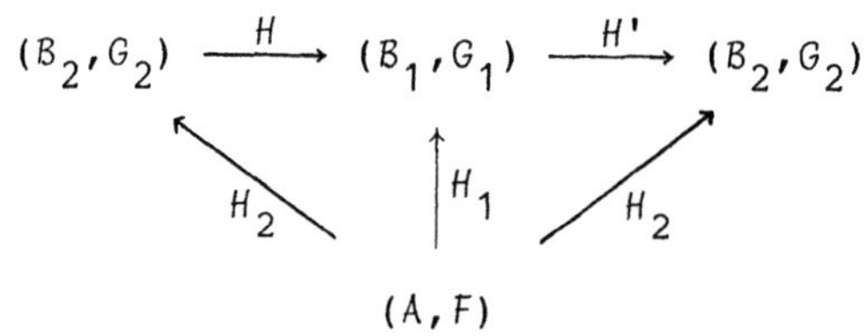

commutes. For each $i \in \{1,2\}$, $H_i(\mathcal{A})$ defined by $|H_i(\mathcal{A})| = \{H_i(A)\colon A \in |\mathcal{A}|\}$ and $\operatorname{Mor} H_i(\mathcal{A}) = \{H_i(f)\colon f \in \operatorname{Mor} \mathcal{A}\}$ is a full subcategory of $\mathcal{B}_i$ which is isomorphic to $\mathcal{A}$ by means of H_i. Then by the commutativity of the above diagram one obtains that $H' \circ H|_{H_2(\mathcal{A})} = I_{\mathcal{B}_2}\big|_{H_2(\mathcal{A})}$. Since H_2 is initially dense, there is a G_2-initial source $(B_2 \xrightarrow{g_i} H_2(A_i))_{i\in I}$ for each $B_2 \in |\mathcal{B}_2|$. Moreover, H and H' are initiality preserving (e.g. H has this property: If $(B^2 \xrightarrow{f_i} B_i^2)_{i\in I}$ is a G_2-initial source in $\mathcal{B}_2$, $(B^1 \xrightarrow{h_i} H(B_i^2))_{i\in I}$ a source in $\mathcal{B}_1$ and $f\colon G_1(B^1) \to G_1(H(B^2))$ with $G_1(H(f_i)) \circ f = G_1(h_i)$, then $(G_1(B^1) \xrightarrow{G_1(h_i)} G_1(H(B_i^2)))_{i\in I} = (G_1(B^1) \xrightarrow{G_1(h_i)} G_2(B_i^2))_{i\in I}$ has a G_2-initial lifting $(\widetilde{B}^2 \xrightarrow{k_i} B_i^2)_{i\in I}$ (hence $G_2(k_i) = G_1(h_i)$ and $G_2(\widetilde{B}^2) = G_1(B^1)$) since $(\mathcal{B}_2,G_2)$ is initially complete. Thus there exists a unique $f^*\colon \widetilde{B}^2 \to B^2$ with $G_2(f^*) = f$ and $f_i \circ f^* = k_i$. With $H(f^*) = \bar{f}\colon B^1 \to H(B^2)$ we have $H(f_i) \circ \bar{f} = h_i$

(note that H is concrete and G_1 is faithful). The uniqueness of $\bar{f}$ is obtained by the uniqueness of f^* since H is full and faithful. Hence $(H(B^2) \xrightarrow{H(f_i)} H(B_i^2))_{i\in I}$ is G_1-initial.). Then

$$((H'\circ H)(B_2) \xrightarrow{(H'\circ H)(g_i)} (H'\circ H)(H_2(B_i)))_{i\in I} =$$

$= ((H'\circ H)(B_2) \xrightarrow{(H'\circ H)(g_i)} H_2(B_i))_{i\in I}$ is G_2-initial. Since $G_2(H'(H(B_2))) = G_2(B_2)$ and $G_2(H'(H(g_i))) = G_2(g_i)$ for each $i \in I$, it follows from the uniqueness of the initial lifting that $(H'\circ H)(B_2) = B_2$. Then for each B_2-morphism f, we have $(H'\circ H)(f) = f$ since G_2 is faithful. Hence $H'\circ H = I_{B_2}$. Correspondingly it is shown that $H\circ H' = I_{B_1}$. Consequently, H is a concrete isomorphism of the desired kind.

6.1.6 Definition. The smallest[40)] (with respect to $\leq$) initial completion of a concrete category (A,F) (over a base category X), if it exists, is called the MacNeille completion of (A,F).

6.1.7. In order to formulate a necessary and sufficient condition for the existence of the MacNeille completion the following explanations are necessary:
If (A,F) is a concrete category over X, then an F-morphism from an object X of X is a pair (f,A) where A is an A-object and $f: X \to F(A)$ is an X-morphism with domain X and codomain $F(A)$. Usually such a morphism is denoted briefly by $X \xrightarrow{f} F(A)$. *If* $F(A) \xrightarrow{g} F(B)$ *is an* X*-morphism for which there exists a (unique)* A*-morphism* $\bar{g}: A \to B$ *with* $F(\bar{g}) = g$, *then sometimes one says that* $A \xrightarrow{g} B$ *is an* A*-morphism (thus not distinguishing between* g *and* $\bar{g}$*)*. An F-source S from X is a pair (X,ξ) where X is an X-object and ξ is a class of F-morphisms (f_i, A_i) from X (especially ξ may be a proper class). Usually such an F-source is denoted by $(X \xrightarrow{f_i} F(A_i))_{i\in I}$.

---

40) by 6.1.5 unique up to isomorphism.

If $S = (X \xrightarrow{f_i} F(A_i))_{i\in I}$ and $T = (Y \xrightarrow{g_j} F(A'_j))_{j\in J}$ are F-sources, then an F-source-morphism $p: S \to T$ is an X-morphism $X \to Y$ such that for each $j \in J$, there exists some $i \in I$ with $(f_i, A_i) = (g_j \circ p, A'_j)$.

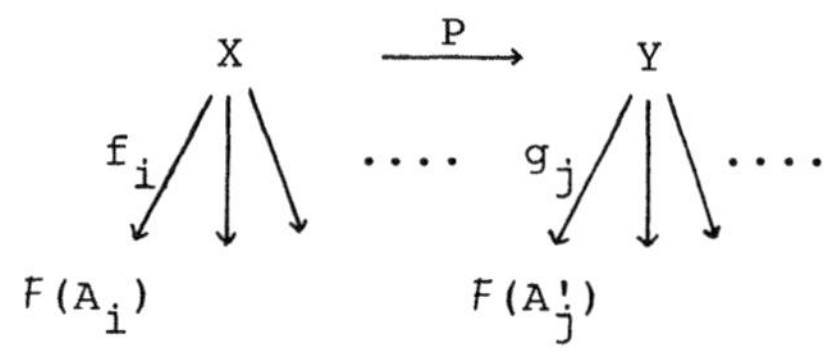

Dually: F-morphism into an X-object, F-sink to X, F-sink-morphism

Since there may be too many F-sources to form a class, one cannot speak of the category of all F-sources (as objects) and all F-source-morphisms. But if we confine ourselves to a "class of sources", then we "have" a category. More exactly:

Agreement. Let B be a conglomerate of F-sources of a concrete category (A,F) over X. Further let B be codable by a class B' (i.e. there is an injection $\varphi: B \to B'$). Then B is considered as a concrete category whose objects are the images of members of B in B' and whose morphisms are all F-source-morphisms between F-sources in B (more exactly: being in a bijective correspondence to them). The underlying functor sends $\varphi((X \xrightarrow{f_i} F(A_i))_{i\in I})$ to X.

An F-source $(X \xrightarrow{f_i} F(A_i))_{i\in I}$ is called closed provided that it contains every F-morphism $X \xrightarrow{f} F(A)$[41] with the following property: For each F-morphism $F(A') \xrightarrow{g} X$ into X such that each $A' \xrightarrow{f_i \circ g} A_i$ is an A-morphism, it follows that $A' \xrightarrow{f \circ g} A$ is an A-morphism.

[41] i.e. there exists some $i \in I$ such that $(f,A) = (f_i,A_i)$.

6.1.8 Theorem. For a concrete category (A,F) over X the following are equivalent:

(1) (A,F) has an initial completion.

(2) (A,F) has a MacNeille completion.

(3) The conglomerate of all closed F-sources in (A,F) is codable by a class.

If these conditions are satisfied, then the MacNeille completion of (A,F) is the category of closed F-sources.

Proof. (2) $\Rightarrow$ (1): trivial.

(3) $\Rightarrow$ (2): α) If A is a small category (i.e. $|A|$ is a set), then all F-sources form a class (since for each $X \in |X|$, all F-sources from X form a set). Let B be the category of closed F-sources and F-source-morphisms (cf. 6.1.7). A functor $H: A \to B$ is defined by

$H(A)$ = F-source from $F(A)$ of all $F(A) \xrightarrow{g} F(A')$ which are A-morphisms $A \xrightarrow{g} A'$,

$H(f) = F(f)$.

(Obviously $H(A)$ is closed; for if $h: F(A) \to F(A')$ is an F-morphism with the required property, then $h \circ 1 = h: A \to A'$ is an A-morphism, i.e. h belongs to $H(A)$ since $g = g \circ 1: A \to A'$ is an A-morphism for each $g: F(A) \to F(A')$ which belongs to $H(A)$.) If an underlying functor $G: B \to X$ is defined by $G((X,\xi)) = X$ for each $(X,\xi) \in |B|$ and $G(p) = p$ for each $p \in \text{Mor } B$, then (B,G) is a concrete category over X and H is a concrete functor.

Moreover H is

1. full. If $\bar{f}: H(A) \to H(A')$ is a B-morphism where $H(A) = (F(A), (F(h_i))_{i\in I})$ and $H(A') = (F(A'), (F(k_j))_{j\in J})$, then $F(1_{A'}) \circ \bar{f} = 1_{F(A')} \circ \bar{f} = \bar{f}: F(A) \to F(A')$ belongs to $(F(k_j))_{j\in J}$, i.e. there exists some A-morphism $f: A \to A'$ with $F(f) = H(f) = \bar{f}$ and

2. an embedding. It suffices to show:

   a) H is faithful: trivial since F is faithful and

   b) $H_1: |A| \to |B|$ defined by $H_1(A) = H(A)$ for each $A \in |A|$

is injective[42]: If $H(A) = H(A') = B$, then by 1. there exist A-morphisms $h\colon A \to A'$ and $k\colon A' \to A$ with $H(h) = H(k) = 1_B$. Thus $F(h) = F(k) = 1_X$ where $X = F(A) = F(A')$. Since F is faithful, h (resp. k) is an isomorphism which must be the identity because F is amnestic. Hence $A = A'$.

In order to show that B is initially complete it suffices to prove that B is finally complete (cf. 6.1.2):

If $((X_i,\xi_i))_{i\in I}$ is any family of B-objects, X any X-object and $(f_i\colon X_i \to X)_{i\in I}$ any family of X-morphisms, then a B-structure ξ on X is defined by:

$X \xrightarrow{a} F(A)$ belongs to ξ if and only if $a\circ f_i$ belongs to ξ_i for each $i \in I$. (The closedness of (X,ξ) results immediately from the closedness of each (X_i,ξ_i).)

Obviously $(f_i\colon (X_i,\xi_i) \to (X,\xi))_{i\in I}$ is G-final.

Further G is initially dense:

If $S = (X,\xi) \in |B|$ with $\xi = \{(f_j,V_j): j \in J\}$, then a G-initial source $(g_i\colon S \to H(W_i))_{i\in I}$ is wanted. Let ξ' be the class of all F-morphisms $X \xrightarrow{g_i} F(W_i)$ such that for each F-morphism $h\colon F(W_i) \to F(V)$ for which $h\colon W_i \to V$ is an A-morphism, it follows that $h\circ g_i$ belongs to ξ. Especially, every F-morphism belonging to ξ has this property since S is closed. By the definition of ξ', $g_i\colon X \to F(W_i)$ is a B-morphism $S \to H(W_i)$ and $(g_i\colon S \to H(W_i))_{i\in I}$ is G-initial (If $(\hat{X} \xrightarrow{1_q} F(U_q))_{q\in Q} \in |B|$ and $f\colon \hat{X} \to X$ is an X-morphism such that $g_i\circ f\colon \hat{X} \to F(W_i)$ is a B-morphism for each $i \in I$, then for each $j \in J$, there is some $i_j \in I$ with $f_j = g_{i_j}$ and some $q \in Q$ with
$1_q = F(1_{W_{i_j}})\circ g_{i_j}\circ f = 1_{F(W_{i_j})}\circ g_{i_j}\circ f = f_j\circ f$, i.e. $f\colon \hat{X} \to X$ is a B-morphism).

In order to prove that H is initiality preserving it suffices to show (cf. 6.1.3) that H is finally dense (since H is full):

If $S = (X,\xi) \in |B|$ with $\xi = \{(f_j,V_j): j \in J\}$, then a G-final sink $(g_i\colon H(W_i) \to S)_{i\in I}$ is wanted. Let ξ' be the class of all F-morphisms

---

42) We say that a functor $H\colon A \to B$ is <u>injective on objects</u>, provided that the property b) is fulfilled.

$F(W_i) \xrightarrow{g_i} X$ such that $f_j \circ g_i\colon W_i \to V_j$ is an A-morphism for each $j \in J$. By the definition of ξ', $g_i\colon F(W_i) \to X$ is a B-morphism $H(W_i) \to S$ for each $i \in I$ and $(g_i\colon H(W_i) \to S)_{i\in I}$ is G-final since S is closed.
Thereby it has been shown that (B,G) is an initial completion of (A,F).
If $H'\colon (A,F) \to (B',G')$ is any initial completion of (A,F), then it remains to show that there is a full concrete embedding $H''\colon (B,G) \to (B',G')$ with $H'' \circ H = H'$. For each $B \in |B|$, let $(f_i\colon B \to H(A_i))_{i\in I}$ be the source of all B-morphisms with codomain in $H(A)$. Then $(G(f_i)\colon G(B) \to G'(H'(A_i)))_{i\in I}$ is the corresponding source in X (note: $G(H(A_i)) = F(A_i) = G'(H'(A_i))$) for which there exists a unique G'-initial source $(b_i\colon B'_B \to H'(A_i))_{i\in I}$ with $G'(B'_B) = G(B)$ and $G'(b_i) = G(f_i)$ for each $i \in I$ since (B',G') is initially complete. Put

$$H''(B) = B'_B \quad \text{for each } B \in |B|$$

and

$$H''(f) = G(f) \quad \text{for each } f \in \text{Mor } B .$$

Thereby a concrete functor $H''\colon B \to B'$ is defined (If $f\colon B \to \hat{B}$ is a B-morphism, then $G(f)\colon G(B) \to G(\hat{B})$ is a B'-morphism $B'_B \to B'_{\hat{B}}$ which is easily verified). If $A \in |A|$ and $(f_i\colon H(A) \to H(A_i))_{i\in I}$ is the source of all B-morphisms whose codomain belongs to $H(A)$, then $(H'(A) \xrightarrow{G(f_i)} H'(A_i))_{i\in I}$ is G'-initial (since the identity $H'(A) \to H'(A)$ occurs) and $G'(H'(A)) = F(A) = G(H(A))$ so that by the unique construction of $H''(H(A))$, it follows that $H'(A) = H''(H(A))$. Moreover since $H''(H(f)) = H'(f)$ for each $f \in \text{Mor } A$ (because of $G'(H''(H(f))) = G'(H'(f)) = F(f)$ and the faithfulness of G'), we have $H'' \circ H = H'$. In order to show that H'' is full we use the fact that H is initially and finally dense:
Now if $f\colon H''(B) \to H''(\hat{B})$ is a B'-morphism, then we consider the family of all B'-morphisms

$H''(H(A)) \xrightarrow{H''(a)} H''(B) \xrightarrow{f} H''(\hat{B}) \xrightarrow{H''(\hat{a})} H''(H(\hat{A}))$ where A and $\hat{A}$ range over $|A|$ independently and a (resp. $\hat{a}$) ranges over all $\mathcal{B}$-morphisms $H(A) \to B$ (resp. $\hat{B} \to H(\hat{A})$). Since $H'' \circ H = H'$ is full, there exists an A-morphism $h: A \to \hat{A}$ with $H''(H(h)) = H''(\hat{a}) \circ f \circ H''(a)$. Hence $G(H(h)) = G(\hat{a}) \circ G'(f) \circ G(a)$ because of $G' \circ H'' = G$. Since $H: (A,F) \to (\mathcal{B},G)$ is initially and finally dense, it follows that $G'(f): G(B) \to G(\hat{B})$ is a $\mathcal{B}$-morphism $B \to \hat{B}$, i.e. there exists a unique $\mathcal{B}$-morphism $\bar{f}: B \to \hat{B}$ with $G(\bar{f}) = G'(f)$. Consequently, $H''(\bar{f}) = f$ since G' is faithful.

Since H'' is full, H'' is also an embedding (cf. the corresponding argumentation with respect to H). Altogether showing that $H: (A,F) \to (\mathcal{B},G)$ is the MacNeille completion of (A,F).

β) If A is not small but the conglomerate of all closed F-sources in (A,F) is still codable by a class, then by 6.1.7 (agreement) one can form the category $\mathcal{B}$ of all closed F-sources and F-source-morphisms and the proof continues as under α).

(1) ⇒ (3): Let $H: (A,F) \to (\mathcal{B},G)$ be an initial completion of (A,F). Without loss of generality let A be a full subcategory of $\mathcal{B}$, i.e. let H be the inclusion functor. For each closed F-source $S = (X \xrightarrow{f_i} F(A_i))_{i \in I}$, there is a unique G-initial source $(B_S \xrightarrow{g_i} A_i)_{i \in I}$ with $G(B_S) = X$ and $G(g_i) = f_i$ for each $i \in I$ since $(\mathcal{B},G)$ is initially complete. It suffices to show that for distinct closed F-sources $S = (X,\xi)$ and $S' = (X',\xi')$, it follows that $B_S \neq B_{S'}$. Then the conglomerate of all closed F-sources in (A,F) is codable by the class $|\mathcal{B}|$.

If $S \neq S'$ and $X = X'$ (the case $X \neq X'$ is trivial), then, without loss of generality, there is some $X \xrightarrow{f} F(A)$ belonging to S' but not to S. Since S is closed, there is some F-morphism $F(A') \xrightarrow{g} X$ such that

(1) $A' \xrightarrow{f_i \circ g} A_i$ is an A-morphism for each $X \xrightarrow{f_i} F(A_i)$ in S

and

(2) $F(A') \xrightarrow{f \circ g} F(A)$ is not an A-morphism.

By (1) and the definition of initiality, $A' \xrightarrow{g} B_S$ is a $\mathcal{B}$-morphism. Then $B_S \neq B_{S'}$, since $A' \xrightarrow{g} B_{S'}$ is not a $\mathcal{B}$-morphism; for if $A' \xrightarrow{g} B_{S'}$ was a $\mathcal{B}$-morphism, then $A' \xrightarrow{f \circ g} A$ would be a $\mathcal{B}$-morphism since $X \xrightarrow{f} \mathcal{F}(A)$ belongs to S' and thus an $\mathcal{A}$-morphism (because $\mathcal{A}$ is a full subcategory of $\mathcal{B}$) in contradiction to (2).

6.1.9 Remarks. (1) It results from the proof of 6.1.8 that *the MacNeille completion* of a concrete category $(\mathcal{A},\mathcal{F})$, if it exists, may be characterized too *as a finally dense initial completion*. Obviously it exists if $\mathcal{A}$ is a small category.

(2) If $\mathcal{H}: (\mathcal{A},\mathcal{F}) \to (\mathcal{B},\mathcal{G})$ is the MacNeille completion of $(\mathcal{A},\mathcal{F})$, then $\mathcal{H}$ is finality preserving (apply the dual statement of 6.1.3 and note that $\mathcal{H}$ is full and initially dense). Moreover, $(\mathcal{B},\mathcal{G})$ is finally complete and $\mathcal{H}$ is an initially dense full concrete embedding. Hence the *MacNeille completion* is an *initially dense final completion*. Since obviously the converse is true, the *MacNeille completion is self-dual*. Then it may be considered too as the *smallest final completion* of $(\mathcal{A},\mathcal{F})$.

6.1.10 Examples. (1) Let $(S,\leq)$ be an ordered set, i.e. $(S,\leq) \in |\underline{Ord}|$. A (small) category $\mathcal{A}$ is defined by

1) $|\mathcal{A}| = S$

and

2) $[s,s']_{\mathcal{A}} = \begin{cases} \{(s,s')\} & \text{if } s \leq s' \\ \emptyset & \text{otherwise} \end{cases}$

If $\mathcal{X}$ denotes the category consisting of exactly one object X and one morphism 1_X and a functor $\mathcal{F}: \mathcal{A} \to \mathcal{X}$ is defined by $\mathcal{F}(s) = X$ for each $s \in |\mathcal{A}| = S$ and $\mathcal{F}(f) = 1_X$ for each $f \in \operatorname{Mor} \mathcal{A}$, then $(\mathcal{A},\mathcal{F})$ is a (small) concrete category over $\mathcal{X}$ having a MacNeille completion by 6.1.7, which may be (equivalently) described as follows:
For each $A \subset S$, let A^* be the set of all upper bounds of A in $(S,\leq)$ and A^+ the set of all lower bounds of A in $(S,\leq)$. If $A = (A^*)^+$, then $A \subset S$ is called a *cut*. If $N(S)$ denotes the set of all cuts in $(S,\leq)$ and $\subset$ the set-theoretic inclusion,

then $(N(S),\subset)$ is a complete lattice and $\mu_S\colon (S,\le) \to (N(S),\subset)$ defined by $\mu_S(s) = (\{s\}^*)^+ = \{x \in S\colon x \le s\}$ is an embedding (i.e. $\mu'_S\colon (S,\le) \to (\mu_S[S],\subset)$ defined by $\mu'_S(s) = \mu_S(s)$ for each $s \in S$ is an Ord-isomorphism).
[This construction is known to be the *MacNeille completion in lattice theory*] . Analogously to the construction of (A,F) we construct a concrete category (B,G) over X from $(N(S),\subset)$. Then μ_S defines a functor $H\colon (A,F) \to (B,G)$ by $H(s) = \mu_S(s)$ for each $s \in S = |A|$ and $H((s,s')) = (\mu_S(s),\mu_S(s'))$ if $s \le s'$. Hence $H\colon (A,F) \to (B,G)$ is a finally dense initial completion of (A,F) , i.e. (up to isomorphism) the MacNeille completion of (A,F) [Note: F-initial source in A (resp. G-initial source in B) means meet in $(S,\le)$ (resp. in $(N(S),\subset)$), initial completeness of (B,G) means completeness of $(N(S),\subset)$, initial denseness of H means meet denseness of μ_S (i.e. each $A \in N(S)$ is the meet of all $\mu_S(s)$ containing A) , final denseness of H means join denseness of μ_S (i.e. each $A \in N(S)$ is the supremum of all $\mu_S(s)$ contained in A)] .

(2) Let (A,F) be the concrete category of compact topological spaces and continuous maps over Set . The monocoreflective (= bicoreflective) hull $R^{co}_{\underline{Top}}\,A$ of A in Top is a topological category, hence an initially complete concrete category over Set . If $I\colon A \to R^{co}_{\underline{Top}}\,A$ denotes the inclusion functor, then I is the MacNeille completion of A ; for I is a full concrete embedding which is initially dense [each space (X,X) of $|R^{co}_{\underline{Top}}\,A|$ is a subspace of a compact space (Y,Y) by means of the one-point compactification, hence the one-element source $((X,X)\hookrightarrow(Y,Y))$ is initial] and finally dense [if $C_i \xrightarrow{j_i} \coprod_{i\in I} C_i \xrightarrow{\omega} X$ is a representation of $X \in |R^{co}_{\underline{Top}}\,A|$ as a quotient object of a coproduct of spaces of $|A|$, then $(C_i \xrightarrow{\omega\circ j_i} X)_{i\in I}$ is a final sink] (and thus by 6.1.3 also initiality preserving), consequently a finally dense initial completion.

(3) a) Let Ω be a proper class. Then $(\Omega\times\{0,1\},\le)$ is an ordered[43)] class provided that

[43)] i.e. $\le$ is reflexive, antisymmetric and transitive.

$$(\alpha,i) < (\beta,j) \Leftrightarrow (i < j \text{ and } \alpha \neq \beta) .$$

Analogously to (1) this one may be considered as a concrete category (A,F) over the category X consisting of exactly one object X and one morphism 1_X . But (A,F) has no initial completion; for it is easily verified, that there is an injection between the conglomerate of all subclasses of Ω and the conglomerate of all closed F-sources, and since the first one is not codable by a class, the latter one cannot be it likewise.

b) As under a) let Ω be a proper class. Then $(\Omega,\leq)$ is an ordered class provided that

$$\alpha \leq \beta \text{ iff } \alpha = \beta .$$

This one may be considered as a concrete category (A,F) over X (cf. a)). Then (A,F) has a MacNeille completion which is not fibre-small[44)] (there is a one-one-correspondence between the conglomerate of all closed F-sources and the class Ω' arising from Ω by adding two elements being not in Ω) .

<u>6.1.11 Remark</u>. Obviously, the MacNeille completion of a concrete category (A,F) , if it exists, is *fibre small*[44)] if and only if for each object X in the base category X , the conglomerate of all closed F-sources from X is codable by a set (i.e. there is an injection into a set) .

<u>6.1.12 Definition</u>. An initial completion $H: (A,F) \to (B,G)$ of a concrete category (A,F) (over a base category X) is called <u>universal</u> provided that for each initially complete category (B',G') over X and each initiality preserving concrete functor $H': (A,F) \to (B',G')$, there is a unique initiality preserving concrete functor $\hat{H}: (B,G) \to (B',G')$ with $\hat{H} \circ H = H'$.

<u>6.1.13 Remark</u>. Obviously the universal initial completion, if it exists, is uniquely determined (up to isomorphism) by its defining property (cf. the corresponding argumentation con-

44) cf. IS_2) in 5.3.1 and replace $|\underline{Set}|$ by $|X|$.

cerning universal maps) .

6.1.14. In order to formulate a necessary and sufficient condition for the existence of the universal initial completion we need the following:

Definition. Let (A,F) be a concrete category over X . An F-source $S = (X \xrightarrow{f_i} F(A_i))_{i \in I}$ is called semi-closed provided that the following are satisfied:

(1) If $X \xrightarrow{f} F(A)$ belongs to S , then $X \xrightarrow{F(g) \circ f} F(A')$ belongs to S for each A-morphism $g: A \to A'$.

(2) If $X \xrightarrow{f} F(A)$ is an F-morphism and $(A \xrightarrow{g_j} A_j^*)_{j \in J}$ is an F-initial source such that each $X \xrightarrow{F(g_j) \circ f} F(A_j^*)$ belongs to S , then $X \xrightarrow{f} F(A)$ belongs to S .

6.1.15 Theorem. For a concrete category (A,F) over X the following are equivalent:

(1) (A,F) has a universal initial completion.

(2) The conglomerate of all semi-closed F-sources is codable by a class.

If these conditions are satisfied, then the universal initial completion is the category of semi-closed F-sources.

Proof. (2) $\Rightarrow$ (1): α) If A is a small category, then there are no problems in forming the category of semi-closed F-sources and F-source-morphisms denoted by B . A functor $H: A \to B$ is defined by

$H(A)$ = F-source from $F(A)$ of all $F(A) \xrightarrow{g} F(A')$ which are A-morphisms $A \xrightarrow{g} A'$

$H(f) = F(f)$

(Obviously $H(A)$ is semi-closed). If an underlying functor $G: B \to X$ is defined by $G((X,\xi)) = X$ for each $(X,\xi) \in |B|$ and $G(p) = p$ for each $p \in \operatorname{Mor} B$, then (B,G) is a concrete category over X and H is a concrete functor which is a full embedding (cf. the corresponding proof concerning the MacNeille completion).

In order to prove the initial completeness of $\mathcal{B}$ it suffices to show the final completeness (see 6.1.2): If $((X_i,\xi_i))_{i\in I}$ is any family of $\mathcal{B}$-objects, X any $\mathcal{X}$-object and $(f_i: X_i \to X)_{i\in I}$ any family of $\mathcal{X}$-morphisms, then a $\mathcal{B}$-structure ξ on X is defined by: $X \xrightarrow{a} \mathcal{F}(A)$ belongs to ξ if and only if $a \circ f_i$ belongs to ξ_i for each $i \in I$. (Since (X_i,ξ_i) is semi-closed for each $i \in I$, it follows immediately that (X,ξ) is semi-closed.) Obviously $(f_i: (X_i,\xi_i) \to (X,\xi))_{i\in I}$ is $\mathcal{G}$-final.

Furthermore, $\mathcal{H}$ is initially dense: If $S = (f_j: X \to \mathcal{F}(V_j))_{j\in J} \in |\mathcal{B}|$, then by condition (1) for semi-closed $\mathcal{F}$-sources, $f_j: S \to \mathcal{H}(V_j)$ is a $\mathcal{B}$-morphism for each $j \in J$ and $(f_j: S \to \mathcal{H}(V_j))_{j\in J}$ is $\mathcal{G}$-initial.

Moreover $\mathcal{H}$ is initiality preserving; for if $(A \xrightarrow{f_i} A_i)_{i\in I}$ is an $\mathcal{F}$-initial source, then $\mathcal{H}(f_i) = \mathcal{F}(f_i): \mathcal{H}(A) \to \mathcal{H}(A_i)$ is a $\mathcal{B}$-morphism for each $i \in I$ by condition (1) for semi-closed $\mathcal{F}$-sources, and $(\mathcal{H}(A) \xrightarrow{\mathcal{H}(f_i)} \mathcal{H}(A_i))_{i\in I}$ is $\mathcal{G}$-initial (for each semi-closed $\mathcal{F}$-source $S = (X \xrightarrow{l_q} \mathcal{F}(U_q))_{q\in Q}$ and each $\mathcal{X}$-morphism $f: X \to \mathcal{F}(A)$ such that $\mathcal{F}(f_i)\circ f: S \to \mathcal{H}(A_i)$ is a $\mathcal{B}$-morphism for each $i \in I$, f belongs to S by condition (2) for semi-closed $\mathcal{F}$-sources and hence $f: S \to \mathcal{H}(A)$ is a $\mathcal{B}$-morphism by condition (1) for semi-closed $\mathcal{F}$-sources).

Consequently, $\mathcal{H}: (\mathcal{A},\mathcal{F}) \to (\mathcal{B},\mathcal{G})$ is an initial completion of $(\mathcal{A},\mathcal{F})$. In order to prove that $\mathcal{H}$ is universal let $(\mathcal{B}',\mathcal{G}')$ be an initially complete category over $\mathcal{X}$ and let $\mathcal{H}': (\mathcal{A},\mathcal{F}) \to (\mathcal{B}',\mathcal{G}')$ be an initiality preserving concrete functor. If $B = (X,\xi) \in |\mathcal{B}|$ with $\xi = \{(f_j,A_j): j \in J\}$, then there exists a unique $\mathcal{G}'$-initial source $(b_j: B'_B \to \mathcal{H}'(A_j))_{j\in J}$ with $\mathcal{G}'(B'_B) = X$ and $\mathcal{G}'(b_j) = f_j$ for each $j \in J$ since $(\mathcal{B}',\mathcal{G}')$ is initially complete. Putting

$$\hat{\mathcal{H}}(B) = B'_B \quad \text{for each } B \in |\mathcal{B}|$$

and

$$\hat{\mathcal{H}}(f) = \mathcal{G}(f) \quad \text{for each } f \in \operatorname{Mor} \mathcal{B}$$

a functor $\hat{\mathcal{H}}: \mathcal{B} \to \mathcal{B}'$ is defined (If $f: B \to \hat{B}$ is a $\mathcal{B}$-morphism, then $G(f)\colon G(B) \to G(\hat{B})$ is obviously a $\mathcal{B}'$-morphism $B'_B \to B'_{\hat{B}})$ which is concrete. Moreover:

$$\hat{H} \circ H = H'$$

(For each $A \in |\mathcal{A}|$, $H(A)$ is the F-source from $F(A)$ of all F-morphisms $F(A) \xrightarrow{g_k} F(A_k)$ for which there exists a unique $\mathcal{A}$-morphism $\bar{g}_k\colon A \to A_k$ with $F(\bar{g}_k) = g_k$.
$(H'(A) \xrightarrow{H'(\bar{g}_k)} H'(A_k))_{k \in K}$ is a G'-initial source (since the identity $H'(A) \to H'(A)$ occurs) with $G'(H'(A)) = F(A)$ and $G'(H'(\bar{g}_k)) = F(\bar{g}_k) = g_k$ for each $k \in K$. Since $(\hat{H}(H(A)) \xrightarrow{b_k} H'(A_k))_{k \in K}$ is the unique G'-initial source with this property, $\hat{H}(H(A)) = H'(A)$. Moreover since $\hat{H}(H(f)) = H'(f)$ for each $\mathcal{A}$-morphism f [because $G'(\hat{H}(H(f))) = G'(H(f)) = F(f)$ and G' is faithful], we have $\hat{H} \circ H = H'$). If $((X,\xi) \xrightarrow{f_i} (X_i,\xi_i))_{i \in I}$ is a G-initial source in $\mathcal{B}$, then ξ is the smallest of those classes ξ' of F-morphisms from X which are $\mathcal{B}$-structures on X (i.e. for which (X,ξ') is semi-closed) and for which all f_i are $\mathcal{B}$-morphisms. Let $\xi = \{(g_j,A_j)\colon j \in J\}$ and $\xi_i = \{(h_{k_i},A_{k_i})\colon k_i \in K_i\}$ for each $i \in I$. Especially all $g_j\colon X \to F(A_j)$ are $\mathcal{B}'$-morphisms $\hat{H}((X,\xi)) = (X,\xi') \to H'(A_j)$ and all $h_{k_i}\colon X_i \to F(A_{k_i})$ are $\mathcal{B}'$-morphisms $\hat{H}((X_i,\xi_i)) = (X_i,\xi'_i) \to H'(A_{k_i})$ by the definition of $\hat{H}$. In order to show that $\hat{H}$ is initiality preserving let ξ_{in} be the initial $\mathcal{B}'$-structure on X with respect to $(X,f_i,(X_i,\xi'_i),I)$. Then

(a) $\xi' \le \xi_{in}$, i.e. $1_X\colon (X,\xi') \to (X,\xi_{in})$ is a $\mathcal{B}'$-morphism (note that $f_i \circ 1_X = f_i\colon (X,\xi') \to (X_i,\xi'_i)$ is a $\mathcal{B}'$-morphism for each $i \in I$ since $\hat{H}$ is a functor).
If ξ_o denotes the class of all F-morphisms $f\colon X \to F(A)$ from X for which $f\colon (X,\xi_{in}) \to H'(A)$ is a $\mathcal{B}'$-morphism, then (X,ξ_o) is semi-closed; for the condition (1) for semi-closed F-sources is fulfilled because H' is a functor and (2) is fulfilled

because H' is initiality preserving. Moreover all $f_i: (X,\xi_o) \to (X_i,\xi_i)$ are B-morphisms since all $h_{k_i} \circ f_i: (X,\xi_{in}) \to H'(A_{k_i})$ are B'-morphisms[as a composite of B'-morphisms]. Since ξ is the smallest class of this kind, it follows that $\xi \subset \xi_o$. Hence $1_X: (X,\xi_o) \to (X,\xi)$ is a B-morphism and thus $\hat{H}((X,\xi_o)) = (X,\xi_o') \xrightarrow{\hat{H}(1_X)=1_X} \hat{H}((X,\xi)) = (X,\xi')$ is a B'-morphism, i.e. $\xi_o' \le \xi'$. On the other hand $1_X: (X,\xi_{in}) \to \hat{H}((X,\xi_o)) = (X,\xi_o')$ is a B'-morphism by the definition of ξ_o and by the initiality of ξ_o', i.e. $\xi_{in} \le \xi_o'$. Consequently,

(b) $\xi_{in} \le \xi'$.

It follows from (a) and (b) that $\xi' = \xi_{in}$. Hence $\hat{H}$ is initiality preserving. The uniqueness of $\hat{H}$ follows immediately from the initial denseness of H.

β) If A is not small but the conglomerate of all semi-closed F-sources in (A,F) is codable by a class, then by 6.1.7 (agreement) one may form the category B of semi-closed F-sources and F-source morphisms and the proof continues as under α).

(1) ⇒ (2): Let $H: (A,F) \to (B,G)$ be a universal initial completion of (A,F), where, without loss of generality, H is considered to be the inclusion functor. For each semi-closed F-source $S = (X,\xi) = (X \xrightarrow{f_i} F(A_i))_{i\in I}$, there is a unique G-initial source $(B_S \xrightarrow{g_i} A_i)_{i\in I}$ with $G(B_S) = X$ and $G(g_i) = f_i$ for each $i \in I$ since (B,G) is initially complete. It suffices to show that the assignment $S \mapsto B_S$ defines an injection from the conglomerate of all semi-closed F-sources to the class $|B|$. For this purpose an initially complete concrete category (B_S,G_S) and an initiality preserving concrete functor $H_S: (A,F) \to (B_S,G_S)$ is defined for each semi-closed F-source $S = (X \xrightarrow{f_i} F(A_i))_{i\in I}$:

$\mathcal{B}_S$-*objects* are triples (B,H,T) where $B \in |\mathcal{B}|$, $H \subset [X,G(B)]_X$ and T is a $\mathcal{B}$-source $(B \xrightarrow{b_l} H(\bar{A}_l))_{l\in L}$ such that the following conditions are satisfied:

(a) For each $h \in H$ and each $l \in L$, $G(b_l)\circ h\colon X \to F(\bar{A}_l)$ belongs to S.

(b) T is maximal with respect to (a), i.e. given any $\mathcal{B}$-morphism $b\colon B \to H(A)$ such that $G(b)\circ h\colon X \to F(A)$ belongs to S for each $h \in H$, then b belongs to T.

(c) H is maximal with respect to (a), i.e. given any X-morphism $\bar{h}\colon X \to G(B)$ such that $G(b_l)\circ\bar{h}\colon X \to F(\bar{A}_l)$ belongs to S for each $l \in L$, then $\bar{h}$ belongs to H.

$\mathcal{B}_S$-*morphisms* $q\colon (B,H,T) \to (B',H',T')$ are $\mathcal{B}$-morphisms $q\colon B \to B'$ such that $t'\circ q$ belongs to T for each t' belonging to T' (under these circumstances $G(q)\circ h \in H'$ for each $h \in H$ since H' is maximal with respect to (a)).

A functor $G_S\colon \mathcal{B}_S \to X$ is defined by $G_S((B,H,T)) = G(B)$ and $G_S(q) = G(q)$ (note that the construction of $\mathcal{B}_S$ is admissible, i.e. the conglomerate of the objects of $\mathcal{B}_S$ is codable by a class since each object (B,H,T) is already uniquely determined by B and H and $|\mathcal{B}|$ as well as the conglomerate of all subsets H is a class). Then $(\mathcal{B}_S,G_S)$ is a concrete category which is initially complete:

Obviously G_S is faithful since G is faithful. Moreover G_S is amnestic since G is amnestic and (c) holds. If one shows the existence of initial $\mathcal{B}_S$-structures (or equivalently: the existence of final $\mathcal{B}_S$-structures) defined correspondingly to Cat top_1) (cf. 1.1.2) where the uniqueness follows from the fact that G_S is amnestic, then G_S is automatically transportable (cf. the corresponding result 1.2.2.7 for topological categories).

Let $Z \in |X|$, $(B_i,H_i,T_i)_{i\in I}$ with $B_i = (X_i,\xi_i)$ for each $i \in I$ be a family of $\mathcal{B}_S$-objects (I is a class!) and $(f_i\colon X_i \to Z)_{i\in I}$ be a family of X-morphisms. If ξ denotes the final $\mathcal{B}$-structure on Z with respect to $(f_i)_{i\in I}$, T the source of all $(Z,\xi) \xrightarrow{b} H(A)$ in $\mathcal{B}$ such that $(X_i,\xi_i) \xrightarrow{b\circ f_i} H(A)$ belongs to T_i for each $i \in I$ and $H = \{h \in [X,Z]_X\colon X \xrightarrow{G(b)\circ h} F(A)$

belongs to S for each b belonging to $T\}$, then (ξ,H,T) is the final $\mathcal{B}_S$-structure on Z with respect to $(f_i)_{i\in I}$ which is easy to check.
A concrete functor $\mathcal{H}_S\colon (\mathcal{A},\mathcal{F}) \to (\mathcal{B}_S,\mathcal{G}_S)$ is defined by $\mathcal{H}_S(A) = (A,H_A,T_A)$ where $H_A = \{h \in [X,\mathcal{F}(A)]_{\mathcal{X}}\colon h$ belongs to $S\}$ and T_A is the source of all $\mathcal{A}$-morphisms starting from A and $\mathcal{H}_S(f) = f$; $\mathcal{H}_S$ is even a full embedding. In order to show that $\mathcal{H}_S$ is initiality preserving let $(A \xrightarrow{g_j} A_j^*)_{j\in J}$ be an $\mathcal{F}$-initial source in $\mathcal{A}$. For the proof of the $\mathcal{G}_S$-initiality of $(\mathcal{H}_S(A) \xrightarrow{g_j} \mathcal{H}_S(A_j^*))_{j\in J}$ consider any $\mathcal{B}_S$-object (B,H,T) and any $\mathcal{X}$-morphism $g\colon \mathcal{G}(B) \to \mathcal{G}(A)$ such that $g_j\circ g\colon (B,H,T) \to \mathcal{H}_S(A_j^*)$ is a $\mathcal{B}_S$-morphism for each $j \in J$. Since $\mathcal{H}$ is initiality preserving, $(A \xrightarrow{g_j} A_j^*)_{j\in J}$ is also $\mathcal{G}$-initial and hence g is a $\mathcal{B}$-morphism because $g_j\circ g$ is a $\mathcal{B}$-morphism for each $j \in J$ (cf. the definition of $\mathcal{B}_S$-morphism). Moreover $a\circ g$ belongs to T for each $\mathcal{A}$-morphism $a\colon A \to A'$; for it suffices to show that for each $h \in H$, $\mathcal{G}(a\circ g)\circ h = \mathcal{F}(a)\circ\mathcal{G}(g)\circ h$ belongs to S which is fulfilled because of the condition (1) for semi-closed sources provided that $\mathcal{G}(g)\circ h$ belongs to S: Since $g_j\circ g$ is a $\mathcal{B}_S$-morphism for each $j \in J$, it follows that $\mathcal{G}(g_j\circ g)\circ h = \mathcal{F}(g_j)\circ\mathcal{G}(g)\circ h \in H_{A_j}$ and thus it belongs to S so that by condition (2) for semi-closed sources, it follows that $\mathcal{G}(g)\circ h$ belongs to S . Therefore it has been shown that $g\colon (B,H,T) \to (A,H_A,T_A)$ is a $\mathcal{B}_S$-morphism.
Since $\mathcal{H}\colon (\mathcal{A},\mathcal{F}) \to (\mathcal{B},\mathcal{G})$ is universal, $\mathcal{H}_S$ can be extended to an initiality preserving concrete functor $\hat{\mathcal{H}}_S\colon (\mathcal{B},\mathcal{G}) \to (\mathcal{B}_S,\mathcal{G}_S)$. Since $T_S = (B_S \xrightarrow{g_i} A_i)_{i\in I}$ with $\mathcal{G}(B_S) = X$ and $\mathcal{G}(g_i) = f_i$ for each $i \in I$ is a $\mathcal{G}$-initial source, it follows that $(\hat{\mathcal{H}}_S(B_S) \xrightarrow{\hat{\mathcal{H}}_S(g_i)} \hat{\mathcal{H}}_S(A_i) = \mathcal{H}_S(A_i))_{i\in I}$ is $\mathcal{G}_S$-initial with $\mathcal{G}_S(\hat{\mathcal{H}}_S(B_S)) = \mathcal{G}(B_S) = X$ and $\mathcal{G}_S(\hat{\mathcal{H}}_S(g_i)) = \mathcal{G}(g_i) = f_i$ for each $i \in I$. It is easy to check that $\hat{\mathcal{H}}_S(B_S) = (B_S,H_S,T_S)$ (note the uniqueness of initial structures) where $H_S = \{h \in [X,X]_{\mathcal{X}}\colon f_i\circ h\colon X \to \mathcal{F}(A_i)$ belongs to S for each $i \in I\}$.

If $S' = (X',\xi') = (X' \xrightarrow{e_k} F(A_k'))_{k\in K}$ is another semi-closed F-source, then there is a unique G-initial source $(B_{S'} \xrightarrow{g_k'} A_k')_{k\in K}$ with $G(B_{S'}) = X'$ and $G(g_k') = e_k$ for each $k \in K$. Moreover if $B_{S'} = B_S$, then $X' = G(B_{S'}) = G(B_S) = X$. Thus $(\hat{H}_S(B_S) \xrightarrow{\hat{H}_S(g_k')} H_S(A_k'))_{k\in K}$ is again G_S-initial with $G_S(\hat{H}_S(B_S)) = G(B_S) = X$ and $G_S(\hat{H}_S(g_k')) = G(g_k') = e_k$ for each $k \in K$. Since $\hat{H}_S(g_k')$ is a B_S-morphism for each $k \in K$ and H_S contains the identity, we have $G(\hat{H}_S(g_k')) = e_k \in H_{A_k'}$ for each $k \in K$, i.e. $(e_k, A_k') \in \xi$ for each $(e_k, A_k') \in \xi'$ so that $\xi' \subset \xi$. By symmetry we get $\xi \subset \xi'$. Thus $\xi' = \xi$ and consequently $S' = S$. Thereby everything has been shown.

6.1.16 Remarks. (1) The universal initial completion $H: (A,F) \to (B,G)$ of a concrete category (A,F) over X, if it exists, is the *largest initial completion of* (A,F). [If $H': (A,F) \to (B',G')$ is any initial completion of (A,F), then a concrete functor $\bar{H}: (B',G') \to (B,G)$ is defined by

$$\bar{H}(B') = (G'(B'), \{(G'(a),A): a: B' \to H'(A) \text{ is a } B'\text{-morphism}\})$$

and

$$\bar{H}(f) = G'(f).$$

[For the proof of condition (2) concerning the semi-closedness of $\bar{H}(B')$ is used that H' is initiality preserving!]
$\bar{H}$ is full; for if $g: \bar{H}(B') \to \bar{H}(\hat{B}')$, then $g: G'(B') \to G'(\hat{B}')$ is an F-source-morphism, i.e. $G'(f_i)\circ g: G'(B') \to F(A_i)$ is a B'-morphism $B' \to H'(A_i)$ for each $i \in I$ provided that $(f_i: \hat{B}' \to H'(A_i))_{i\in I}$ is the source of all B'-morphisms with co-domain in $H'(A)$ (this source is G'-initial since H' is initially dense), and thus $g: B' \to \hat{B}'$ is a B'-morphism. Then $\bar{H}$ is also an embedding (cf. the corresponding argumentation concerning the construction of the MacNeille completion $H: (A,F) \to (B,G)$). Moreover $\bar{H}\circ H' = H$ since H' is full. Consequently, $H' \le H$.]

(2) Dualizing the construction of the universal initial completion $H: (A,F) \to (B,G)$ one obtains the *universal final completion* $\bar{H}: (A,F) \to (\bar{B},\bar{G})$, i.e. a final completion of (A,F) such that for each finally complete category (B',G') over X and each finality preserving concrete functor $H': (A,F) \to (B',G')$, there is a unique finality preserving concrete functor $\hat{H}: (\bar{B},\bar{G}) \to (B',G')$ with $\hat{H} \circ \bar{H} = H'$. If the universal final completion exists, then it is the category of semi-closed F-sinks and simultaneously the largest final completion of (A,F) . In the following it is shown by means of an example (6.1.17 (1)) that the universal initial completion is not self-dual(in contrast to the MacNeille completion), i.e. that it generally does not coincide with the universal final completion.

<u>6.1.17 Examples</u>. (1) According to 6.1.10 (1) every ordered set $(S,\leq)$ may be considered to be a small category (A,F) over the category X with exactly one object X and exactly one morphism 1_X . (A,F) has a MacNeille completion as well as a universal initial (and a universal final) completion. Let

$H: (A,F) \to (B,G)$ be the universal initial completion of (A,F) .	$\bar{H}: (A,F) \to (\bar{B},\bar{G})$ be the universal final completion of (A,F) .

If one identifies the semi-closed

F-sources	F-sinks

with those subsets M of S satisfying the following properties

1) $s \in M$ and $s \leq s'$ implies $s' \in M$	1') $s \in M$ and $s' \leq s$ implies $s' \in M$
2) $s \in S$ and $s = \inf_S N$ with $N \subset M$ implies $s \in M$,	2') $s \in S$ and $s = \sup_S N$ with $N \subset M$ implies $s \in M$,

then

$(\mathcal{B},\mathcal{G})$ | $(\bar{\mathcal{B}},\bar{\mathcal{G}})$

can be considered to be the set

UI(S) | UF(S)

of all subsets of S determined in this way ordered by

"$\supset$" | "$\subset$"

(note:

"$B \leq B'$ | $\bar{B} \leq' \bar{B}'$

if and only if there exists a

$\mathcal{B}$-morphism $f: B \to B'$ | $\bar{\mathcal{B}}$-morphism $\bar{f}: \bar{B} \to \bar{B}'$"

defines an order relation since

$\mathcal{G}: \mathcal{B} \to \mathcal{X}$ | $\bar{\mathcal{G}}: \bar{\mathcal{B}} \to \mathcal{X}$

is faithful and amnestic.) and

$\mathcal{H}: (\mathcal{A},\mathcal{F}) \to (\mathcal{B},\mathcal{G})$ | $\bar{\mathcal{H}}: (\mathcal{A},\mathcal{F}) \to (\bar{\mathcal{B}},\bar{\mathcal{G}})$

yields an embedding

$\Psi_S: (S,\leq) \to (UI(S),\supset)$ | $\bar{\Psi}_S: (S,\leq) \to (UF(S),\subset)$

from $(S,\leq)$ to the complete lattice

$(UI(S),\supset)$ | $(UF(S),\subset)$

defined as follows

$\Psi_S(s) = \{x \in S: x \geq s\}$ | $\bar{\Psi}_S(s) = \{x \in S: x \leq s\}$ [Construction by Derdêrian and Ringleb (1969)]

If $(S,\leq)$ has the diagram

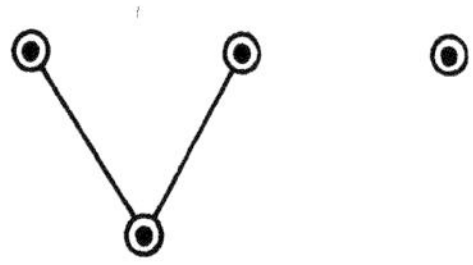

then the following diagrams show that the MacNeille completion, the universal initial completion and the universal final completion are pairwise distinct:

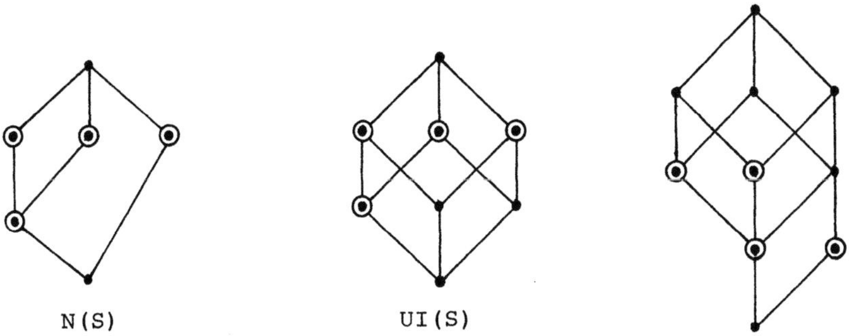

N(S) UI(S) UF(S)

(2) a) The category Top of topological spaces (and continuous maps) is the universal initial completion as well as the MacNeille completion of the category A of T_0-spaces (and continuous maps) [If $F: A \to \underline{Set}$ denotes the forgetful functor, then an F-morphism $e: X \to F(A)$ is surjective if and only if e is *semi-universal*, i.e. if for any F-initial source $(A' \xrightarrow{h_i} A_i)_{i \in I}$, any source $(A \xrightarrow{f_i} A_i)_{i \in I}$ in A and any F-morphism $X \xrightarrow{f} F(A')$ with $F(h_i) \circ f = F(f_i) \circ e$, there exists a unique A-morphism $g: A \to A'$ such that the diagram

$$\begin{array}{ccc} X & \xrightarrow{\quad e \quad} & F(A) \\ {\scriptstyle f}\downarrow & \swarrow {\scriptstyle F(g)} & \downarrow {\scriptstyle F(f_i)} \\ F(A') & \xrightarrow[\quad]{F(h_i)} & F(A_i) \end{array}$$

commutes for each $i \in I$ ("⇒": Choose $h: F(A) \to X$ such that $e \circ h = 1_{F(A)}$, then $f \circ h$ completes the above square commutatively and from the F-initiality of $(A' \xrightarrow{h_i} A_i)_{i \in I}$, it follows that there exists some $g: A \to A'$ with the desired property. The uniqueness of g follows from the faithfulness of F and the fact that e is surjective. "⇐": Let $X \xrightarrow{e} F(A) =$ $= X \xrightarrow{g} F(B) \xrightarrow{F(m)} F(A)$ be the (epi,mono)-factorization in

Set with $(m: B \to A)$ $\mathcal{F}$-initial in $\mathcal{A}$ (cf. 5.3.3. IS_1')). Then there exists a unique $\mathcal{A}$-morphism $f: A \to B$ such that the diagram

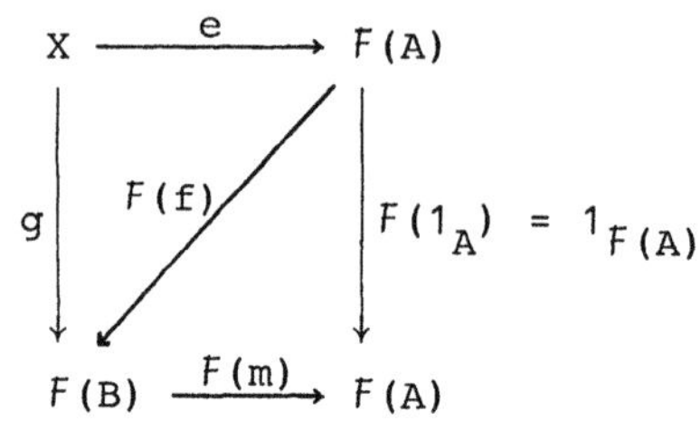

commutes. $\mathcal{F}(m)$ is surjective because of $\mathcal{F}(m) \circ \mathcal{F}(f) = 1_{\mathcal{F}(A)}$. Thus $e = \mathcal{F}(m) \circ g$ is likewise surjective.) If an equivalence relation $\sim$ on the class of all tripels (X,e,B), where $X \in |\underline{Set}|$ and $e: X \to \mathcal{F}(B)$ is surjective (= semi-universal), is defined by

$(X,e,B) \sim (X',e',B')$ iff $X = X'$ and there exists an isomorphism $h: B \to B'$ with $\mathcal{F}(h) \circ e = e'$

and if one assigns to each equivalence class $[(X,e,B)]$ the $\mathcal{F}$-source of all $\mathcal{F}$-morphisms $f: X \to \mathcal{F}(A)$ such that there exists some $k: B \to A$ with $\mathcal{F}(k) \circ e = f$, then this assignment is well-defined (i.e. independent of the representative selected) and yields a bijection in the conglomerate of all semi-closed $\mathcal{F}$-sources (the condition (2) for semi-closed $\mathcal{F}$-sources results from the semi-universality of the surjective map $e: X \to \mathcal{F}(B)$; the assignment is surjective since by IS_1') for the initial structured category $(\mathcal{A},\mathcal{F})$, there is an (epi,mono-source)-factorization $X \xrightarrow{f_i} \mathcal{F}(A_i) = X \xrightarrow{e} \mathcal{F}(B) \xrightarrow{\mathcal{F}(g_i)} \mathcal{F}(A_i)$ for each semi-closed source $(X \xrightarrow{f_i} \mathcal{F}(A_i))_{i\in I}$ where $(g_i: B \to A_i)_{i \in I}$ is $\mathcal{F}$-initial). Choosing a system of representatives $|\mathcal{C}|$ with respect to $\sim$ a category $\mathcal{C}$ is obtained by choosing, as morphisms from (X,e,B) to (X',e',B'), all pairs (f,g) with $f: X \to X'$, $g: B \to B'$ and $e' \circ f = \mathcal{F}(g) \circ e$ and defining the composition componentwise. The above bijection can be extended to an isomorphism between $\mathcal{C}$ and the universal initial completion of $\mathcal{A}$. On the other hand the objects of $\mathcal{C}$ are in a bijective correspondence to the objects of Top (there is assigned to each topological space $(X,\mathcal{X})$ the representative of the equivalence class containing the $\mathcal{A}$-reflection of $(X,\mathcal{X})$; the inverse assignment is obtained by assigning to each $(X,e,B) \in |\mathcal{C}|$ the topological space $(X,\mathcal{X}_e)$, where $\mathcal{X}_e$ is the initial topology on X with respect to (X,e,B)). This correspondence can be extended to an isomorphism between $\mathcal{C}$ and Top.

Similar results are obtained for the MacNeille completion of $\mathcal{A}$ by replacing "semi-universal" by "semi-final" and showing that an $\mathcal{F}$-morphism $e: X \to \mathcal{F}(A)$ is surjective if and only if e is *semi-final*, i.e. if there is an $\mathcal{F}$-sink $(\mathcal{F}(A_i) \xrightarrow{f_i} X)_{i\in I}$ for which there exists an $\mathcal{A}$-sink $(A_i \xrightarrow{g_i} A)_{i\in I}$ with $e\circ f_i = \mathcal{F}(g_i)$ for each $i \in I$ such that for any $\mathcal{A}$-sink $(A_i \xrightarrow{\tilde{g}_i} \tilde{A})_{i\in I}$ and any $\mathcal{F}$-morphism $X \xrightarrow{\tilde{e}} \mathcal{F}(\tilde{A})$ with $\tilde{e}\circ f_i = \mathcal{F}(\tilde{g}_i)$ for each $i \in I$, there exists a unique $\mathcal{A}$-morphism $g: A \to \tilde{A}$ with $\mathcal{F}(g)\circ e = \tilde{e}$, i.e. such that the diagram

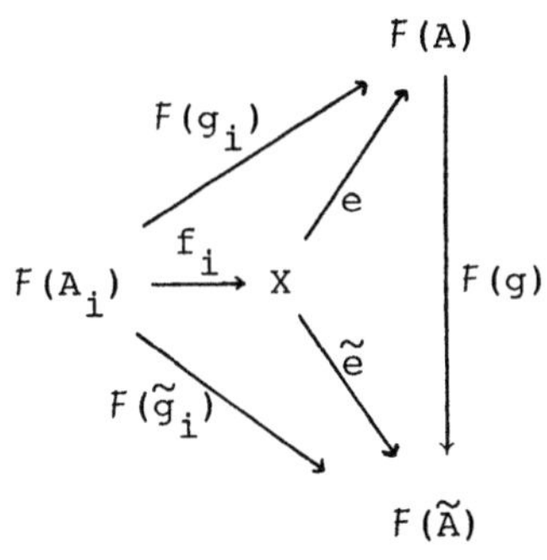

commutes for each $i \in I$ ("⇒": Choose $f: \mathcal{F}(A) \to X$ such that $e\circ f = 1_{\mathcal{F}(A)}$. Let $(\mathcal{F}(A_i) \xrightarrow{f_i} X)_{i\in I}$ be the $\mathcal{F}$-sink of all $\mathcal{F}$-morphisms $\mathcal{F}(A_k) \xrightarrow{f_k} X$ such that $e\circ f_k$ is constant together with f.

A map $h\colon \mathcal{F}(A) \to \mathcal{F}(\tilde{A})$ is definable provided that for each pair $(x,y) \in X \times X$ the equality $e(x) = e(y)$ implies $\tilde{e}(x) = \tilde{e}(y)$ [then $h = \mathcal{F}(g)$ with $g \in [A,\tilde{A}]_{\mathcal{A}}$ since the identity $1_{\mathcal{F}(A)}$ appears in the above diagram; g is uniquely determined because e is surjective and $\mathcal{F}$ is faithful]. If $\tilde{e}(x) \neq \tilde{e}(y)$, then without loss of generality there would be an open neighbourhood O of $\tilde{e}(x)$ not containing $\tilde{e}(y)$. Further let us choose some $A_i \in |\mathcal{A}|$ being not discrete as well as some subset $M \subset \mathcal{F}(A_i)$ being not open. Then $f_i\colon A_i \to X$ defined by

$$f_i(t) = \begin{cases} x & \text{for } t \in M \\ y & \text{for } t \notin M \end{cases}$$

would have the property that $\tilde{e} \circ f_i$ is continuous [since $e \circ f_i$ is constant] which is impossible because $(\tilde{e} \circ f_i)^{-1}[O] = M$ is not open.

"$\Leftarrow$": It suffices to show that "semi-final" implies "semi-universal" which is easy to verify.)].

b) The category Unif of uniform spaces (and uniformly continuous maps) is the universal initial completion as well as the MacNeille completion of the category $\mathcal{A}$ of separated uniform spaces (and uniformly continuous maps) [analogously to a)].

(3) Choose $(\mathcal{A},\mathcal{F})$ as under 6.1.10 (3) b). Then $(\mathcal{A},\mathcal{F})$ has no universal initial completion (resp. universal final completion) [there is a 1-1-correspondence between the conglomerate of all semi-closed $\mathcal{F}$-sources (resp. $\mathcal{F}$-sinks) and the conglomerate of all subclasses of the class Ω].

(4) Let $(\mathcal{A},\mathcal{F})$ be the concrete category of compact Hausdorff spaces (and continuous maps) over the category Set. Then the category Prox considered as a concrete category over Set is the universal initial completion of $(\mathcal{A},\mathcal{F})$ (without proof).

6.1.18 Remark. The MacNeille completion of the category of compact Hausdorff spaces (and continuous maps) is the category CGUnif of compactly generated uniform spaces (a uniform space is called compactly generated provided that a pseudometric on this space is uniformly continuous if and only if it is uniformly continuous on each compact subset). Simultaneously CGUnif is the cartesian closed topological hull of the category of compact spaces; for each nontrivial (i.e. there exists at least one $A \in |\mathcal{A}|$ with $\mathcal{F}(A) \neq \emptyset$) concrete category $(\mathcal{A},\mathcal{F})$ over Set having finite concrete products (i.e. $\mathcal{A}$ has finite products preserved by $\mathcal{F}$) and constant morphisms (i.e. each constant map $f\colon \mathcal{F}(A) \to \mathcal{F}(B)$ is an $\mathcal{A}$-morphism $f\colon A \to B$), this hull is defined as a cartesian closed topological category $(\mathcal{B},\mathcal{G})$ containing $(\mathcal{A},\mathcal{F})$ as a full concrete subcategory (i.e. the inclusion functor $\mathcal{I}\colon \mathcal{A} \to \mathcal{B}$ is a full concrete embedding) such that the following are satisfied:

(1) $(\mathcal{A},\mathcal{F})$ is finally dense in $(\mathcal{B},\mathcal{G})$ (i.e. $\mathcal{I}\colon (\mathcal{A},\mathcal{F}) \to (\mathcal{B},\mathcal{G})$ is finally dense).

(2) The power-objects[45] B^A of $\mathcal{A}$-objects A,B are initially dense in $\mathcal{B}$.

It is an open question under which conditions to a concrete category $(\mathcal{A},\mathcal{F})$ its MacNeille completion is cartesian closed.

6.2. Completion of nearness spaces

6.2.1 Definitions. Let (X,μ) be a nearness space and ξ the corresponding set of all near collections in (X,μ).

1) A non-empty[46] subset $\mathcal{A}$ of $\mathcal{P}(X)$ is called a cluster provided that $\mathcal{A}$ is a maximal element of the set ξ, ordered by inclusion.

2) A point $x \in X$ is called an adherencepoint of a subset $\mathcal{A}$ of $\mathcal{P}(X)$ provided that $x \in \bigcap_{A\in\mathcal{A}} \bar{A}$ (where the closure is formed in the underlying topological space $(X,\mathcal{X}_\mu)$).

---

45) cf. 4.1.1.(2).

46) Superfluous if $X \neq \emptyset$.

3) (X,μ) is called complete provided that every cluster has an adherencepoint.

6.2.2 Remark. (1) As well-known a uniform space is said to be complete iff every Cauchy filter converges. If a nearness space (X,μ) is called separated provided that for each near Cauchy system $\mathcal{A}$ in (X,μ), the collection $\mathcal{B} = \{B \subset X: \mathcal{A} \cup \{B\}$ is near in $(X,\mu)\}$ is near in (X,μ), then especially *every uniform space is separated* [If (X,μ) is a uniform nearness space, $\mathcal{A}$ a near Cauchy system, $\mathcal{B} = \{B \subset X: \mathcal{A} \cup \{B\}$ is near in $(X,\mu)\}$ and $\mathcal{U} \in \mu$, then there exists some $\mathcal{V} \in \mu$ with $\mathcal{V} *\!< \mathcal{U}$. Since $\mathcal{A}$ is a Cauchy system, there exist $A \in \mathcal{A}$ and $V \in \mathcal{V}$ with $A \subset V$. Furthermore, there is some $U \in \mathcal{U}$ with $St(V,\mathcal{V}) \subset U$. Then $\{A, X \setminus U\}$ is not near. Thus $U \cap B \neq \emptyset$ for each $B \in \mathcal{B}$; for if $U \cap B = \emptyset$, i.e. $B \subset X \setminus U$, for some $B \in \mathcal{B}$, then $\{A,B\}$ would not be near in contradiction to $\{A,B\} << \mathcal{A} \cup \{B\}$. Consequently, $\mathcal{B}$ is near in (X,μ).]. Moreover *a separated nearness space* X *is complete if and only if every Cauchy filter converges.* [1. "⇒": If $\mathcal{F}$ is a Cauchy filter on X, then by 3.2.3.11 and 3.2.3.12 $\mathcal{A} = \sec \mathcal{F}$ is a near grill. Because of $\mathcal{F} \subset \sec \mathcal{F}$ ($\mathcal{F}$ is a filter!) $\mathcal{F}$ is also near so that $\mathcal{A} = \sec \mathcal{F}$ is a Cauchy system. Furthermore, $\mathcal{B} = \{B \subset X: \mathcal{A} \cup \{B\}$ is near in $X\}$ is near in X (since X is separated) and we have $\mathcal{A} \subset \mathcal{B}$. Obviously $\mathcal{B}$ is a maximal near collection. Since $\mathcal{A}$ is a near Cauchy system, $\mathcal{B}$ is not empty. Hence $\mathcal{B}$ is a cluster. By assumption $\mathcal{B}$ has an adherencepoint $x \in X$. Especially x is an adherencepoint of $\mathcal{A} = \sec \mathcal{F}$. Consequently, $\mathcal{U}(x) \subset \sec \mathcal{A} = \sec^2 \mathcal{F} = \mathcal{F}$ ($\mathcal{F}$ is a filter!), i.e. $\mathcal{F}$ converges to x.

2. "⇐": Let $\mathcal{B}$ be a cluster in X. Then $\mathcal{B}$ is a near grill (cf. the last part of 3.2.3.16) and by 3.2.3.11 and 3.2.3.12 $\sec \mathcal{B}$ is a Cauchy filter converging to some $x \in X$ by assumption. Thus $\mathcal{B}$ has the adherencepoint x.].

Therefore the concept of completeness defined above is a suitable generalization of the concept of completeness for uniform spaces.

(2) If (X,μ) is a topological nearness space, then $\mathcal{A} \subset P(X)$ is near if and only if $\bigcap_{A\in\mathcal{A}} \bar{A} \neq \emptyset$, i.e. if $\mathcal{A}$ has an adherencepoint. Thus, *every topological nearness space is complete.*

<u>6.2.3 Theorem</u>. Let (X,μ) be a nearness space. Put $X^* = X \cup X'$, where X' is the set of all clusters in X without an adherencepoint. If μ^* denotes the set of all covers $\mathcal{U}^*$ of X^* for which there exists some $\mathcal{U} \in \mu$ with $o(\mathcal{U}) < \mathcal{U}^*$ where $o(\mathcal{U}) = \{U^o \cup \{x^* \in X': (X \setminus U) \notin x^*\}: U \in \mathcal{U}\}$, then (X^*,μ^*) is a complete nearness space containing (X,μ) as a dense subspace.

<u>Proof</u>. Put for each $A \subset X$,

$$o(A) = A^o \cup \{x^* \in X': (X \setminus A) \notin x^*\} .$$

(1) (X^*,μ^*) is a nearness space.
We have $o(X) = X^*$ and from $\{X\} \in \mu$, it follows that $o(\{X\}) = \{X^*\} \in \mu^*$. Thus μ^* is a non-empty set of non-empty covers.

N_1) holds by definition.

N_2) If $\mathcal{U}^*, \mathcal{W}^* \in \mu^*$, then there exist $\mathcal{U}, \mathcal{W} \in \mu$ with $o(\mathcal{U}) < \mathcal{U}^*$ and $o(\mathcal{W}) < \mathcal{W}^*$. Hence

$$o(\mathcal{U}) \wedge o(\mathcal{W}) < \mathcal{U}^* \wedge \mathcal{W}^* ,$$

so that $\mathcal{U}^* \wedge \mathcal{W}^* \in \mu^*$, if it can be shown that the following are satisfied:

(a) $o(\mathcal{U}) \wedge o(\mathcal{W}) = o(\mathcal{U} \wedge \mathcal{W})$

(b) $o(\mathcal{W}) \in \mu^*$ for each $\mathcal{W} \in \mu$.

In order to prove (a) we show that for arbitrary subsets A and B of X , $o(A \cap B) = o(A) \cap o(B)$. Let $x \in X$. Then $x \in o(A \cap B)$ if and only if $x \in (A \cap B)^o = A^o \cap B^o$, i.e. $x \in o(A) \cap o(B)$. Now let $x^* \in X'$. Then $x^* \in o(A \cap B)$ if and only if $X \setminus (A \cap B) = (X \setminus A) \cup (X \setminus B) \notin x^*$. Since x^* is a grill (cf. the last part of 3.2.3.16), this assertion is equivalent to the fact, that $X \setminus A \notin x^*$ and $X \setminus B \notin x^*$, i.e. $x^* \in o(A) \cap o(B)$.

In order to prove (b) it suffices to show that $o(\mathcal{W})$ is a cover of X^*. Since $\mathcal{W} \in \mu$, $\{\mathrm{int}_\mu W: W \in \mathcal{W}\} \in \mu$ by N_3). Hence $\{\mathrm{int}_\mu W: W \in \mathcal{W}\}$ is a cover of X. Thus for each $x \in X$, there is some $W \in \mathcal{W}$ with $x \in \mathrm{int}_\mu W = W^o \subset o(W)$. If $x^* \in X'$, then $\{X \setminus W: W \in \mathcal{W}\}$ is not near (because of $\mathcal{W} \in \mu$) and hence not a subset of X^*, i.e. there is some $W \in \mathcal{W}$ with $X \setminus W \notin x^*$. Consequently $x^* \in o(W)$.

N_3) If $\mathcal{U}^* \in \mu^*$, then there exists some $\mathcal{U} \in \mu$ with $o(\mathcal{U}) < \mathcal{U}^*$. In order to prove that $\{\mathrm{int}_{\mu^*} U^*: U^* \in \mathcal{U}^*\} \in \mu^*$ is valid it suffices to show that $o(\mathcal{U}) = \{\mathrm{int}_{\mu^*} o(U): U \in \mathcal{U}\}$. We even show

$$o(U) = \mathrm{int}_{\mu^*}\, o(U) \quad \text{for each} \quad U \subset X .$$

If $x \in o(U)$ and $x \in X$, then $x \in U^o = \mathrm{int}_\mu U$ so that $\mathcal{U} = \{U, X \setminus \{x\}\} \in \mu$. Thus $\mathcal{U}^* = \{X^* \setminus \{x\}, o(U)\} \in \mu^*$, i.e. $x \in \mathrm{int}_{\mu^*} o(U)$ since $o(\mathcal{U}) < \mathcal{U}^*$. If $x^* \in o(U)$ and $x^* \in X'$, then $X \setminus U \notin x^*$. Since x^* is maximal, it follows that $x^* \cup \{X \setminus U\}$ is not near in (X,μ) so that $\mathcal{U} = \{X \setminus A: A \in x^*\} \cup \{U\} \in \mu$. Then $\mathcal{U}^* = \{X^* \setminus \{x^*\}, o(U)\} \in \mu^*$, i.e. $x^* \in \mathrm{int}_{\mu^*} o(U)$ because of $o(\mathcal{U}) < \mathcal{U}^*$ (note that for each $A \in x^*$, we have $o(X \setminus A) \subset X^* \setminus \{x^*\}$ because $x^* \in o(X \setminus A)$ implies $A \notin x^*$ by definition).

(2) (X,μ) is a subspace of (X^*,μ^*), i.e. $\mu = \{\mathcal{U}^* \wedge \{X\}: \mathcal{U}^* \in \mu^*\}$.

If $\mathcal{U}^* \in \mu^*$, then there exists some $\mathcal{U} \in \mu$ with $o(\mathcal{U}) < \mathcal{U}^*$, where without loss of generality $\mathcal{U}$ is considered to be an open cover (note N_3) and the definition of o). Thus $\mathcal{U} < \{U^* \cap X: U^* \in \mathcal{U}^*\}$ since for each $U \in \mathcal{U}$, there exists some $U^* \in \mathcal{U}^*$ with $U \subset o(U) \subset U^*$. Hence $\mathcal{U}^* \wedge \{X\} \in \mu$. If $\mathcal{W} \in \mu$, then $o(\mathcal{W}) < \mathcal{W}^* = \{V \cup X': V \in \mathcal{W}\}$, i.e. $\mathcal{W}^* \in \mu^*$. Moreover $\mathcal{W} = \mathcal{W}^* \wedge \{X\}$.

(3) X is dense in (X^*,μ^*).

We show that $\bar{A}^{X^*} = \bar{A}^X \cup \{x^* \in X': A \in x^*\}$ for each $A \subset X$ (then especially $\bar{X}^{X^*} = X \cup \{x^* \in X': X \in x^*\} = X \cup X' = X^*$). We have $\bar{A}^{X^*} = X^* \setminus \mathrm{int}_{\mu^*}(X^* \setminus A) = X^* \setminus \mathrm{int}_{\mu^*}((X \setminus A) \cup X') = X^* \setminus o(X \setminus A) =$

$= X^* \setminus ((X \setminus A)^{o} \cup \{x^* \in X': A \notin x^*\}) =$
$= (X \setminus (X \setminus A)^{o}) \cup \{x^* \in X': A \in x^*\} = \bar{A}^{X} \cup \{x^* \in X': A \in x^*\}$
using the fact that for each $U \subset X$ holds

$$o(U) = \text{int}_{\mu^*} (U \cup X') .$$

(If $x \in \text{int}_{\mu^*}(U \cup X')$, then $\mathcal{U}^* = \{X^* \setminus \{x\} , U \cup X'\} \in \mu^*$. Hence there exists some $\mathcal{W} \in \mu$ with $o(\mathcal{W}) < \mathcal{U}^*$. In order to show that $x \in \text{int}_{\mu^*} o(U) = o(U)$ (i.e. $\{X^* \setminus \{x\} , o(U)\} \in \mu^*$) it suffices to prove that $o(\mathcal{W}) < \{X^* \setminus \{x\} , o(U)\}$ is valid. If $V \in \mathcal{W}$ and $x \notin o(V)$, then $o(V) \subset X^* \setminus \{x\}$. If $x \in o(V)$, then obviously $o(V) \subset U \cup X'$ because of $o(\mathcal{W}) < \mathcal{U}^*$, so that $V^{o} \subset U$. Hence $V^{o} \subset U^{o}$ and thus $\overline{X \setminus U} = X \setminus U^{o} \subset X \setminus V^{o} = \overline{X \setminus V}$. Furthermore, if $x^* \in X'$ with $(X \setminus V) \notin x^*$, then $(X \setminus U) \notin x^*$ [otherwise $\overline{X \setminus V} \in x^*$ and thus $(X \setminus V) \in x^*$[47)]]. Therefore $o(V) \subset o(U)$. Consequently, $\text{int}_{\mu^*}(U \cup X') \subset \text{int}_{\mu^*} o(U) = o(U)$. The converse is trivial since $o(U) \subset U \cup X'$.)

(4) (X^*,μ^*) is complete.

(a) At first we show: $\mathcal{A}^* \subset \mathcal{P}(X^*)$ is near in (X^*,μ^*) if and only if $\mathcal{A} = \{A \subset X$: there exists some $A^* \in \mathcal{A}^*$ with $A^* \subset \bar{A}^{X^*}\}$ is near in (X,μ) . By definition $\mathcal{A}^*$ is near in (X^*,μ^*) if and only if $\{X^* \setminus A^*: A^* \in \mathcal{A}^*\} \notin \mu^*$, i.e. $\mathcal{U} = \{U \subset X$: there exists some $A^* \in \mathcal{A}^*$ with $o(U) \subset X^* \setminus A^*\} \notin \mu$ or equivalently $\mathcal{B} = \{X \setminus U: U \in \mathcal{U}\}$ is near in (X,μ) . $A \subset X$ belongs to $\mathcal{B}$ if and only if $X \setminus A \in \mathcal{U}$, i.e. if there exists some $A^* \in \mathcal{A}^*$ with $o(X \setminus A) = \text{int}_{\mu^*}((X \setminus A) \cup X') =$ $= \text{int}_{\mu^*} (X^* \setminus A) \subset X^* \setminus A^*$. This is equivalent to $A^* \subset (X^* \setminus \text{int}_{\mu^*}(X^* \setminus A)) = \bar{A}^{X^*}$. Hence $\mathcal{B} = \mathcal{A}$.

(b) Let $\mathcal{A}^*$ be a cluster in (X^*,μ^*) . Then by (a), $\mathcal{A} = \{A \subset X$: there exists some $A^* \in \mathcal{A}^*$ with $A^* \subset \bar{A}^{X^*}\} =$ $= \{A \subset X: \bar{A}^{X^*} \in \mathcal{A}^*\} = \{A \subset X: A \in \mathcal{A}^*\}$[47)] $= \mathcal{A}^* \cap \mathcal{P}(X)$ is a cluster in (X,μ) ; for $\mathcal{A}$ is near and if $\mathcal{B} \supset \mathcal{A}$ is near, then $\mathcal{C} = \mathcal{A} \cup \{B\}$ is near for each $B \in \mathcal{B}$ and consequently $\mathcal{A}^* \cup \{B\} = \mathcal{D}^*$ is near since $\mathcal{D} = \{D \subset X$: there exists some $D^* \in \mathcal{D}^*$ with $D^* \subset \bar{D}^{X^*}\}$ is near (we have $\bar{\mathcal{D}} = \{\bar{D}^{X}: D \in \mathcal{D}\} << \mathcal{C}$ because for each $D \in \mathcal{D}$,

---

47) If $\mathcal{A}$ is a cluster in (X,μ) and $\bar{B} \in \mathcal{A}$, then $\mathcal{A} \cup \{\bar{B}\}$ and $\{\bar{A}: A \in \mathcal{A}\} \cup \{\bar{B}\}$ are near and by N_3) $\mathcal{A} \cup \{B\}$ is near so that $B \in \mathcal{A}$ since $\mathcal{A}$ is maximal.

there exists some $D^* \in \mathcal{D}^*$ with $D^* \subset \bar{D}^{X^*}$:

case 1: $D^* \in \mathcal{A}^*$. Then $D \in \mathcal{A} \subset \mathcal{C}$ and $D \subset \bar{D}^X$.

case 2: $D^* = B$. Then $B \in \mathcal{C}$ is contained in $\bar{D}^X$ because of $B \subset X$.

Then by N_1) $\bar{\mathcal{D}}$ is near so that $\mathcal{D}$ is near because of N_3)); hence $B \in \mathcal{A}^*$ (since $\mathcal{A}^*$ is maximal), i.e. $B \in \mathcal{A}$ (because of $B \subset X$) so that $\mathcal{B} = \mathcal{A}$. If $\mathcal{A}$ has an adherencepoint x in (X,μ) , then $\mathcal{M} = \{\bar{A}: A \in \mathcal{A}\} \cup \{\overline{\{x\}}\}$ is near since for each $\mathcal{U} \in \mu$, there exists some $U \in \mathcal{U}$ with $x \in U$, i.e. U meets all elements of $\mathcal{M}$. Then by N_3) $\mathcal{A} \cup \{\{x\}\}$ is near. Since $\mathcal{A}$ is maximal we have $\{x\} \in \mathcal{A}$ and thus $\{x\} \in \mathcal{A}^*$. Therefore x is an adherencepoint[48] of $\mathcal{A}^*$ in (X^*,μ^*) . If $\mathcal{A}$ has not an adherencepoint in (X,μ) , then $\mathcal{A} = x^* \in X'$. By (a) we obtain that $\mathcal{A}^* \cup \{\{x^*\}\}$ is near in (X^*,μ^*) ; for $\{A \subset X\colon$ there exists some $A^* \in \mathcal{A}^*$ with $A^* \subset \bar{A}^{X^*}\} \cup \{B \subset X\colon x^* \in \bar{B}^{X^*}\} = \mathcal{A} \cup \{B \subset X\colon B \in x^*\} = \mathcal{A} \cup \mathcal{A} = \mathcal{A}$ is near in (X,μ) . Since $\mathcal{A}^*$ is maximal, it follows that $\{x^*\} \in \mathcal{A}^*$. Thus x^* is an adherencepoint of $\mathcal{A}^*$.

6.2.4 Definition. Let (X,μ) be a nearness space and (X^*,μ^*) the complete nearness space constructed in 6.2.3. Then the inclusion map $j_X\colon (X,\mu) \to (X^*,\mu^*)$ is called the canonical completion of (X,μ) . Occasionally (X^*,μ^*) is already called the canonical completion of (X,μ) .

6.2.5 Remarks. (1) *If* (X,μ) *is a uniform* N_1*-space, then* (X^*,μ^*) *is also a uniform* N_1*-space* [1. Let $x^* \in X^*$. If $x^* = x \in X$, then $X \setminus \{x\}$ is open in (X,X_μ) and $o(X \setminus \{x\}) = (X \setminus \{x\}) \cup \{x^* \in X'\colon \{x\} \notin x^*\} = (X \setminus \{x\}) \cup X' = X^* \setminus \{x\}$ is open in (X^*,X_{μ^*}) , i.e. $\{x\}$ is closed in (X^*,X_{μ^*}) . If $x^* \in X'$, then

48) If $\mathcal{B}$ is near in some nearness space (Y,ν) and $\{y\} \in \mathcal{B}$ with $y \in Y$, then for each $B \in \mathcal{B}$, $\{B,\{y\}\} \subset \mathcal{B}$ is near by N_1), i.e. $y \in \bar{B}$. Thus y is an adherencepoint of $\mathcal{B}$.

$$\bigcap_{A \in x^*} \bar{A}^{X^*} = \bigcap_{A \in x^*} \bar{A}^{X} \cup \bigcap_{A \in x^*} \{y^* \in X' : A \in y^*\} =$$

$= \emptyset \cup \{y^* \in X' : x^* \subset y^*\} = \emptyset \cup \{x^*\} = \{x^*\}$ is closed in (X^*, X_{μ^*}) [as an intersection of sets which are closed in (X^*, X_{μ^*})]. Consequently, (X^*, μ^*) is an N_1-space.

2. For each $\mathcal{U}^* \in \mu^*$, there is some $\mathcal{U} \in \mu$ with $o(\mathcal{U}) < \mathcal{U}^*$ as well as some $\mathcal{V} \in \mu$ with $\mathcal{V} *< \mathcal{U}$. If we can show that $o(\mathcal{V}) *< o(\mathcal{U})$, then (X^*, μ^*) is uniform. If $V \in \mathcal{V}$, then there exists some $U \in \mathcal{U}$ with $St(V, \mathcal{V}) \subset U$. If $V' \in \mathcal{V}$ with $o(V') \cap o(V) \neq \emptyset$, then $V' \cap V \neq \emptyset$; for either there exists some $x \in V^o \cap V'^o \subset V \cap V'$, i.e. $V' \cap V \neq \emptyset$, or there exists some $x^* \in X'$ with $(X \setminus V') \notin x^*$ and $(X \setminus V) \notin x^*$, hence (because x^* is a grill) $(X \setminus V') \cup (X \setminus V) = X \setminus (V \cap V') \notin x^*$ so that $V \cap V' \neq \emptyset$ since $X \in x^*$ because X is dense in (X^*, μ^*) [note: $\mathcal{A}^* = \{\{x^*\}, X\}$ is near in (X^*, μ^*); hence $\mathcal{A} = \{A \subset X$: there exists some $A^* \in \mathcal{A}^*$ with $A^* \subset \bar{A}^{X^*}\}$ is near (X, μ) and thus $x^* \cup \{X\} \subset \mathcal{A}$ so that $X \in x^*$ since x^* is maximal]. Then $V' \subset St(V, \mathcal{V}) \subset U$ and thus $o(V') \subset o(U)$ so that $St(o(V), o(\mathcal{V})) \subset o(U)$. Consequently, $o(\mathcal{V}) *< o(\mathcal{U})$.).

Therefore the canonical completion (X^*, μ^*) *of a uniform* N_1-*space* (X, μ) *(= separated uniform space* $(X, \mathcal{W}_\mu)$) *is nothing else but the Hausdorff completion (= complete hull) of* $(X, \mathcal{W}_\mu)$ *in the sense of A. Weil* (up to isomorphism) [note also 6.2.2 ①].

② As well-known the canonical completion of a uniform N_1-space is an epireflection with respect to the subcategory of complete uniform N_1-spaces. If the category of separated N_1-spaces (and uniformly continuous maps) is denoted by $\underline{\text{SepNear}}_1$, then the full subcategory $\underline{\text{CSepNear}}_1$ of complete separated N_1-spaces is still epireflective in $\underline{\text{SepNear}}_1$ but the epireflection is not obtained by the canonical completion but by the simple completion which is constructed as follows: Let $\tilde{X}$ be the set of all clusters in $(X, \mu) \in |\underline{\text{SepNear}}_1|$ and let $e: X \to \tilde{X}$ be defined by $e(x) = \{A \subset X : x \in \bar{A}^X\}$ for each $x \in X$. A cover $\mathcal{A}$ of $\tilde{X}$ belongs to $\tilde{\mu}$ if and only if the following are satisfied:

(1) $e^{-1}\mathcal{A} \in \mu$

(2) For each $\tilde{x} \in \tilde{X}$ there exists some $A \in \mathcal{A}$ such that $\tilde{x} \in A$ and $e^{-1}[A]$ meets each element of $\tilde{x}$.

Then $(\tilde{X},\tilde{\mu})$ is a complete separated N_1-space and $e: (X,\mu) \to (\tilde{X},\tilde{\mu})$ is a dense embedding in <u>Near</u> hence an epimorphism in <u>$SepNear_1$</u> (note that the underlying topological space of a separated N_1-space is a T_2-space [= Hausdorff space]!) and an extremal monomorphism in <u>Near</u> . The proof of these assertions is left to the interested reader. In order to show that $e: (X,\mu) \to (\tilde{X},\tilde{\mu})$ is a reflection let (Y,ν) be a complete separated N_1-space and $f: (X,\mu) \to (Y,\nu)$ a uniformly continuous map. If $\tilde{x} \in \tilde{X}$, then $f\tilde{x} = \{f[A]: A \in \tilde{x}\}$ is a near Cauchy system having precisely one adherencepoint $g(\tilde{x})$ in (Y,ν) since (Y,ν) is a complete separated N_1-space. It is easy to check that the map $g: \tilde{X} \to Y$ defined in this way is uniformly continuous and that we have $g \circ e = f$. In general $(\tilde{X},\tilde{\mu})$ is not isomorphic to (X^*,μ^*) which is shown by the following

example: Let Y be the closed unit interval $[0,1]$ with the usual nearness structure (= topological, = uniform structure). Further let X be the nearness subspace of Y given by the set $[0,1] \setminus \{\frac{1}{n}: n = 1,2,...\}$. Then $X^* \cong Y$ is even a compact Hausdorff space which is especially proximal but $\tilde{X}$ is neither uniform nor contigual (although X has these properties!).

Now in the following a class of nearness spaces containing the class of all uniform N_1-spaces is studied such that the canonical completion becomes still an epireflection.

<u>6.2.6 Definition</u>. A nearness space (X,μ) is called <u>regular</u> iff it satisfies the following condition:

(R) For each $\mathcal{U} \in \mu$, there is some (refinement) $\mathcal{V} \in \mu$ such that for each $V \in \mathcal{V}$, there exists some $U \in \mathcal{U}$ with $\{X \setminus V, U\} \in \mu$.

<u>6.2.7 Remark</u>. (1) If a seminearness space (X,μ) fulfills the condition (R) , then (X,μ) is already a nearness space (namely, if $\mathcal{U} \in \mu$, then there exists some $\mathcal{V} \in \mu$ such that (R) is

satisfied. Thus $\mathcal{V} < \{\mathrm{int}_\mu U : U \in \mathcal{U}\}$ so that the assertion follows.) .

(2) a) *Every uniform nearness space* (X,μ) *is regular* (If $\mathcal{U} \in \mu$, then there exists some $\mathcal{V} \in \mu$ with $\mathcal{V} \ {}^*\!\!< \mathcal{U}$, i.e. for each $V \in \mathcal{V}$, there exists some $U \in \mathcal{U}$ with $\mathrm{St}(V,\mathcal{V}) \subset U$. Hence $\mathcal{V} \prec \{X \setminus V, U\}$ and thus $\{X \setminus V, U\} \in \mu$.).

b) *Every topological nearness space* (X,μ) *which is regular as topological space is regular as nearness space* (If $\mathcal{U} \in \mu$, then there exists an open cover $\mathcal{O}$ of $(X,\mathcal{X}_\mu)$ with $\mathcal{O} < \mathcal{U}$. Thus for each $x \in X$, there exists some $O_x \in \mathcal{O}$ with $x \in O_x \in \mathring{\mathcal{U}}(x)$. Since $(X,\mathcal{X}_\mu)$ is a regular space, it follows that O_x contains some closed neighbourhood V_x of x. Then $\mathcal{V} = \{V_x : x \in X\}$ belongs to μ because it is refined by $\{V_x^{\circ} : x \in X\}$. Moreover for each $V_x \in \mathcal{V}$, there exists some $U \in \mathcal{U}$ with $V_x \subset O_x \subset U$ and $\{X \setminus V_x, U\}$ belongs to μ because it is refined by the open cover $\{X \setminus V_x, O_x\}$.) .

(3) *If* (X,μ) *is a regular nearness space, then* $(X,\mathcal{X}_\mu)$ *is a regular topological space* (Let $x \in X$ and $U \in \mathcal{U}(x)$. Then $x \in \mathrm{int}_\mu U$, i.e. $\mathcal{U} = \{X \setminus \{x\}, U\} \in \mu$. By assumption there exists some $\mathcal{V} \in \mu$ such that (R) is satisfied. By N_3) there exists some $V \in \mathcal{V}$ with $x \in \mathrm{int}_\mu V$, i.e. $V \in \mathcal{U}(x)$. Further there is some $W \in \mathcal{U}$ with $\{X \setminus V, W\} \in \mu$ where obviously $W = U$. If $y \in X \setminus U$, then $\{X \setminus V, U\} < \{X \setminus \{y\}, X \setminus V\}$ and hence $\{X \setminus \{y\}, X \setminus V\} \in \mu$, i.e. $y \in \mathrm{int}_\mu (X \setminus V)$. Thus $X \setminus U \subset (X \setminus V)^{\circ}$ or equivalently $\bar{V} \subset U$.).

(4) *Every regular nearness space* (X,μ) *is separated.* (If $\mathcal{A}$ is a near Cauchy system in (X,μ), then we have to show that $\mathcal{B} = \{B \subset X : \mathcal{A} \cup \{B\}$ is near in $(X,\mu)\}$ is near in (X,μ). Let $\mathcal{U} \in \mu$. Then there exists some $\mathcal{V} \in \mu$ such that (R) is satisfied. Since $\mathcal{A}$ is a Cauchy system, there exist $A \in \mathcal{A}$ and $V \in \mathcal{V}$ with $A \subset V$. Moreover there is some $U \in \mathcal{U}$ with $\mathcal{W} = \{X \setminus V, U\} \in \mu$. For each $B \in \mathcal{B}$, $\mathcal{A} \cup \{B\}$ is near; hence there is some $W \in \mathcal{W}$ with $A \cap W \neq \emptyset$ for each $A \in \mathcal{A}$ and $B \cap W \neq \emptyset$. Thus $W = U$ because of $A \subset V$. Consequently, $\mathcal{B}$ is near in (X,μ) .)

⑤ *The category* <u>RegNear</u> *of regular nearness spaces and uniformly continuous maps is bireflective in* <u>Near</u> . (It suffices to show that for a given source $(f_i:(X,\mu) \to (X_i,\mu_i))_{i\in I}$ in <u>Near</u> such that (X_i,μ_i) is regular for every $i \in I$ and μ is the initial <u>Near</u>-structure on X with respect to $(f_i)_{i\in I}$ it follows that (X,μ) is regular. Let $\mathcal{U} \in \mu$. Then there exists a finite set $\{i_1,\ldots,i_n\} \subset I$ such that for each $j \in \{1,\ldots,n\}$ there is some $\mathcal{U}_{i_j} \in \mu_{i_j}$ with $f_{i_1}^{-1}\,\mathcal{U}_{i_1} \wedge\ldots\wedge f_{i_n}^{-1}\,\mathcal{U}_{i_n} < \mathcal{U}$. By assumption, for each $\mathcal{U}_{i_j}$ there is some $\mathcal{V}_{i_j} \in \mu_{i_j}$ such that the condition (R) is satisfied. Thus $\mathcal{V} = f_{i_1}^{-1}\,\mathcal{V}_{i_1} \wedge\ldots\wedge f_{i_n}^{-1}\,\mathcal{V}_{i_n} \in \mu$ and we are going to show that $\mathcal{V}$ refines $\mathcal{U}$ such that (R) is satisfied. Let $V \in \mathcal{V}$. Then for each $j \in \{1,\ldots,n\}$ there is some $V_{i_j} \in \mathcal{V}_{i_j}$ with $V = \bigcap_{j=1}^{n} f_{i_j}^{-1}[V_{i_j}]$. Further there are $U_{i_j} \in \mathcal{U}_{i_j}$ such that $\mathcal{W}_{i_j} = \{X \setminus V_{i_j}, U_{i_j}\} \in \mu_{i_j}$ for each $j \in \{1,\ldots,n\}$. Consequently $\mathcal{W} = f_{i_1}^{-1}\,\mathcal{W}_{i_1} \wedge\ldots\wedge f_{i_n}^{-1}\,\mathcal{W}_{i_n} \in \mu$. Furthermore there exists some $U \in \mathcal{U}$ with $\bigcap_{j=1}^{n} f_{i_j}^{-1}[U_{i_j}] \subset U$. In order to show that $\{X \setminus V, U\} \in \mu$ it suffices to prove that $St(V,\mathcal{W}) \subset U$ (because this inclusion implies that $\{V, X \setminus U\}$ is not near). Let $W = \bigcap_{j=1}^{n} f_{i_j}^{-1}[W_{i_j}]$ be an element of $\mathcal{W}$ such that $V \cap W \neq \emptyset$. Thus for each $j \in \{1,\ldots,n\}$, $f_{i_j}^{-1}[V_{i_j} \cap W_{i_j}] = f_{i_j}^{-1}[V_{i_j}] \cap f_{i_j}^{-1}[W_{i_j}] \neq \emptyset$. Therefore $V_{i_j} \cap W_{i_j} \neq \emptyset$ for each $j \in \{1,\ldots,n\}$ which implies $W_{i_j} = U_{i_j}$ for each $j \in \{1,\ldots,n\}$. Consequently $W \subset U$. Hence $St(V,\mathcal{W}) \subset U$.)

<u>6.2.8 Proposition</u>. If (X,μ) is a regular nearness space, then (X^*,μ^*) is regular.

Proof. If $\mathcal{U}^* \in \mu^*$, then there exists some $\mathcal{U} \in \mu$ with $o(\mathcal{U}) < \mathcal{U}^*$ and some $\mathcal{V} \in \mu$ such that (R) is satisfied. It suffices to show that for each $V \in \mathcal{V}$, there exists some $U \in \mathcal{U}$ with $\{X^* \setminus o(V), o(U)\} \in \mu^*$ (for then there is also some $U^* \in \mathcal{U}^*$ with $\{X^* \setminus o(V), U^*\} \in \mu^*$ because of $o(\mathcal{U}) < \mathcal{U}^*$, hence $o(\mathcal{V})$ is the desired uniform cover of X^* such that (R) is satisfied): For each $V \in \mathcal{V}$, there exists some $U \in \mathcal{U}$ with $\{X \setminus V, U\} \in \mu$; hence $\{o(X \setminus V), o(U)\} \in \mu^*$. The set $o(X \setminus V) \cap o(V)$ is empty since it is an open set in X^* which does not contain any element of X (X is dense in X^*!). Hence $o(X \setminus V) \subset X^* \setminus o(V)$ and thus $\{X^* \setminus o(V), o(U)\} \in \mu^*$.

6.2.9 Proposition. Let (X,μ) be a nearness space, (A,μ_A) a dense subspace of (X,μ) and (Y,ν) a complete, regular N_1-space. Then each uniformly continuous map $f: (A,\mu_A) \to (Y,\nu)$ has a unique uniformly continuous extension $\bar{f}: (X,\mu) \to (Y,\nu)$.

Proof. For each $x \in X$, $\mathcal{B}_x = \{B \subset X: x \in \bar{B}^X\}$ is near (because $\bar{\mathcal{B}}_x = \{\bar{B}^X: B \in \mathcal{B}_x\}$ is near) and a Cauchy system ($\mathcal{U} \in \mu$ implies the existence of some $U \in \mathcal{U}$ with $x \in \text{int}_\mu U$ and since $A \in \mathcal{B}_x$, we have $\tilde{A} = A \cap \text{int}_\mu U \in \mathcal{B}_x$ with $\tilde{A} \subset U$). Since $f: (A,\mu_A) \to (Y,\nu)$ is uniformly continuous, it follows that $f\mathcal{B}_x = \{f[B]: B \in \mathcal{B}_x\}$ is near (cf. 3.2.3.4) and obviously a Cauchy system. Since (Y,ν) is separated (as regular nearness space), $\mathcal{C} = \{C \subset Y: f\mathcal{B}_x \cup \{C\}$ is near in $(Y,\nu)\}$ is a cluster which has an adherencepoint $y \in Y$ by the completeness of (Y,ν). Especially, y is an adherencepoint of $f\mathcal{B}_x$. Since (Y,ν) is an N_1-space, y is the unique adherencepoint of $f\mathcal{B}_x$ (If y' is an adherencepoint of $f\mathcal{B}_x$, then y' is also an adherencepoint of $f\mathcal{B}_x \cup \{\{y'\}\}$ and thus $f\mathcal{B}_x \cup \{\{y'\}\}$ is near [$\mathcal{V} \in \nu$ implies the existence of some $V \in \mathcal{V}$ with $y' \in \text{int}_\nu V$ and hence V meets every element of $f\mathcal{B}_x \cup \{\{y'\}\}$] so that $\{y'\} \in \mathcal{C}$ and hence $y \in \overline{\{y'\}}^Y$, i.e. $y = y'$). By putting $\bar{f}(x) = y$ a map $\bar{f}: X \to Y$ is defined. If $x \in A$, then x is an adherencepoint of $\mathcal{B}_x$ so that $f(x)$ is an adherencepoint of $f\mathcal{B}_x$ and thus $\bar{f}(x) = f(x)$, i.e. $\bar{f}$ is an extension of f. As well-known a continuous map f from a dense subspace A of a topological space X to a regular Hausdorff space Y has a unique continuous extension g provided that $\lim_{\substack{z \to x \\ z \in A}} f(z)$ exists for each $x \in X$; then especially $g(x) = \lim_{\substack{z \to x \\ z \in A}} f(z)$. In the present case this limit exists (note that (Y,ν) is complete and separated [cf. 6.2.2 ① and 6.2.7 ④] and argue analogously to uniform spaces) and coincides with $\bar{f}(x)$ which is easy to verify. It remains to show that $\bar{f}$ is uniformly continuous. If $\mathcal{U} \in \nu$, then there exists some $\mathcal{V} \in \nu$ such that (R) is satisfied. Moreover $f^{-1}\mathcal{V} \in \mu_A$, hence there exists some open uniform cover $\mathcal{W}$ of (X,μ) with $\mathcal{W} \wedge \{A\} < f^{-1}\mathcal{V}$. It suffices to show that $\mathcal{W} < \bar{f}^{-1}\mathcal{U}$. For each $W \in \mathcal{W}$, there exists some $V \in \mathcal{V}$ with $W \cap A \subset f^{-1}[V] \subset \bar{f}^{-1}[V]$ and some $U \in \mathcal{U}$ with $\{Y \setminus \bar{V}, U\} \in \nu$. Now we show that $W \subset \bar{f}^{-1}[U]$. If $c \in W$, then $c \in \overline{W \cap A}^X$ because W is open and A is dense in (X,μ). Since $\bar{f}$ is continuous, it follows that $\bar{f}(c) \in \overline{\bar{f}[W \cap A]}^Y \subset \bar{V}^Y$. Because of $\{Y \setminus V, U\} \in \nu$ we have $\bar{V}^Y \subset U$, so that $\bar{f}(c) \in U$, i.e. $c \in \bar{f}^{-1}[U]$. This completes the proof.

<u>6.2.10 Theorem</u>. The category <u>$\mathrm{CRegNear}_1$</u> of complete regular N_1-spaces (and uniformly continuous maps) is epireflective in the category <u>$\mathrm{RegNear}_1$</u> of regular N_1-spaces (and continuous maps). For each $X \in |\mathrm{RegNear}_1|$, the canonical completion $j_X\colon X \to X^*$ is an epireflection with respect to <u>$\mathrm{CRegNear}_1$</u> .

<u>Proof</u>. If (X,μ) is a regular N_1-space, then (X^*,μ^*) is also a regular N_1-space (cf. 6.2.8 and 6.2.5 ① 1.) which is complete by construction. Each uniformly continuous map $f\colon (X,\mu) \to (Y,\nu)$ from (X,μ) to a complete regular N_1-space (Y,ν) has a unique uniformly continuous extension $\bar{f}\colon (X^*,\mu^*) \to (Y,\nu)$ by 6.2.9. Thus the inclusion map $j_X\colon (X,\mu) \to (X^*,\mu^*)$ is a reflection. It is even an epireflection since X is dense in the Hausdorff space $(X^*,\mathcal{X}_{\mu^*})$.

6.2.11 Remark. If $j_X: (X,\mu) \to (X^*,\mu^*)$ is the canonical completion, then $j_X: (X,\mathcal{X}_\mu) \to (X^*,\mathcal{X}_{\mu^*})$ is an extension of $(X,\mathcal{X}_\mu)$, i.e. a dense embedding in the topological sense (note that initial nearness structures induce initial topological structures). If (X,μ) is contigual, then (X^*,μ^*) is also contigual (If $\beta \subset \mu$ is a base consisting of finite covers of X, then $\{o(\mathcal{U}): \mathcal{U} \in \beta\}$ is a base for μ^* consisting of finite covers of X^*). Thus, for each contigual nearness space (X,μ), (X^*,μ^*) is complete and contigual and hence $(X^*,\mathcal{X}_{\mu^*})$ is compact (If $\mathcal{A}^*$ is near in (X^*,μ^*) and $X \neq \emptyset$, then there exists some cluster $\mathcal{B}^*$ containing $\mathcal{A}^*$ which was shown in 3.2.3.16 during the proof of the fact that every contigual nearness space is grill-determined. Since (X^*,μ^*) is complete, $\mathcal{B}^*$ [and thus $\mathcal{A}^*$] has an adherencepoint $x^* \in X^*$. If $\mathcal{U}^* \subset P(X^*)$ such that $X^* = \bigcup_{U^* \in \mathcal{U}^*} \mathrm{int}_{\mu^*} U^*$, then $\{X^* \setminus U^*: U^* \in \mathcal{U}^*\}$ does not have any adherencepoint. Hence it is not near in (X^*,μ^*), i.e. $\mathcal{U}^* \in \mu^*$. Thus (X^*,μ^*) is a topological nearness space so that $(X^*,\mathcal{X}_{\mu^*})$ is compact by 3.1.3.4.). If (X,μ) is a proximal N_1-space, then $j_X: (X,\mathcal{X}_\mu) \to (X^*,\mathcal{X}_{\mu^*})$ is a (Hausdorff) compactification of $(X,\mathcal{X}_\mu)$ (cf. 6.2.5 (1), 6.2.7 (2) a) and 6.2.7 (3)). In the following some important (Hausdorff) compactifications and extensions are given by means of the canonical completion and finally it is shown that every regular Hausdorff extension and every Hausdorff compactification (up to equivalence) may be obtained by the canonical completion.

6.2.12 Examples. (1) Let $(X,\mathcal{X})$ be a T_1-space. Then $T(j_X): T((X,(\mu_\mathcal{X})_c)) = (X,\mathcal{X}) \to T((X^*,(\mu_\mathcal{X})_c{}^*))$ is a compact T_1-extension of $(X,\mathcal{X})$ [note that $T((Y,\nu)) = T((Y,\nu_c))$ for each nearness space (Y,ν)], the so-called Wallman extension of $(X,\mathcal{X})$.

(2) Let $(X,\mathcal{X})$ be a Tychonoff space. Further let $(f_i: (X,\mathcal{X}) \to (\mathbb{R}, \text{usual topology}))_{i\in I}$ be the family of all continuous maps from $(X,\mathcal{X})$ to $(\mathbb{R}$, usual top.) and let $(f_i: (X,\mu) \to \mathbb{R}_u)_{i\in I}$ be initial. Then $T(j_X): T((X,\mu)) = (X,\mathcal{X}) \to T((X^*,\mu^*))$ is a realcompact extension

of $(X,\mathcal{X})$, namely the *Hewitt realcompactification* of $(X,\mathcal{X})$.

③ Let $(X,\mathcal{X})$ be a Tychonoff space. Then $T(j_X): T((X,(\mu_{\mathcal{X}})_p)) = (X,\mathcal{X}) \to T((X^*,(\mu_{\mathcal{X}})_p{}^*))$ is a compactification of $(X,\mathcal{X})$, namely the *Stone-Čech compactification* of $(X,\mathcal{X})$.

④ Let $(X,\mathcal{X})$ be a locally compact, non-compact Hausdorff space. Further let

$$\mu = \{\mathcal{U} \subset P(X): X = \bigcup_{U\in\mathcal{U}} U^o \text{ and there exists some } V \in \mathcal{U} \text{ such that } \overline{X\setminus V}^X \text{ is compact in } (X,\mathcal{X})\}.$$

Then $T(j_X): T((X,\mu)) = (X,\mathcal{X}) \to T((X^*,\mu^*))$ is a compactification of $(X,\mathcal{X})$, namely the *Alexandroff compactification* of $(X,\mathcal{X})$.

<u>6.2.13 Theorem</u>. If $f: (X,\mathcal{X}) \to (Y,\mathcal{Y})$ is a regular Hausdorff extension[49)] (resp. Hausdorff compactification) of a topological space $(X,\mathcal{X})$, then there is a regular N_1-structure (resp. a unique proximal N_1-structure) μ on X such that $T((X,\mu)) = (X,\mathcal{X})$ and there exists a homeomorphism $h: T((X^*,\mu^*)) \to (Y,\mathcal{Y})$ with $h\circ T(j_X) = f$ where $j_X: (X,\mu) \to (X^*,\mu^*)$ denotes the canonical completion of (X,μ).

<u>Proof</u>. (1) Let $f: (X,\mathcal{X}) \to (Y,\mathcal{Y})$ be a regular Hausdorff extension of $(X,\mathcal{X})$ and ν the <u>Near</u>-structure on Y induced by $\mathcal{Y}$. Further let $f: (X,\mu) \to (Y,\nu)$ be initial. Then $f: (X,\mu) \to (Y,\nu)$ is an embedding and (X,μ) is regular (note that regularity is hereditary and that (Y,ν) is regular by 6.2.7 ② b)). Since initial <u>Near</u>-structures induce initial topological structures, we have $T((X,\mu)) = (X,\mathcal{X})$ so that (X,μ) is an N_1-space. Since (Y,ν) is a topological nearness space, it follows that (Y,ν) is complete. Hence $f: (X,\mu) \to (Y,\nu)$ is a dense embedding in a complete regular N_1-space and thus an epireflection of $(X,\mu) \in |\underline{\text{RegNear}_1}|$ with respect to $\underline{\text{CRegNear}_1}$. Consequently there is an isomorphism $h: (X^*,\mu^*) \to (Y,\nu)$ with $h\circ j_X = f$ (note 6.2.10 as well as 2.1.6 and 2.2.2). Then the homeomorphism induced by h fulfills

49) i.e. a dense embedding into a regular Hausdorff space.

the desired property.

(2) If $f: (X,\mathcal{X}) \to (Y,\mathcal{Y})$ is a Hausdorff compactification of $(X,\mathcal{X})$, then ν and μ are chosen as under (1). Then (X,μ) is a proximal N_1-space (because (X,μ) is isomorphic to a subspace of the proximal N_1-space (Y,ν)). Since every compact Hausdorff space is also regular, there is a homeomorphism $h: T((X^*,\mu^*)) \to (Y,\mathcal{Y})$ with $h \circ T(j_X) = f$ (cf. (1)). Now let $(X,\bar{\mu})$ be a proximal N_1-space with $T((X,\bar{\mu})) = (X,\mathcal{X})$ such that there exists a homeomorphism $\bar{h}: T((X^*,\bar{\mu}^*)) \to (Y,\mathcal{Y})$ with $\bar{h} \circ T(\bar{j}_X) = f$ where $\bar{j}_X: (X,\bar{\mu}) \to (\bar{X}^*,\bar{\mu}^*)$ is the canonical completion of $(X,\bar{\mu})$. Then $\hat{h} = \bar{h}^{-1} \circ h: T((X^*,\mu^*)) \to T((\bar{X}^*,\bar{\mu}^*))$ is a homeomorphism with $\bar{j}_X = \hat{h} \circ j_X$. Since (X^*,μ^*) and $(\bar{X}^*,\bar{\mu}^*)$ are topological nearness spaces (note that (X^*,μ^*) and $(\bar{X}^*,\bar{\mu}^*)$ are contigual and complete and cf. 6.2.11), it follows that $\hat{h}: (X^*,\mu^*) \to (\bar{X}^*,\bar{\mu}^*)$ is an isomorphism. Then the following diagrams

$$
\begin{array}{ccc}
(X,\mu) & \xrightarrow{1_X} & (X,\bar{\mu}) \\
{\scriptstyle j_X}\downarrow & & \downarrow{\scriptstyle \bar{j}_X} \\
(X^*,\mu^*) & \xrightarrow{\hat{h}} & (\bar{X}^*,\bar{\mu}^*)
\end{array}
\qquad\qquad
\begin{array}{ccc}
(X,\bar{\mu}) & \xrightarrow{1_X} & (X,\mu) \\
{\scriptstyle \bar{j}_X}\downarrow & & \downarrow{\scriptstyle j_X} \\
(\bar{X}^*,\bar{\mu}^*) & \xrightarrow{\hat{h}^{-1}} & (X^*,\mu^*)
\end{array}
$$

are commutative. Since μ (resp. $\bar{\mu}$) is initial with respect to j_X (resp. $\bar{j}_X$), we get that $1_X: (X,\mu) \to (X,\bar{\mu})$ and $1_X: (X,\bar{\mu}) \to (X,\mu)$ are uniformly continuous so that $\mu \le \bar{\mu}$ and $\bar{\mu} \le \mu$, i.e. $\mu = \bar{\mu}$.

CHAPTER VII

COHOMOLOGY AND DIMENSION OF NEARNESS SPACES

It is a well-known fact that cohomology theory leads to better results in dimension theory than homology theory. The beautiful results characterizing finite-dimensional compact metric spaces by means of homology resp. cohomology (cf. Hurewicz and Wallman [48]) may be generalized in a slightly modified form to compact Hausdorff spaces provided Lebesgue's covering dimension is considered (cf. Nagata [64]). But already for the wider class of paracompact Hausdorff spaces a corresponding homological characterization of covering dimension is not valid. Nevertheless a cohomological characterization of finite-dimensional paracompact Hausdorff spaces is known. In 1952 C.H. Dowker [24] has shown that Čech's cohomology theory (and homology theory) may be defined for structures which - as we know today - include nearness structures. H.L. Bentley [11] and D. Czarcinski [21] have proved that these theories satisfy a variant of the Eilenberg-Steenrod axioms. During this chapter it is expected that the reader is acquainted with simplicial cohomology and classical Čech cohomology.

For uniform spaces, Isbell [50] examined two dimension functions, namely the uniform dimension and the large dimension respectively. Their generalizations for nearness spaces are denoted by dim and Dim respectively. For proximity spaces they coincide with the δ-dimension of Smirnov. For normal topological R_0-spaces dim is identical with Lebesgue's covering dimension. Thus, at first a concept of normality for nearness spaces is introduced in such a way that all uniform spaces and all normal topological R_0-spaces are normal. It is shown that for normal nearness spaces Urysohn's lemma as well as the extension theorem of Tietze-Urysohn are valid. Furthermore, normal nearness spaces may be characterized by means of the existence of a partition of unity for each finite uniform cover. Via a uniform version of the classical theorem of Borsuk a characterization of dim for normal nearness spaces by means of an extension property of maps into spheres is obtained.

The aim of this chapter is to give a cohomological characterization of the dimension dim of a finite-dimensional normal nearness space. An essential tool in this direction is Hopf's extension theorem. C.H. Dowker [23] has pointed out that this theorem is valid for normal topological spaces provided the Čech cohomology groups are based on finite open covers. Therefore we introduce Čech cohomology based on finite uniform covers and show that this theory satisfies the above mentioned variant of the Eilenberg-Steenrod axioms. Since the usual (topological) Čech cohomology based on finite open covers does not satisfy the homotopy axiom one may conclude that the category of topological spaces is not the right category for considering Čech cohomology groups based on finite coverings. The category of nearness spaces is a better one. One reobtains Dowker's result (corollary 7.3.2) from the generalization of Hopf's extension theorem (theorem 7.3.1). The cohomological characterization theorem 7.3.3 for dim contains a well-known result of Kodama [56] for normal topological spaces and is applicable to uniform spaces and proximity spaces (corollaries 7.3.4, 7.3.5, 7.3.6).

7.1 Cohomology theories for nearness spaces

7.1.1. In order to give an exact definition of a cohomology theory for nearness spaces we introduce at first the category $\underline{\text{Near}}_2$ of pairs of nearness spaces:

Objects of $\underline{\text{Near}}_2$ are pairs $((X,\mu),(Y,\mu_Y))$ - shortly (X,Y) - where (X,μ) is a nearness space, Y a subset of X and $\mu_Y = \{\alpha \wedge \{Y\}: \alpha \in \mu\}$, i.e. (Y,μ_Y) is a subspace of (X,μ).

Morphisms $f: (X,Y) \to (X',Y')$ are uniformly continuous maps $f: X \to X'$ such that $f[Y] \subset Y'$.

Definition. Let G be a fixed abelian group. A cohomology theory for nearness spaces with coefficients G is a pair (H^*,δ^*) where $H^* = (H^q)_{q\in\mathbb{Z}}$ is a family of contravariant functors $H^q: \underline{\text{Near}}_2 \to \underline{\text{Ab}}$ from the category $\underline{\text{Near}}_2$ into the category $\underline{\text{Ab}}$ of abelian groups (and homomorphisms) for each integer q and $\delta^* = (\delta^q)_{q\in\mathbb{Z}}$ is a family of natural transformations $\delta^q: H^q \circ T \to H^{q+1}$ with a functor $T: \underline{\text{Near}}_2 \to \underline{\text{Near}}_2$ defined by $T(X,Y) = (Y,\emptyset)$ and $T(f) = f|_Y$ for each $f: (X,Y) \to (X',Y')$ such that the following are satisfied:

1) *Exactness axiom.* For any pair (X,Y) with inclusion maps $i: (Y,\emptyset) \to (X,\emptyset)$ and $j: (X,\emptyset) \to (X,Y)$ there is an exact sequence

$$\cdots \xrightarrow{\delta^{q-1}(X,Y)} H^q(X,Y) \xrightarrow{H^q(j)} H^q(X,\emptyset) \xrightarrow{H^q(i)} H^q(Y,\emptyset) \xrightarrow{\delta^q(X,Y)} H^{q+1}(X,Y) \to \cdots$$

2) *Homotopy axiom.* If $g: (X,Y) \to (Z,W)$ and $h: (X,Y) \to (Z,W)$ are uniformly homotopic (i.e. there exists a uniformly continuous map $F: (X\times I, Y\times I) \to (Z,W)$ such that $F(\cdot,0) = g$, $F(\cdot,1) = h$ where I denotes the unit interval $[0,1]$ with its usual uniform (= topological) structure) then $H^q(g) = H^q(h)$ for each integer q.

3) *Excision axiom.* If Y and U are subspaces of (X,μ) such that $\{X \setminus U, Y\} \in \mu$ then the inclusion map $i: (X\setminus U, Y\setminus U) \to (X,Y)$ induces isomorphisms

$$H^q(i): H^q(X,Y) \to H^q(X\setminus U, Y\setminus U) \quad \text{for each integer } q.$$

4) *Dimension axiom.* If P is a nearness space with a single point then

$$H^q(P,\emptyset) = \begin{cases} 0 & \text{for } q \neq 0 \\ G & \text{for } q = 0 \end{cases}$$

7.1.2. 1) Let (K,L) be a simplicial pair (i.e. K is a simplicial complex and L a subcomplex, possibly empty), and let G be an abelian group. Then the group $C^q(K,L;G)$ of q-dimensional cochains is defined in the usual way (cf. Eilenberg and Steenrod [26;VI,4]). Thus the homology groups of the cochain complex $\{C^q(K,L;G),\delta\}$ may be defined and are called the cohomology groups of the pair (K,L) (notation: $H^q(K,L;G)$). The coboundary homomorphism $\delta\colon H^{q-1}(L;G) \to H^q(K,L;G)$ and the homomorphism $f^*\colon H^q(K',L';G) \to H^q(K,L;G)$ for a simplicial map $f\colon (K,L) \to (K',L')$ are defined in the usual way.

2) Let $(X,Y) \in |\underline{\mathrm{Near}}_2|$. For every uniform cover α of X let (X_α,Y_α) be the following simplicial pair: X_α is the nerve of the covering α (i.e. the vertices of X_α are the non-empty elements of α and the simplexes of X_α are those finite non-empty sets of vertices of X_α whose intersection is non-empty) and Y_α is a subcomplex of X_α which is described as follows: The vertices of Y_α are the elements of $\alpha' = \{A \in \alpha\colon A \cap Y \neq \emptyset\}$; a simplex of Y_α is a finite set of elements of α' whose intersection meets Y. Thus Y_α is the nerve of $\alpha \wedge \{Y\}$ (up to an isomorphism).

3) If $\beta > \alpha$ (covering β is a refinement of covering α) then any projection $\Pi_\alpha^\beta\colon (X_\beta,Y_\beta) \to (X_\alpha,Y_\alpha)$ defines a homomorphism $\Pi_\alpha^{\beta *}\colon H^q(X_\alpha,Y_\alpha;G) \to H^q(X_\beta,Y_\beta;G)$ which is independent of the choice of the projection Π_α^β. There results a direct spectrum $\{H^q(X_\alpha,Y_\alpha;G);\Pi_\alpha^{\beta *}\}$ whose limit group is designated by $\check{H}^q(X,Y;G)$ and called the <u>q-dimensional Čech cohomology group</u> of the pair of nearness spaces (X,Y). Using the same method as Eilenberg and Steenrod [26;IX,4] one can show that any $\underline{\mathrm{Near}}_2$-morphism $f\colon (X,Y) \to (X',Y')$ induces homomor-

phisms $\check{H}^q(f): \check{H}^q(X',Y';G) \to \check{H}^q(X,Y;G)$. Thus we obtain contravariant functors $\check{H}^q$: $\underline{\text{Near}}_2 \to \underline{\text{Ab}}$, the so-called <u>Čech cohomology functors</u>. The coboundary operator $\delta^q_{(X,Y)}: \check{H}^q(Y,\emptyset;G) \to \check{H}^{q+1}(X,Y;G)$ is defined in the usual way (cf. Eilenberg and Steenrod [26;IX,7]).

<u>7.1.3 Theorem</u>. For each integer q , let $\check{H}^q$: $\underline{\text{Near}}_2 \to \underline{\text{Ab}}$ be the Čech cohomology functor and $\delta^q = (\delta^q_{(X,Y)})_{(X,Y)\in|\underline{\text{Near}}_2|}$ the corresponding family of coboundary operators. Then $((\check{H}^q)_{q\in\mathbb{Z}}, (\delta^q)_{q\in\mathbb{Z}})$ is a cohomology theory for nearness spaces with coefficients G .

For the <u>proof</u> the reader is referred to Bentley [11].

<u>7.1.4 Remarks</u>. ① The above definitions and results may also be formulated for merotopic spaces instead of nearness spaces.

② If X is a topological nearness space (i.e. a topological R_o-space) and Y is a closed subspace, then $\check{H}^q(X,Y;G)$ is isomorphic with the usual q-dimensional Čech cohomology group of the closed pair (X,Y) of topological spaces [26;IX,8] (obviously the directed set of all open coverings of X is a cofinal subset of the directed set of all uniform covers of X).

<u>7.1.5</u>. Let C: $\underline{\text{Near}} \to \underline{\text{C-Near}}$ denote the contigual bireflector from the category <u>Near</u> of nearness spaces (and uniformly continuous maps) into the full subcategory <u>C-Near</u> of contigual nearness spaces. This functor can be extended to a functor $\tilde{C}$: $\underline{\text{Near}}_2 \to \underline{\text{Near}}_2$ defined by $\tilde{C}(X,Y) = (C(X),C(Y))$.

<u>Theorem</u>. Let (H^*,δ^*) be a cohomology theory for nearness spaces with coefficients G . Then $(\tilde{H}^*,\tilde{\delta}^*)$ is again a cohomology theory for nearness spaces with coefficients G provided that $\tilde{H}^* = (H^q \circ \tilde{C})_{q\in\mathbb{Z}}$ and $\tilde{\delta}^* = (\tilde{\delta}^q)_{q\in\mathbb{Z}}$ where $\tilde{\delta}^q: H^q\circ\tilde{C}\circ T \to H^{q+1}\circ\tilde{C}$ is defined by $\tilde{\delta}^q_{(X,Y)} = \delta^q_{(C(X),C(Y))}$.

Proof. The assertion follows immediately from the following facts:

1. C preserves extremal monomorphisms (= embeddings).
2. a) $C(X) = X$
   b) $C(X \times Y) = C(X) \times C(Y)$ } provided X is contigual.

Since 1. and 2. a) are rather trivial, it suffices to show 2. b): Evidently every uniform cover of $X \times C(Y)$ is a uniform cover of $C(X \times Y)$. Conversely, if $\mathcal{A}$ is a uniform cover of $C(X \times Y)$, then $\mathcal{A}$ is refined by some finite uniform cover $\mathcal{B}$ of $X \times Y$. Hence there exist a finite uniform cover $\mathcal{C}$ of X and a uniform cover $\mathcal{D}$ of Y such that $\{C \times D : C \in \mathcal{C}, D \in \mathcal{D}\}$ refines $\mathcal{B}$. For each $B \in \mathcal{B}$ and each $C \in \mathcal{C}$ define $E(B,C) = \{y \in Y : C \times \{y\} \subset B\}$. Then for each $C \in \mathcal{C}$ $\mathcal{F}_C = \{E(B,C) : B \in \mathcal{C}\}$ is refined by $\mathcal{D}$ and finite, hence a uniform cover of Y. By N_2), $\mathcal{F} = \mathcal{F}_{C_1} \wedge \ldots \wedge \mathcal{F}_{C_n}$ is also a uniform cover of Y provided that $\mathcal{C} = \{C_1, \ldots, C_n\}$. Since $\{C \times F : C \in \mathcal{C}, F \in \mathcal{F}\}$ refines $\mathcal{B}$, and hence $\mathcal{A}$, one may conclude that $\mathcal{A}$ is a uniform cover of $X \times C(Y)$.

7.1.6. Combining the Čech cohomology functors $\check{H}^q = \check{H}^q(-;G)$ with the functor $\tilde{C}$ (corresponding to the above theorem) we obtain cohomology functors $\check{H}^q_f(-;G)$: $\underline{\mathrm{Near}}_2 \to \underline{\mathrm{Ab}}$. For each pair of nearness spaces (X,Y) the groups $\check{H}^q_f(X,Y;G) = \check{H}^q(C(X),C(Y);G)$ are called the q-dimensional Čech cohomology groups of (X,Y) based on finite uniform covers (Since the directed set $(\mu^f_C, <)$ of all finite uniform covers of $C(X)$ is a cofinal subset of the directed set $(\mu_C, <)$ of all uniform covers of $C(X)$ the limit $\check{H}^q(C(X),C(Y);G)$ may be taken to be based on $(\mu^f_C, <)$ which is identical with the directed set $(\mu^f, <)$ of all finite uniform covers of X. Thus $\check{H}^q_f(X,Y;G)$ may be constructed in the same way as $\check{H}^q(X,Y;G)$ provided only finite uniform covers of X are considered.).

7.1.7 Remark. If X is a topological nearness space (i.e. a topological R_0-space) and Y is a closed subspace, then $\check{H}^q_f(X,Y;G)$ is isomorphic to the usual q-dimensional Čech cohomology group of the closed pair (X,Y) of topological spaces based on finite open coverings [26;IX,8] (obviously the directed set of all finite open coverings of X is a cofinal subset of the directed set of all finite uniform covers of X). Though $\check{H}_f^*$ satisfies the "uniform" homotopy axiom, the "topological" homotopy axiom is not satisfied; namely C.H. Dowker [23] has shown that $\check{H}_f^1(\mathbb{R}) = \check{H}_f^1(\mathbb{R},\emptyset;\mathbb{Z})$ has an infinite number of elements, especially $\check{H}_f^1(\mathbb{R}) \neq 0$ (note: $\mathbb{R}$ is homotopically equivalent to a one-point space P and $\check{H}_f^1(P) = 0$). Thus *for considering Čech cohomology groups with respect to finite covers the theory of nearness spaces is more suitable than the theory of topological spaces.*

7.1.8 Convention. We will write $\check{H}^q(X,Y)$ (resp. $\check{H}_f^q(X,Y)$) instead of $\check{H}^q(X,Y;G)$ (resp. $\check{H}_f^q(X,Y;G)$) if the group G coincides with the group $\mathbb{Z}$ of integers. Furthermore, $\check{H}^q(X)$ (resp. $\check{H}_f^q(X)$) denotes $\check{H}^q(X,\emptyset)$ (resp. $\check{H}_f^q(X,\emptyset)$).

7.2 Normality and dimension of nearness spaces

7.2.1 Definition. A nearness space (X,μ) is called normal provided that $C(X)$ is regular where C denotes the contigual bireflector.

7.2.2 Remark. Uniform nearness spaces (= uniform spaces) and proximal nearness spaces (= proximity spaces) are normal nearness spaces (cf. 3.1.3.8 ③ and 6.2.7. ② a)). *A topological nearness space (= topological R_0-space) is normal if and only if it is a normal topological space in the usual sense.*
("⇐". Let (X,μ) be a topological nearness space which is normal in the usual sense. In order to show that $C((X,\mu)) = (X,\mu_c)$ is regular, let $\mathcal{A} \in \mu_c$. Then there exists a finite open cover $\mathcal{B} = \{B_i : i \in \{1,\dots,n\}\}$ of X such that $\mathcal{B} < \mathcal{A}$. By normality there exists an open cover $\mathcal{C} = \{C_i : i \in \{1,\dots,n\}\} \in \mu_c$ with $\bar{C}_i \subset B_i$ for each

$i \in \{1,\dots,n\}$ (cf. Willard [82;15.10]). Hence $\{(X \setminus C_i)^{\circ}, B_i\} \in \mu_c$ and therefore $\{X \setminus C_i, B_i\} \in \mu_c$ for each $i \in \{1,\dots,n\}$, i.e. (X,μ_c) is regular.

"$\Rightarrow$". Let A and B be closed disjoint subsets of X. Then $\mathcal{U} = \{X \setminus A, X \setminus B\}$ is an open cover of X, i.e. $\mathcal{U} \in \mu_c$. By regularity of (X,μ_c) there is some $\mathcal{V} \in \mu_c$ such that for each $V \in \mathcal{V}$ there is some $U \in \mathcal{U}$ with $\{X \setminus V, U\} \in \mu_c$. Furthermore there is some finite open cover $\mathcal{C}$ with $\mathcal{C} < \mathcal{V}$. Thus there exists some two-element open cover $\mathcal{D} = \{D_1, D_2\}$ such that $\mathcal{D} < \mathcal{U}$ and $\{X \setminus D_1, X \setminus A\} \in \mu_c$ as well as $\{X \setminus D_2, X \setminus B\} \in \mu_c$. By $N_3)$ we may conclude that $\{(X \setminus D_1)^{\circ}, X \setminus A\} \in \mu_c$ and $\{(X \setminus D_2)^{\circ}, X \setminus B\} \in \mu_c$, i.e. $\bar{D}_1 \subset X \setminus A$ and $\bar{D}_2 \subset X \setminus B$. Hence $X \setminus \bar{D}_1, X \setminus \bar{D}_2$ is a separation of A and B.)

<u>7.2.3 Definition</u>. A uniform cover $\mathcal{B} = \{B_i : i \in I\}$ of a nearness space (X,μ) is called a <u>shrinking</u> of a uniform cover $\mathcal{A} = \{A_i : i \in I\}$ of (X,μ) provided that $\{X \setminus B_i, A_i\} \in \mu$ for each $i \in I$.

<u>7.2.4 Proposition</u>. A nearness space (X,μ) is normal if and only if every finite uniform cover of X has a shrinking.

<u>Proof</u>. 1) "$\Rightarrow$". Let $\mathcal{U} = \{U_1,\dots,U_n\}$ be a finite uniform cover of X. Since $C((X,\mu)) = (X,\mu_c)$ is regular and $\mathcal{U} \in \mu_c$ there is some finite $\mathcal{W} \in \mu$ such that for each $W \in \mathcal{W}$ there may be chosen some $i(W) \in \{1,\dots,n\}$ with $\{X \setminus W, U_{i(W)}\} \in \mu$. Thus for each $i \in \{1,\dots,n\}$, $\mathcal{V}_i = \{W \in \mathcal{W} : i(W) = i\}$ is a finite set of elements of $\mathcal{W}$. For each $W \in \mathcal{V}_i$, $\{X \setminus U_i, W\}$ is not near. Therefore $\{X \setminus U_i, V_i\}$ is not near, where $V_i = \bigcup_{W \in \mathcal{V}_i} W$.

Since $\mathcal{W} < \mathcal{V} = \{V_1,\dots,V_n\}$, $\mathcal{V}$ is a uniform cover of X such that $\{X \setminus V_i, U_i\} \in \mu$ for each $i \in \{1,\dots,n\}$. Thus $\mathcal{V}$ is a shrinking of $\mathcal{U}$.

2) "$\Leftarrow$". In order to show that (X,μ_c) is regular let $\mathcal{U} \in \mu_c$. Then there exists a finite $\mathcal{V} = \{V_1,\dots,V_n\} \in \mu$ with $\mathcal{V} < \mathcal{U}$. By assumption $\mathcal{V}$ has a shrinking

$\mathcal{W} = \{W_1,\dots,W_n\} \in \mu$. Since $\mathcal{W}$ is finite it follows that $\mathcal{W} \in \mu_c$. Obviously, for each $W \in \mathcal{W}$ there is some $U \in \mathcal{U}$ with $\{X \setminus W, U\} \in \mu_c$, i.e. (X,μ_c) is regular.

<u>7.2.5 Definition</u>. Let (X,μ) be a nearness space and A,U subsets of X with $\{X \setminus A, U\} \in \mu$. Then U is called a <u>uniform neighborhood</u> of A and we write $A <_\mu U$.

<u>7.2.6 Remark</u>. Every normal nearness space (X,μ) has the following property: Whenever $\{A,B\} \subset P(X)$ is not near there are uniform neighborhoods U and V of A and B respectively such that $\{U,V\}$ is not near. (This is an immediate consequence of the fact that each two-element uniform cover has a shrinking.)

<u>7.2.7 Theorem</u> (Urysohn's lemma). Let (X,μ) be a normal nearness space. Whenever $\{A,B\} \subset P(X)$ is not near there is a uniformly continuous map $f\colon (X,\mu) \to [0,1]$ from (X,μ) into the unit interval $[0,1]$ (endowed with the usual uniformity) such that $f[A] \subset \{0\}$ and $f[B] \subset \{1\}$.

<u>Proof</u>. Let A,B be subsets of X such that $\{X \setminus A, X \setminus B\} \in \mu$. By 7.2.6 there is a uniform neighborhood $U_{\frac{1}{2}}$ of A with $U_{\frac{1}{2}} <_\mu X \setminus B$. Now we have $\{X \setminus A, U_{\frac{1}{2}}\} \in \mu$ and $\{X \setminus U_{\frac{1}{2}}, X \setminus B\} \in \mu$. Hence there are uniform neighborhoods $U_{\frac{1}{4}}$ and $U_{\frac{3}{4}}$ of A and $U_{\frac{1}{2}}$ respectively such that

$$A <_\mu U_{\frac{1}{4}} <_\mu U_{\frac{1}{2}} <_\mu U_{\frac{3}{4}} <_\mu X \setminus B .$$

Suppose sets $U_{\frac{k}{2^n}}$, $k = 1,\dots,2^n - 1$ have been defined in such a way that

$$A <_\mu U_{\frac{1}{2^n}} <_\mu \dots <_\mu U_{\frac{2^n-1}{2^n}} <_\mu X \setminus B$$

By induction, then, for each rational of the form $r = \frac{k}{2^n}$ for some $n > 0$ and $k = 1,\dots,2^n - 1$ we have defined a set $U_r \subset X$. Further put $U_o = A$, $U_1 = X \setminus B$ and $U_t = X$ if $t > 1$. Thus the index set $D \subset [0,\infty)$ of the family $(U_t)_{t \in D}$ is a dense subset of $[0,\infty)$ such that the following are satisfied:

(a) $t < s$ implies $U_t <_\mu U_s$ whenever $t,s \in D$.

(b) For each near collection $\mathcal{C} \subset \mathcal{P}(X)$ there is some $t \in D$ such that $\mathcal{C} \cup \{U_t\}$ is near in (X,μ).

For every $x \in X$ let $f(x) = \inf \{r \in D: x \in U_r\}$. This defines a map $f: (X,\mu) \to [0,\infty)$ with $f[A] \subset \{0\}$ and $f[B] \subset \{1\}$. Obviously $f[X] \subset [0,1]$. In order to prove that f is uniformly continuous it suffices to show that for each near collection $\mathcal{C}$ in (X,μ) there exists some real number $t \geq 0$ such that for every $C \in \mathcal{C}$, $t \in \overline{f[C]}$ (i.e. for each $\varepsilon > 0$ there is some $x_C \in C$ with $t - \varepsilon < f(x_C) < t + \varepsilon$). [Evidently, this implies that $f\mathcal{C}$ is near.]

Let $\mathcal{C}$ be a near collection in (X,μ) and let $D' = \{r \in D: \mathcal{C} \cup \{U_r\}$ is near in $(X,\mu)\}$. By (a), D' has the following property: Every $s \in D$ with $s \geq r$ for some $r \in D'$ belongs to D'. Let $t = \inf D'$ (by (b), D' is non-void!). Further let $\varepsilon > 0$ and $C \in \mathcal{C}$.

<u>Case 1</u>: $t = 0$. Thus there are real numbers $r,s \in D'$ such that $0 < r < s < \varepsilon$. By (a), we have $U_r <_\mu U_s$, i.e. $\{X \setminus U_r, U_s\} \in \mu$. Since $\mathcal{C} \cup \{U_r\}$ is near, we may conclude that $C \cap U_s \neq \emptyset$, i.e. there is some $x_C \in C$ with $x_C \in U_s$. Hence $0 \leq f(x_C) \lesssim s < \varepsilon$.

<u>Case 2</u>: $t > 0$. Then there exist real numbers $p,q \in D$ and $r,s \in D'$ with $t - \varepsilon < p < q < t < r < s < t + \varepsilon$. It suffices to show that there is some $x_C \in C$ with $x_C \in U_s \setminus U_p$. Since $\mathcal{D}_1 = \{X \setminus U_p, U_q\}$ and $\mathcal{D}_2 = \{X \setminus U_r, U_s\}$ belong to μ, we obtain $\mathcal{D}_1 \wedge \mathcal{D}_2 \in \mu$ and therefore $\mathcal{D} = \{U_s \setminus U_p, X \setminus U_r, U_q\} \in \mu$. Thus there is some $D \in \mathcal{D}$ which meets every element of $\mathcal{C} \cup \{U_r\}$.

First assumption: $D = U_s \setminus U_p$. Then $C \cap (U_s \setminus U_p) \neq \emptyset$, i.e. there is some $x_C \in C$ with $x_C \in U_s \setminus U_p$.
Second assumption: $D = X \setminus U_r$. This is a contradiction since $U_r \cap (X \setminus U_r) = \emptyset$.
Third assumption: $D = U_q$. Since $q < t$, $\mathcal{C} \cup \{U_q\}$ is not near. Thus there is some uniform cover $\mathcal{F}$ of (X,μ) such that for each $F \in \mathcal{F}$ there exists some $E \in \mathcal{C} \cup \{U_q\}$ with $F \cap E = \emptyset$. Let $\mathcal{G} = \mathcal{D}_2 \wedge \mathcal{F} \in \mu$. Since $r \in D'$, there is some $G \in \mathcal{G}$ which meets every element of $\mathcal{C} \cup \{U_r\}$. Therefore $G = U_s \cap F_o$ where $F_o \in \mathcal{F}$. Obviously $F_o \cap U_q = \emptyset$ and consequently $F_o \cap U_p = \emptyset$. Because of $G \cap C \neq \emptyset$ there is some $x_C \in C$ with $x_C \in G$, i.e. $x_C \in U_s \setminus U_p$.
This completes the proof.

7.2.8 Definitions. Let (X,μ) be a nearness space.

1) A partition of unity on (X,μ) is a family $(f_i)_{i\in I}$ of uniformly continuous maps from (X,μ) into the unit interval $[0,1]$ endowed with the usual uniformity such that

$$\sum_{i\in I} f_i(x) = 1 \quad \text{for every} \quad x \in X .$$

(This equality means that for a fixed $x_o \in X$ at most countably many members of the family $(f_i)_{i\in I}$ have values different from zero at the point x_o and that the series $\sum_{j=1}^{\infty} f_{i_j}(x_o)$, where $\{i_1,i_2,\ldots\} = \{i \in I\colon f_i(x_o) \neq 0\}$, is convergent and its sum is equal to 1 . Since the series under consideration is absolutely convergent, the arrangement of terms is of no importance.)

2) If $\mathcal{U} = \{U_i\colon i \in I\}$ is a given uniform cover of X , we say that a partition $(f_i)_{i\in I}$ of unity is subordinated to $\mathcal{U}$ iff each f_i vanishes outside the set U_i (i.e. $f_i^{-1}[(0,1]] \subset U_i$ for each $i \in I$).

3) A partition of unity $(f_i)_{i\in I}$ is called an equiuniformly continuous partition of unity provided that for each $\varepsilon > 0$ there is some $\mathcal{A} \in \mu$ such that $\mathcal{A} < f_i^{-1}\mathcal{A}_\varepsilon$ for every $i \in I$

where $A_\varepsilon = \{U(x,\varepsilon): x \in [0,1]\}$ and $U(x,\varepsilon) = \{y \in [0,1]: |x-y| < \varepsilon\}$.

7.2.9 Proposition. Let (X,μ) be a nearness space. Then the following are equivalent:

(1) (X,μ) is normal.

(2) Every finite uniform cover of X has a partition of unity subordinated to it.

(3) Every finite uniform cover of X has an equiuniformly continuous partition of unity subordinated to it.

Proof. (1) ⇒ (2). Let $\mathcal{A} = \{A_1,\dots,A_n\}$ be a finite uniform cover of X. By 7.2.4 there is a shrinking $\mathcal{B} = \{B_1,\dots,B_n\}$ of $\mathcal{A}$. Further, by Urysohn's lemma, there are uniformly continuous maps $f_i: (X,\mu) \to [0,1]$ with $f_i[X \setminus A_i] \subset \{0\}$ and $f_i[B_i] \subset \{1\}$ for each $i \in \{1,\dots,n\}$. Since all f_i are bounded, $f = f_1 + \dots + f_n$ is uniformly continuous. Then $\frac{1}{f}$ is also uniformly continuous (note: $1 \le f_1(x) + \dots + f_n(x)$ for every $x \in X$). For each $i \in \{1,\dots,n\}$ let us define $g_i: (X,\mu) \to [0,1]$ by $g_i(x) = \dfrac{f_i(x)}{f_1(x)+\dots+f_n(x)}$ for every $x \in X$. Then $(g_i)_{i\in\{1,\dots,n\}}$ is a partition of unity subordinated to $\mathcal{A}$ (cf. also exercise 70.).

(2) ⇒ (1). Let $\mathcal{A} = \{A_1,\dots,A_n\}$ be a finite uniform cover of X. By assumption there is a partition $(g_i)_{i\in\{1,\dots,n\}}$ of unity subordinated to it. If ρ denotes the usual uniformity on $[0,1]$, we obtain $[\frac{1}{n},1] <_\rho [\frac{1}{n+1},1]$ and therefore $g_i^{-1}[[\frac{1}{n},1]] <_\mu g_i^{-1}[[\frac{1}{n+1},1]]$ for every $i \in \{1,\dots,n\}$ since each g_i is uniformly continuous. Put $B_i = g_i^{-1}[[\frac{1}{n+1},1]]$ for each $i \in \{1,\dots,n\}$. Since $\mathcal{C}_i = \{X \setminus g_i^{-1}[[\frac{1}{n},1]], B_i\} \in \mu$ for every $i \in \{1,\dots,n\}$ we get $\mathcal{C} = \mathcal{C}_1 \wedge \dots \wedge \mathcal{C}_n \in \mu$. Then, for each $i \in \{1,\dots,n\}$, $\mathrm{St}(g_i^{-1}[[\frac{1}{n},1]],\mathcal{C}) \subset B_i = \mathrm{St}(g_i^{-1}[[\frac{1}{n},1]],\mathcal{C}_i)$. Further $\{\mathrm{St}(x,\mathcal{C}): x \in X\} \in \mu$ because it is refined by $\mathcal{C}$. Thus $\mathcal{B} = \{B_i: i \in \{1,\dots,n\}\} \in \mu$ because it is refined by

$\{St(x,\mathcal{C}) : x \in X\}$ (For each $x \in X$ there is some $j \in \{1,\dots,n\}$ such that $g_j(x) \geq \frac{1}{n}$ since $\sum_{i=1}^{n} g_i(x) = 1$. Hence $St(x,\mathcal{C}) \subset B_j$ since $St(g_j^{-1}[[\frac{1}{n},1]],\mathcal{C}) \subset B_j$ and $x \in g_j^{-1}[[\frac{1}{n},1]]$). Additionally $\mathcal{B}$ is a shrinking of $\mathcal{A}$:
Let $i \in \{1,\dots,n\}$. Since $\{\{0\}, [\frac{1}{n+1},1]\}$ is not near in $([0,1],\rho)$ we obtain that $\{g_i^{-1}(0),B_i\}$ is not near in (X,μ). Further $X \setminus A_i \subset g_i^{-1}(0)$. Thus $\{X \setminus A_i, B_i\}$ is not near in (X,μ), i.e. $B_i <_\mu A_i$.

(2) ⇔ (3). Let $\mathcal{A} = \{A_1,\dots,A_n\}$ be a finite uniform cover of X and $(g_i)_{i\in\{1,\dots,n\}}$ a partition of unity subordinated to it. Then (g_i) is equiuniformly continuous since $\mathcal{A}$ is finite (note that, for each $\varepsilon > 0$, $g_1^{-1}\mathcal{A}_\varepsilon \wedge\dots\wedge g_n^{-1}\mathcal{A}_\varepsilon < g_i^{-1}\mathcal{A}_\varepsilon$ for every $i \in \{1,\dots,n\}$).

7.2.10 Remarks. ① *Every subspace* (in Near) *of a normal nearness space is normal* (since C preserves embeddings and 6.2.7.⑤ is valid). Thus a nearness subspace of a normal topological space is normal though a topological subspace of a normal topological space is not normal in general.

② *The product* (in Near) *of a proximal nearness space with a normal nearness space is normal* (cf. 6.2.7. ⑤ and part 2. of the proof of 7.1.5.).

7.2.11 Definitions. A) Let $(K,\mathcal{K})$ be a finite simplicial complex, i.e. $\mathcal{K}$ has only finitely many elements.

1) The uniform realization (K_u,μ_u) of $(K,\mathcal{K})$ is a metrizable uniform nearness space (= metrizable uniform space) defined as follows:

$$K_u = \{p \in \mathbb{R}^K : p(v) \geq 0 \text{ for all } v \in K \text{ and, for some } S \in \mathcal{K},\ p(v) = 0 \text{ for all } v \in K \setminus S \text{ and } \sum_{v\in S} p(v) = 1\}.$$

μ_u is induced by the metric d given by

$$d(p,q) = \max\{|p(v)-q(v)| : v \in K\} \text{ for all } p,q \in K_u .$$

2) For each $v \in K$, the <u>star</u> of v in (K_u,μ_u) is defined by

$$St(v)_u = \{p \in K_u : p(v) \neq 0\}$$

B) Let (X,μ) be a nearness space and $\mathcal{A}$ a finite uniform cover of X. A uniformly continuous map $f\colon (X,\mu) \to \mathcal{A}_u$ of (X,μ) into the uniform realization of the nerve of the covering $\mathcal{A}$ is called a <u>canonical mapping</u> provided that $f^{-1}[St(A)_u] \subset A$ for each vertex A of the nerve of $\mathcal{A}$.

<u>7.2.12 Theorem</u>. A nearness space (X,μ) is normal if and only if for each finite uniform cover $\mathcal{A}$ of X there is a canonical mapping f of (X,μ) into (the uniform realization of) the nerve of $\mathcal{A}$ [50].

<u>Proof</u>. 1) "⇒". Let $\mathcal{A}$ be a finite uniform cover of X. Since (X,μ) is normal there is an equiuniformly continuous partition $(f_A)_{A\in\mathcal{A}}$ of unity subordinated to $\mathcal{A}$ (cf. 7.2.9). For each $x \in X$, let $f(x)$ be a map which assigns to each non-void element A of $\mathcal{A}$ the value $f_A(x)$. Thus a map f of (X,μ) into $\mathcal{A}_u$ is defined (note that for each $x \in X$ we have

1) $f(x)(A) = f_A(x) = 0$ for each non-void $A \in \mathcal{A}$ which does not contain x

and 2) $\sum_{\substack{A\in\mathcal{A}\\ x\in A}} f_A(x) = 1$).

Since $(f_A)_{A\in\mathcal{A}}$ is equiuniformly continuous, f is uniformly continuous. In order to show that f is a canonical mapping let A be a non-void element of $\mathcal{A}$ and $x \in f^{-1}[St(A)_u]$. Thus $f(x)(A) = f_A(x) \neq 0$. Since f_A vanishes outside the set A we obtain $x \in A$.

2) "⇐". Let $\mathcal{A}$ be a finite uniform cover of X. By assumption there is a canonical mapping $f\colon (X,\mu) \to \mathcal{A}_u$. For each non-void element A of $\mathcal{A}$ let us define $f_A\colon X \to [0,1]$

[50] Sometimes we do not make a notational distinction between a finite simplicial complex and its uniform realization.

by $f_A(x) = f(x)(A)$ for each $x \in X$; if $A \in \mathcal{A}$ is empty put $f_A(x) = 0$ for each $x \in X$. Since f is uniformly continuous $(f_A)_{A\in\mathcal{A}}$ is equiuniformly continuous. Furthermore $(f_A)_{A\in\mathcal{A}}$ is a partition of unity subordinated to $\mathcal{A}$. This follows immediately from the definitions. By 7.2.9, (X,μ) is normal.

7.2.13 Theorem (Tietze, Urysohn). Let (X,μ) be a normal nearness space, and $A \subset X$. Every uniformly continuous map $f\colon (A,\mu_A) \to [0,1]$ of the nearness subspace (A,μ_A) of (X,μ) into the unit interval $[0,1]$ (endowed with its usual uniformity) has a uniformly continuous extension $F\colon (X,\mu) \to [0,1]$.

Proof. In order to simplify the proof we consider a uniformly continuous map $f\colon (A,\mu_A) \to [-1,1]$. Let $A_1 = f^{-1}[[\frac{1}{3},1]]$ and $B_1 = f^{-1}[[-1,-\frac{1}{3}]]$. Now $\{A_1,B_1\}$ is not near in (A,μ_A) , and therefore in (X,μ) , so by Urysohn's lemma there is a uniformly continuous $f_1\colon (X,\mu) \to [-\frac{1}{3},\frac{1}{3}]$ such that $f_1[A_1] \subset \{\frac{1}{3}\}$ and $f_1[B_1] \subset \{-\frac{1}{3}\}$. Evidently, for each x in A , $|f(x) - f_1(x)| \leq \frac{2}{3}$, so that $f - f_1$ is a map of A into $[-\frac{2}{3},\frac{2}{3}]$. Now we repeat the process with $f - f_1 = g_1$. That is, divide $[-\frac{2}{3},\frac{2}{3}]$ into thirds (at $-\frac{2}{9}$ and $\frac{2}{9}$) and let $A_2 = g_1^{-1}[[\frac{1}{3}\cdot(\frac{2}{3})^1, (\frac{2}{3})^1]]$, $B_2 = g_1^{-1}[[-(\frac{2}{3})^1, -\frac{1}{3}(\frac{2}{3})^1]]$. Then there is a Urysohn function $f_2\colon (X,\mu) \to [-\frac{1}{3}(\frac{2}{3})^1, \frac{1}{3}(\frac{2}{3})^1]$ such that $f_2[A_2] \subset \{\frac{1}{3}\cdot(\frac{2}{3})^1\}$ and $f_2[B_2] \subset \{-\frac{1}{3}\cdot(\frac{2}{3})^1\}$. Evidently, $|(f-f_1) - f_2| \leq (\frac{2}{3})^2$ on A . Thus we define inductively a sequence $(f_n)_{n\in\mathbb{N}}$ of uniformly continuous functions on A such that

$$|f - \sum_{k=1}^{n} f_k| \leq (\frac{2}{3})^n \quad .$$

Put $F(x) = \sum_{i=1}^{\infty} f_i(x)$ for each $x \in X$. Certainly $F(x) = f(x)$ for each $x \in A$, so it remains only to show that F is uniformly continuous.

Let $\varepsilon > 0$ be given. Pick $N > 0$ so that $\sum_{n=N+1}^{\infty} (\frac{2}{3})^n < \frac{\varepsilon}{2}$.

Since each f_i is uniformly continuous, $f_i^{-1} A^i_{\frac{\varepsilon}{2N}} \in \mu$ where

$A^i_{\frac{\varepsilon}{2N}} = \{U(x_i,\frac{\varepsilon}{2N}) : x_i \in [-\frac{1}{3}(\frac{2}{3})^{i-1}, \frac{1}{3}(\frac{2}{3})^{i-1}]\}$ with

$U(x_i,\frac{\varepsilon}{2N}) = \{y \in [-\frac{1}{3}(\frac{2}{3})^{i-1}, \frac{1}{3}(\frac{2}{3})^{i-1}] : |x_i - y| < \frac{\varepsilon}{2N}\}$. Thus

$\mathcal{A} = f_1^{-1} A^1_{\frac{\varepsilon}{2N}} \wedge \ldots \wedge f_N^{-1} A^N_{\frac{\varepsilon}{2N}} \in \mu$. Furthermore $\mathcal{A} < F^{-1} \mathcal{A}_\varepsilon$ so that $F^{-1}\mathcal{A}_\varepsilon \in \mu$.

Namely, let $U \in \mathcal{A}$, i.e. $U = f_1^{-1}[U(x_1,\frac{\varepsilon}{2N})] \cap \ldots \cap f_N^{-1}[U(x_N,\frac{\varepsilon}{2N})]$.

Put $x_0 = \sum_{i=1}^{N} x_i$. Then $x_0 \in [-1,1]$. It remains to show that

$U \subset F^{-1}[V(x_0,\varepsilon)]$ with $V(x_0,\varepsilon) = \{z \in [-1,+1] : |x_0 - z| < \varepsilon\}$.

Let $c \in U$. Then $|F(c) - x_0| \le \left|\left(\sum_{i=1}^{N} f_i(c)\right) - x_0\right| + \left|\sum_{i=N+1}^{\infty} f_i(c)\right| <$

$< \sum_{i=1}^{N} |f_i(c) - x_i| + \frac{\varepsilon}{2} < N \cdot \frac{\varepsilon}{2N} + \frac{\varepsilon}{2} = \varepsilon$, i.e. $F(c) \in V(x_0,\varepsilon)$.

7.2.14 Theorem (Borsuk's Theorem). Let (X,μ) be a normal nearness space, and $A \subset X$. Further, let $f,g: (A,\mu_A) \to S^n$ be uniformly continuous maps of the nearness subspace (A,μ_A) of (X,μ) into an n-sphere S^n (endowed with its usual uniformity). If f is uniformly homotopic to g and f has a uniformly continuous extension $F: (X,\mu) \to S^n$ then g has a uniformly continuous extension $G: (X,\mu) \to S^n$ which is uniformly homotopic to F.

Proof. Since f and g are uniformly homotopic, there is a uniformly continuous map $h: A \times [0,1] \to S^n$ such that $h(\cdot,0) = f$ and $h(\cdot,1) = g$. Put $L = (A \times [0,1]) \cup (X \times \{0\}) \subset X \times [0,1]$ and define $k: L \to S^n$ by

$$k(x,t) = \begin{cases} h(x,t) & \text{for each } (x,t) \in A \times [0,1] \\ F(x) & \text{for each } (x,t) \in X \times \{0\} \end{cases} .$$

(1) k is uniformly continuous.

Let $\mathcal{V}$ be a uniform cover of S^n and $\mathcal{W}$ a uniform star-refinement of $\mathcal{V}$. Since F is uniformly continuous, $F^{-1}\mathcal{W} \in \mu$. Furthermore $h^{-1}\mathcal{W}$ is a uniform cover of $A \times [0,1]$, i.e. there are uniform covers $\mathcal{A}'$ and $\mathcal{B}'$ of A and $[0,1]$ respectively such that $\mathcal{A}' \otimes \mathcal{B}' := \{A' \times B' : A' \in \mathcal{A}' \text{ and } B' \in \mathcal{B}'\} < h^{-1}\mathcal{W}$. Especially there is some $\mathcal{A}'' \in \mu$ with $\mathcal{A}' = \mathcal{A}'' \wedge \{A\}$. Put $\mathcal{U} = F^{-1}\mathcal{W} \wedge \mathcal{A}''$. Then it is easily verified that $(\mathcal{U} \otimes \mathcal{B}') \wedge \{L\} < k^{-1}\mathcal{V}$.

(2) Let us define a uniformly continuous map $h'\colon X \times [0,1] \to S^n$ such that $h'(\cdot,0) = F$ and $h'(\cdot,1)|_A = g$. Since $k\colon L \to S^n$ is uniformly continuous and L is a subspace of the normal space $X \times [0,1]$ (cf. 7.2.10.(2)), there is a uniformly continuous extension j of k over a uniform neighborhood N of L. (Note that S^n may be identified with the boundary of $[0,1]^{n+1}$ endowed with the usual uniformity. Thus k may be considered to be a mapping into $[0,1]^{n+1}$. Then $p_l \circ k$ is uniformly continuous for each $l \in \{1,\dots,n+1\}$ where $p_l\colon [0,1]^{n+1} \to [0,1]_l$ with $[0,1]_l = [0,1]$ denotes the projection. By 7.2.13 there are uniformly continuous extensions $q_l\colon X \times [0,1] \to [0,1]$ of $p_l \circ k$ for each $l \in \{1,\dots,n+1\}$. Thus $q\colon X \times [0,1] \to [0,1]^{n+1}$ defined by $p_l \circ q = q_l$ for each $l \in \{1,\dots,n+1\}$ is a uniformly continuous extension of k. Then $N = q^{-1}[\mathrm{St}(S^n, \mathcal{A}_{\frac{1}{4}})]$ is a uniform neighborhood of L because the uniform cover $q^{-1}\mathcal{A}_{\frac{1}{4}}$ refines $\{(X \times [0,1]) \setminus L, N\}$.

If p denotes the central projection of $[0,1]^{n+1} \setminus \{\text{centre}\}$ onto S^n then $j\colon N \to S^n$, defined by $j(z) = p(q(z))$ for each $z \in N$, is the desired extension of k.) Thus there are some $\varepsilon > 0$ and some $\mathcal{B} \in \mu$ such that $\mathcal{B} \otimes \mathcal{A}_\varepsilon < \{(X \times [0,1]) \setminus L, N\}$ and therefore $\mathrm{St}(L, \mathcal{B} \otimes \mathcal{A}_\varepsilon) \subset N$. Without loss of generality we may assume that $\mathrm{St}(L, \mathcal{B} \otimes \mathcal{A}_\varepsilon) = N$. Furthermore for all $x \in X$ and all $t, t' \in [0,1]$ it follows from $(x,t) \in N$ and $t' \le t$ that $(x,t') \in N$.

By Urysohn's lemma there is a uniformly continuous map $e: X \times [0,1] \to [0,1]$ with $e[L] \subset \{1\}$ and $e[X \setminus N] \subset \{0\}$. Then define $h': X \times [0,1] \to S^n$ by $h'(x,t) = j(x, t \cdot e(x,t))$ Because of the shape of N and the size of e, h' is well-defined. Uniform continuity of h' is clear if we note that the functions being multiplied are bounded. Obviously, $h'(\cdot,0) = F$ and $h'(\cdot,1)|_A = g$.

(3) $G = h'(\cdot,1)$ is the desired uniformly continuous extension of g.

7.2.15 Definitions. Let (X,μ) be a nearness space.

(1) The large dimension $\mathrm{Dim}(X,\mu)$ of (X,μ) is said to be $\leq n$ provided every uniform cover $\mathcal{U}$ of X has a refinement $\mathcal{V} \in \mu$ of order $\leq n+1$ (i.e. each $x \in X$ is contained in at most $n+1$ elements of $\mathcal{V}$). The precise number $\mathrm{Dim}(X,\mu)$ is the smallest such n, or -1 for the special case that X is empty; and we write $\mathrm{Dim}(X,\mu) = \infty$ if there is no such n.

(2) The small dimension $\dim(X,\mu)$ of (X,μ) is defined to be the large dimension of (X,μ_c) where $1_X: (X,\mu) \to (X,\mu_c)$ denotes the contigual bireflection of (X,μ). Especially, $\dim(X,\mu) \leq n$ iff every finite uniform cover $\mathcal{U}$ of X has a (finite) refinement $\mathcal{V} \in \mu$ of order $\leq n+1$.

7.2.16 Remarks. (1) For uniform spaces Isbell [50] has introduced the uniform dimension δd and the large dimension Δd. The first one coincides with dim and the latter one is identical with Dim provided uniform spaces are considered. For proximal nearness spaces Dim and dim coincide with the δ-dimension of Smirnov [79]. For normal topological R_0-spaces dim is identical with the Lebesgue covering dimension.

(2) Obviously, $\dim(X,\mu) \leq \mathrm{Dim}(X,\mu)$ for each nearness space (X,μ). If (A,μ_A) is a (nearness) subspace of any nearness space (X,μ) we obtain

a) $\mathrm{Dim}(A,\mu_A) \leq \mathrm{Dim}(X,\mu)$

b) $\dim(A,\mu_A) \leq \dim(X,\mu)$.

7.2.17 Theorem. Let (X,μ) be a normal nearness space. Then $\dim(X,\mu) \leq n$ if and only if every uniformly continuous map of any subspace (A,μ_A) of (X,μ) into an n-sphere S^n has a uniformly continuous extension over (X,μ).

Proof. 1) "⇒". Let (A,μ_A) be a subspace of (X,μ) and let $f: (A,\mu_A) \to S^n$ be uniformly continuous. Further let $\varepsilon = \frac{1}{4}$. Then there are $z_1,\ldots,z_r \in S^n$ such that $\mathcal{V} = \{U(z_i,\varepsilon): i \in \{1,\ldots,r\}\}$ is a uniform cover of S^n where $U(z_i,\varepsilon) = \{x \in S^n: \| x-z_i \| < \varepsilon\}$. $f^{-1}\mathcal{V} = \{A_1,\ldots,A_r\}$ is a uniform cover of A with $A_i = f^{-1}[U(z_i,\varepsilon)]$ for each $i \in \{1,\ldots,r\}$. Since (A,μ_A) is normal (cf. 7.2.10.(1)), $\{A_1,\ldots,A_r\}$ has a shrinking $\{B_1,\ldots,B_r\}$. Thus $B_i <_\mu A_i \cup (X \setminus A)$ for each $i \in \{1,\ldots,r\}$. Put $C_i = A_i \cup (X \setminus A)$. Then $\mathcal{C} = \{C_1,\ldots,C_r\}$ is a uniform cover of X with $C_i \cap A = A_i$ for each $i \in \{1,\ldots,r\}$ (Since (X,μ) is normal, it follows by 7.2.6 that there is some $D_i \subset X$ with $B_i <_\mu D_i <_\mu C_i$ for each $i \in \{1,\ldots,r\}$. Hence $A \subset \bigcup_{i=1}^{r} B_i <_\mu \bigcup_{i=1}^{r} D_i$. Consequently $X \setminus (\bigcup_{i=1}^{r} D_i) <_\mu X \setminus A \subset C_1$.

Furthermore C_1 is a uniform neighbourhood of $D_1 \cup (X \setminus (\bigcup_{i=1}^{r} D_i))$.

Since $D_1 \cup (X \setminus \bigcup_{i=1}^{r} D_i) \cup D_2 \cup \ldots \cup D_r = X$ we obtain $\{C_1,\ldots,C_r\} \in \mu$). It follows from $\dim(X,\mu) \leq n$ that there is some uniform cover $\mathcal{U} = \{U_1,\ldots U_r\}$ of X of order $\leq n+1$ with $\mathcal{U} < \mathcal{C}$.

A map $k: X \to [0,1]^{n+1}$ is defined as follows:

Let $i \in \{1,\ldots,r\}$. If $U_i \cap A \neq \emptyset$, put $x_i = f(a_i)$ for some $a_i \in U_i \cap A$. Otherwise choose some $x_i \in [0,1]^{n+1}$. Since (X,μ) is normal, $\mathcal{U}$ has a shrinking $\{V_1,\ldots,V_r\}$. By Urysohn's lemma there are uniformly continuous maps $g_i: (X,\mu) \to [0,1]$, $i \in \{1,\ldots,r\}$, such that $g_i[V_i] \subset \{1\}$ and $g_i[X \setminus U_i] \subset \{0\}$.

Then $\frac{g_i}{\sum_{i=1}^{r} g_i} : (X,\mu) \to [0,1]$ is a uniformly continuous map (cf. exercise 70.). For each $x \in X$ let

$k(x) = \sum_{i=1}^{r} \left(\frac{g_i(x)}{\sum_{i=1}^{r} g_i(x)}\right) \cdot x_i$. Thus a uniformly continuous map $k: (X,\mu) \to [0,1]^{n+1}$ is defined.

$k[X]$ is nowhere dense in $[0,1]^{n+1}$, i.e. $(\overline{k[X]})^{\circ} = \emptyset$.

Namely, if $x \in X$, there are s elements $x_{i_1},\ldots,x_{i_s} \in \{x_1,\ldots,x_r\}$, $s \le n+1$, such that $x \in U_{i_1} \cap\ldots\cap U_{i_s}$. By definition of k , $k(x)$ belongs to the convex hull of $\{x_{i_1},\ldots,x_{i_s}\}$ which is an at most n-dimensional subspace of $[0,1]^{n+1}$. Thus $k[X]$ is contained in a finite union of such subspaces (since $\mathcal{U}$ is finite) which is a nowhere dense subset.

If the centre c of $[0,1]^{n+1}$ belongs to $k[X]$, choose some $d \in [0,1]^{n+1} \setminus k[X]$ such that $\text{dist}(d,c) \le \frac{1}{4}$. Otherwise put $d = c$. Let z be the central projection from d . Then $g = z \circ k: (X,\mu) \to S^n$ is uniformly continuous (note that S^n may be considered to be the boundary of $[0,1]^{n+1}$) and for each $x \in A$ we obtain $g(x) \in St(f(x),\mathcal{V})$. Namely, let $x \in A$. Then there are $U_{i_1},\ldots,U_{i_s} \in \mathcal{U}$, $s \le n+1$, such that $x \in \bigcap_{j=1}^{s} U_{i_j}$. By definition, for each $j \in \{1,\ldots,s\}$, there is some $a_{i_j} \in U_{i_j} \cap A$ with $f(a_{i_j}) = x_{i_j}$. Since $\mathcal{U} \wedge \{A\} < f^{-1}\mathcal{V}$ we get $x_{i_j} \in St(f(x),\mathcal{V})$. By the choice of d the central projection of the convex hull of $\{x_{i_1},\ldots,x_{i_s}\}$ is also contained in $St(f(x),\mathcal{V})$. Since $k(x)$ belongs to the convex hull of $\{x_{i_1},\ldots,x_{i_s}\}$ it follows that $g(x) = z(k(x)) \in St(f(x),\mathcal{V})$. Now it is easily seen that $g|_A$ and f are uniformly homotopic (the composition of

$h: A \times [0,1] \to [0,1]^{n+1}$, defined by $h(x,t) = t\cdot g|_A(x)+(1-t)\cdot f(x)$, and the central projection is a uniform homotopy!). By Borsuk's theorem one obtains a uniformly continuous extension of f .

2) "$\Leftarrow$". Suppose every uniformly continuous map of any subspace (A,μ_A) of (X,μ) into S^n can be extended over X . Now let us prove that the same is true for mappings $f: A \to S^{n+1}$. Such a mapping can be described by means of its "latitude" $f_o: A \to [0,1]$ and its "longitude" $f_1: A \to S^n$ where the longitude is undefined at the poles, however. Let P and Q be small polar caps on S^{n+1}, and let $A_1 = (A \setminus f^{-1}[P]) \setminus f^{-1}[Q]$. By 7.2.13 f_o has a uniformly continuous extension g_o over X . Furthermore, by assumption, there is a uniformly continuous map $g_1: X \to S^n$ extending the well-defined uniformly continuous map $f_1: A_1 \to S^n$. For each $x \in X$, let $g(x)$ be the point at latitude $g_o(x)$ and longitude $g_1(x)$. Thus a uniformly continuous map $g: X \to S^{n+1}$ is defined. For x in A , $g(x)$ has the same latitude as $f(x)$ and unless the latitude is high, the longitude is also the same. Therefore $g|_A$ and f are uniformly homotopic. By Borsuk's Theorem we get a uniformly continuous extension of f .

Consequently, mappings of subspaces of X into S^n or any higher dimensional sphere can be extended. Without loss of generality we may assume that (X,μ) is compact (note: For every nearness space (X,μ) , the contigual nearness space (X,μ_c) is densely embedded in the compact nearness space $(X^*,(\mu_c)^*)$ [cf. 6.2.11] and $\dim(X,\mu) = \dim(X,\mu_c) \leq \dim(X^*,(\mu_c)^*)$; furthermore, if we say that a nearness space (X,μ) has the extension property $P(k)$ provided that every uniformly continuous map of any subspace of (X,μ) into S^k has a uniformly continuous extension over X , we get:

(1) (X,μ) has the extension property $P(k)$ iff (X,μ_c) has the extension property $P(k)$.

(2) If (X,μ) is a normal nearness space and (D,μ_D) is a dense subspace of (X,μ) , then (X,μ) has the extension property $P(k)$ iff (D,μ_D) has the extension property $P(k)$.

[Since (1) is easy to check, it suffices to prove (2):

1. "⇒". obvious.
2. "⇐". Let (A,μ_A) be a subspace of (X,μ) and $f: (A,\mu_A) \to S^k$ uniformly continuous. Then there is a uniform neighborhood N of A and a uniformly continuous extension $g: (N,\mu_N) \to S^k$ of f (use the same idea as in part (2) of the proof of 7.2.14 where a similar extension over a uniform neighborhood was constructed via the extension theorem of Tietze-Urysohn). Since $A \subset \text{int}_\mu N$ and $\bar{D} = X$ we have $N \cap D \neq \emptyset$. Thus, by assumption, $g|_{N\cap D}$ has a uniformly continuous extension $h: (D,\mu_D) \to S^k$. By 6.2.9 there is a uniformly continuous extension $l: (X,\mu) \to S^k$ of h. Since $N \cap D$ is dense in N and $l|_{N\cap D} = h|_{N\cap D} = g|_{N\cap D}$ we obtain $l|_N = g$. Thus l is the desired extension of f.]).

Now let $\mathcal{A}$ be any finite open cover of (X,μ), and φ a canonical mapping of (X,μ) into the nerve of $\mathcal{A}$. Suppose B is a simplex of the nerve of $\mathcal{A}$ of maximum dimension m such that $m > n$. Let S denote the boundary of B, $A = \varphi^{-1}[S]$, $Y = \varphi^{-1}[B]$. Then the mapping $\varphi': A \to S$ defined by $\varphi'(x) = \varphi(x)$ for every $x \in A$ can be extended over X. Especially, it has a continuous extension φ_1 over Y. Let $\bar{\varphi}$ be a mapping from (X,μ) into the nerve of $\mathcal{A}$ defined by $\bar{\varphi}(x) = \varphi'(x)$ for every $x \in X \setminus Y$ and $\bar{\varphi}(x) = \varphi_1(x)$ for every $x \in Y$. Then $\bar{\varphi}$ is continuous and canonical. Continuing this process we get a canonical mapping φ^* from (X,μ) into the n-skeleton[51] of the nerve of $\mathcal{A}$. Obviously the stars of the vertices in $\mathcal{A}_u$ form a uniform cover $\mathcal{B}$ [50; p. 59]. Then $\varphi^{*-1}\mathcal{B}$ is an open covering of order $\leq n+1$ such that $\varphi^{*-1}\mathcal{B} < \mathcal{A}$.

<u>7.2.18 Corollary</u>. Let (X,μ) be a normal topological nearness space (= normal topological R_0-space). Then the following are equivalent:

(1) $\dim(X,\mu) \leq n$.

[51] The <u>n-skeleton</u> of a simplicial complex $(K,\mathcal{K})$ is defined to be the simplicial complex consisting of all m-simplexes of $(K,\mathcal{K})$ (i.e. simplexes of $(K,\mathcal{K})$ containing exactly m+1 vertices) for $m \leq n$.

(2) Every uniformly continuous map of any (nearness) subspace of (X,μ) into an n-sphere S^n has a continuous (= uniformly continuous) extension over (X,μ) .
(3) Every continuous map of any closed (topological) subspace of (X,μ) into an n-sphere S^n has a continuous extension over (X,μ) .

Proof. The equivalence of (1) and (2) is obvious since 7.2.17 is valid and (X,μ) is topological. The equivalence of (2) and (3) follows from 6.2.9 and the fact that for any closed subset A of X the structure μ_A of the nearness subspace (A,μ_A) of (X,μ) is induced by the structure $(X_\mu)_A$ of the topological subspace $(A,(X_\mu)_A)$ of (X,X_μ) .

7.3 A cohomological characterization of dimension

7.3.1 Theorem (Hopf's extension theorem). Let (X,μ) be a normal nearness space with $\dim(X,\mu) \leq n+1$, and $U \subset X$. Let $g\colon U \to S^n$ be a uniformly continuous map. Then g can be extended to a uniformly continuous map $G\colon X \to S^n$ if and only if $\check{H}^n_f(g)\,[\check{H}^n_f(S^n)] \subset \check{H}^n_f(i)\,[\check{H}^n_f(X)]$ where $i\colon U \to X$ denotes the inclusion map.

Proof. 1) "⇒". $G \circ i = g$ implies $\check{H}^n_f(i) \circ \check{H}^n_f(G) = \check{H}^n_f(g)$. Thus $\check{H}^n_f(g)(e) = \check{H}^n_f(i)(\check{H}^n_f(G)(e)) \in \check{H}^n_f(i)[\check{H}^n_f(X)]$ for every $e \in \check{H}^n_f(S^n)$.

2) "⇐". S^n is regarded as an oriented elementary n-sphere [52] as well as its uniform realization. Let τ be the covering of S^n by the stars of its vertices. Since the stars of the vertices of S^n form a uniform covering (cf. [50 ; p. 59]), τ is uniform. It is shown in [50 ; p. 61] that the nerve of the covering of a simplicial complex K which consists of the stars of the vertices is isomorphic with K . Thus, we may identify the simplicial complexes S^n

[52] i.e. the complex consisting of all the oriented faces of $\dim \leq n$ of an oriented $(n+1)$-simplex.

and $(S^n)_\tau$. Let $e_o^n = [y_o^n]$ be an element of $H^n((S^n)_\tau)$ represented by an oriented[53] n-simplex y_o^n of S^n. Since $g: U \to S^n$ is uniformly continuous $\alpha' = g^{-1}\tau = \{g^{-1}[V]: V \in \tau\}$ is a (finite) uniform cover. Mapping $g^{-1}[V] \in \alpha'$ onto $V \in \tau$ we obtain a simplicial map of $U_{\alpha'}$ into $(S^n)_\tau$ which is also denoted by g. Then $H^n(g)(e_o^n) = d_o^n$ is an element of $H^n(U_{\alpha'})$. By the assumption for the element $[d_o^n] = [H^n(g)(e_o^n)] = \check{H}_f^n(g)(e^n)$ with $e^n = [e_o^n] \in \check{H}_f^n(S^n)$ there exists a finite uniform cover γ of X and an element $b_\gamma^n \in H^n(X_\gamma)$ such that $[d_o^n] = \check{H}_f^n(i)([b_\gamma^n])$. The inclusion map $i: U \to X$ induces a map $i^*: H^n(X_\gamma) \to H^n(U_{\gamma'})$ where $\gamma' = i^{-1}\gamma$. Put $i^*(b_\gamma^n) = d_\gamma^n$; then $[d_o^n] = [d_\gamma^n]$. Therefore a finite uniform cover β' of U exists with $\beta' < \alpha' \wedge \gamma'$ and $\prod_{\alpha'}^{\beta'*}(d_o^n) = \prod_{\gamma'}^{\beta'*}(d_\gamma^n)$. Denote by α (resp. β) a finite uniform cover of X with $i^{-1}\alpha = \alpha'$ (resp. $i^{-1}\beta = \beta'$) and let be $\beta^* = \alpha \wedge \beta \wedge \gamma$. Since X has $\dim X \le n+1$ there is a finite uniform cover δ of X such that $\delta < \beta^*$ and $\dim X_\delta \le n+1$. Obviously, $\prod_{\alpha'}^{\delta'*}(d_o^n) =$ $= \prod_{\beta'}^{\delta'*} \circ \prod_{\alpha'}^{\beta'*}(d_o^n) = \prod_{\beta'}^{\delta'*} \circ \prod_{\gamma'}^{\beta'*}(d_\gamma^n) = \prod_{\gamma'}^{\delta'*}(d_\gamma^n) = \prod_{\gamma'}^{\delta'*} \circ i^*(b_\gamma^n) =$ $= i^* \circ \prod_\gamma^{\delta*}(b_\gamma^n)$, where $\delta' = i^{-1}\delta$ and the map from $H^n(X_\delta)$ into $H^n(U_{\delta'})$ induced by the inclusion map $i: U \to X$ is also denoted by i^*. This implies $i^*(\prod_\gamma^{\delta*}(b_\gamma^n)) = \prod_{\alpha'}^{\delta'*}(g^*([y_o^n])) = (g \circ \prod_{\alpha'}^{\delta'})^*([y_o^n]) =$ $= [(g \circ \prod_{\alpha'}^{\delta'})^n(y_o^n)]$, where $(g \circ \prod_{\alpha'}^{\delta'})^n: C^n(S^n) \to C^n(U_{\delta'})$ is induced by $g \circ \prod_{\alpha'}^{\delta'}: U_{\delta'} \to S^n$. Thus, $(g \circ \prod_{\alpha'}^{\delta'})^n(y_o^n)$ is the restriction of an n-cocycle of X_δ. Applying [64; VIII. 3 E_n)] $g \circ \prod_{\alpha'}^{\delta'}$ can be extended to a mapping $\prod$ of X_δ into S^n which is uniformly continuous with respect to the uniform realizations of the considered complexes. By Theorem 7.2.12 the canonical mapping $\varphi: X \to X_\delta$ is uniformly continuous. Let f be the restriction of $\prod \circ \varphi$ to U. Since φ is canonical, for

[53] During the proof we use oriented simplicial cohomology groups.

every $x \in U$, $\varphi(x)$ belongs to the closed simplex of $U_{\delta'}$ determined by x (i.e. whose vertices correspond to those elements of δ' which contain x). This closed simplex is mapped by $\prod|_{U_{\delta'}} = g \circ \prod_{\alpha}^{\delta'}$ into the closure of the simplex of S^n containing $g(x)$. Therefore $f(x) = \prod(\varphi(x))$ is in the same closed simplex as $g(x)$. Hence f and g are uniformly homotopic, for we can join $g(x)$ to $f(x)$ by a straight-line segment and move $f(x)$ to $g(x)$ along this segment. Since f is extendable over X, by theorem 7.2.14 we can extend g over X.

7.3.2 Corollary (cf. Dowker [23]). Let X be a normal topological R_0-space with $\dim X \leq n+1$, and let $U \subset X$ be a closed (topological) subspace of X. Let g be a continuous map of U into the n-sphere S^n. Then g can be extended to a continuous map $G: X \to S^n$ if and only if $\check{H}_f^n(g)[\check{H}_f^n(S^n)] \subset \check{H}_f^n(i)[\check{H}_f^n(X)]$, where $i: U \to X$ denotes the inclusion map.

7.3.3 Theorem. Let (X,μ) be a normal nearness space of finite small dimension. Then the following are equivalent:

(1) $\dim(X,\mu) \leq n$.

(2) $\check{H}_f^m(X,A) = 0$ for every integer $m \geq n+1$ and every subspace A of X.

(3) For every integer $m \geq n$ and every subspace A of X the homomorphism

$$\check{H}_f^m(i): \check{H}_f^m(X) \to \check{H}_f^m(A)$$

induced by the inclusion map $i: A \to X$ is an onto mapping.

Proof. "(1) ⇒ (2)". It suffices to show that $\check{H}_f^{n+1}(X,A) = 0$. Since X has $\dim X \leq n$ the directed set of all finite uniform covers α of X with $\dim X_\alpha \leq n$ is a cofinal subset of the directed set of all finite uniform covers of X. Combining this with the well-known fact from simplicial cohomology that $H^{n+1}(X_\alpha, A_\alpha) = 0$ we obtain the expected formula.

"(2) ⇒ (3)". It follows from the exactness axiom of the used cohomology theory that the homomorphism $\check{H}^m_f(i)\colon \check{H}^m_f(X) \to \check{H}^m_f(A)$ induced by the inclusion map $i\colon A \to X$ is an onto mapping for every integer $m \geq n$ and every subspace A of X.

"(3) ⇒ (1)". Let $\dim X = m+1 > n$. Then $\dim X \nleq m$, and by 7.2.17 there is a subspace (A,μ_A) of (X,μ) and a uniformly continuous map $g\colon A \to S^m$ which cannot be extended over X. Thus by 7.3.1 there is some $e \in \check{H}^m_f(S^m)$ such that $\check{H}^m_f(g)(e) \in \check{H}^m_f(A)$ does not belong to $\check{H}^m_f(i)[\check{H}^m_f(X)]$, i.e. the homomorphism $\check{H}^m_f(i)$ is not onto.

<u>7.3.4 Corollary</u> (Kodama [56]). Let X be a normal topological R_o-space of finite covering dimension. Then the following are equivalent:

(1) $\dim X \leq n$.

(2) $\check{H}^m_f(X,C) = 0$ for every integer $m \geq n+1$ and every closed subspace C of X.

(3) For every integer $m \geq n$ and every closed subspace C of X the homomorphism

$$\check{H}^m_f(i)\colon \check{H}^m_f(X) \to \check{H}^m_f(C)$$

induced by the inclusion map $i\colon C \to X$ is an onto mapping.

<u>Proof</u>. 1. A closed (topological) subspace is a nearness subspace. Thus "(1) ⇒ (2)" follows immediately from 7.3.3. By the exactness property, (2) implies (3).

2. "(3) ⇒ (1)" results from 7.2.18 and 7.3.2.

<u>7.3.5 Corollary</u>. Let (X,μ) be a uniform space of finite large dimension. Then the following are equivalent:

(1) $\mathrm{Dim}(X,\mu) \leq n$.

(2) $\check{H}^m_f(X,A) = 0$ for every integer $m \geq n+1$ and every subspace A of X.

(3) For every integer $m \geq n$ and every subset A of X the homomorphism

$$\check{H}^m_f(i): \check{H}^m_f(X) \to \check{H}^m_f(A)$$

induced by the inclusion map $i: A \to X$ is an onto mapping.

Proof. Since Isbell [50] has shown that $\dim(X,\mu) = \mathrm{Dim}(X,\mu)$ provided that (X,μ) has finite large dimension, 7.3.5 follows immediately from 7.3.3.

7.3.6 Corollary. Let (X,μ) be a proximal nearness space (= proximity space) of finite δ-dimension (cf. 7.2.16 (1)). Then the following are equivalent:

(1) $\dim(X,\mu) \leq n$.

(2) $\check{H}^m(X,A) = 0$ for every integer $m \geq n+1$ and every subspace A of X .

(3) For every integer $m \geq n$ and every subspace A of X the homomorphism

$$\check{H}^m(i): \check{H}^m(X) \to \check{H}^m(A)$$

induced by the inclusion map $i: A \to X$ is an onto mapping.

Proof. Since a proximal nearness space is contigual, $\check{H}^*_f$ may be replaced by $\check{H}^*$. Thus 7.3.6 follows immediately from 7.3.3.

A P P E N D I X
REPRESENTABLE FUNCTORS

<u>A.1.</u> Let C be a category. For each $X \in |C|$, there is a covariant hom-functor

$$H^X : C \longrightarrow \underline{\text{Set}}$$

defined as follows:

(1) $H^X(Y) = [X,Y]_C$ for each $Y \in |C|$.

(2) If $f : Y \rightarrow Z$ is a C-morphism, then
$H^X(f) : H^X(Y) \rightarrow H^X(Z)$

is defined by $H^X(f)(g) = f \circ g$ for each $g \in H^X(Y)$.

<u>A.2 Theorem.</u> If $F : A \rightarrow C$ is a diagram, $L \in |C|$ and $l_A : L \rightarrow F(A)$ is a C-morphism for each $A \in |A|$, then the following are equivalent:

(1) $(L,(l_A))$ is a limit of F.

(2) For each $X \in |C|$, $(H^X(L),(H^X(l_A)))$ is a limit of $H^X \circ F$.

<u>Proof.</u> (1) $\Rightarrow$ (2): If $(L,(l_A))$ is a limit of F and $X \in |C|$, then for each A-morphism $f : A \rightarrow A'$, we have $F(f) \circ l_A = l_{A'}$ so that $H^X(F(f)) \circ H^X(l_A) = H^X(F(f) \circ l_A) = H^X(l_{A'})$ since H^X is a functor. Thus $(H^X(L),(H^X(l_A)))$ is a lower bound of $H^X \circ F$. If $(S,(s_A))$ is any lower bound of $H^X \circ F$ and $q \in S$ ($S \in |\underline{\text{Set}}|$!), then $(X,(s_A(q))_{A \in |A|})$ is a lower bound of F; for if $f : A \rightarrow A'$ is an A-morphism, then $H^X(F(f)) \circ s_A = s_{A'}$ so that $H^X(F(f))(s_A(q)) = F(f) \circ s_A(q) = s_{A'}(q)$. Hence there is a unique morphism $f_q : X \rightarrow L$ such that the diagram

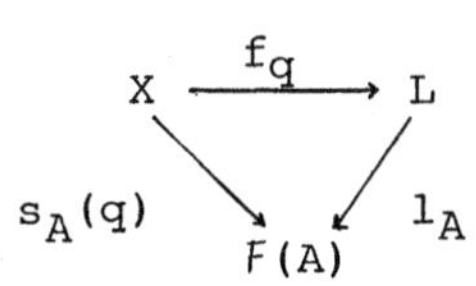

commutes for each $A \in |A|$. Then $f : S \rightarrow H^X(L)$ defined by $f(q) = f_q$ for each $q \in S$ is the uniquely determined map for which the diagram

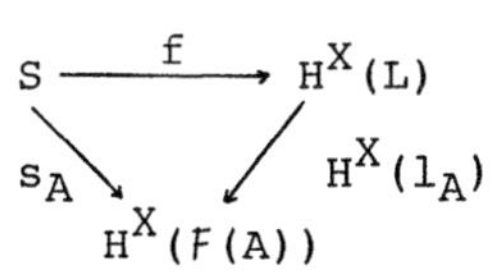

commutes; for we have $H^X(l_A)(f(q)) = l_A \circ f_q = s_A(q)$ for each $q \in S$, and $(H^X(l_A)\circ f')(q) = s_A(q)$ for each $q \in S$ implies $l_A \circ f'(q) = s_A(q)$ for each $q \in S$ so that $f'(q) = f_q = f(q)$ for each $q \in S$ by the uniqueness of f_q.

(2) $\Rightarrow$ (1): Put $X = L$. If $f : A \longrightarrow A'$ is an $\mathcal{A}$-morphism, then $H^L(F(f))\circ H^L(l_A) = H^L(l_{A'})$. Hence $H^L(F(f))(H^L(l_A)(1_L)) = H^L(l_{A'})(1_L)$ so that $F(f)\circ l_A = F(f)\circ l_A \circ 1_L = l_{A'}\circ 1_L = l_{A'}$. Thus $(L,(l_A))$ is a lower bound of F. If $(S,(s_A))$ is a lower bound of F, then $(H^S(S),(H^S(s_A)))$ is a lower bound of $H^S\circ F$. Then by assumption, there exists a unique map $\bar{g} : H^S(S) \longrightarrow H^S(L)$ such that the diagram

$$\begin{array}{ccc} H^S(S) & \xrightarrow{\ \bar{g}\ } & H^S(L) \\ {\scriptstyle H^S(s_A)}\searrow & & \swarrow{\scriptstyle H^S(l_A)} \\ & H^S(F(A)) & \end{array}$$

commutes for each $A \in |\mathcal{A}|$. Putting $g = \bar{g}(1_S) : S \longrightarrow L$, we have $l_A \circ g = s_A$ for each $A \in |\mathcal{A}|$. If $g' : S \longrightarrow L$ with $l_A \circ g' = s_A$, then $H^S(l_A)\circ H^S(g') = H^S(s_A)$. Thus $H^S(g') = \bar{g}$ by the uniqueness of $\bar{g}$. Consequently, $g' = g'\circ 1_S = H^S(g')(1_S) = \bar{g}(1_S) = g$.

<u>A.3 Theorem.</u> If $F, G : \mathcal{B} \longrightarrow \mathcal{C}$ are functors and $\eta_B : F(B) \longrightarrow G(B)$ is a $\mathcal{C}$-morphism for each $B \in \mathcal{B}$, then the following are equivalent:

(1) $\eta = (\eta_B) : F \longrightarrow G$ is a natural transformation.

(2) For each $X \in |\mathcal{C}|$, $H^X(\eta) = (H^X(\eta_B)) : H^X\circ F \longrightarrow H^X\circ G$ is a natural transformation.

<u>Proof.</u> (1) $\Rightarrow$ (2): Trivial since H^X is a functor.

(2) $\Rightarrow$ (1): If $f : B \longrightarrow B'$ is a $\mathcal{B}$-morphism, then $H^X(\eta_{B'}\circ F(f)) = H^X(\eta_{B'})\circ H^X(F(f)) = H^X(G(f))\circ H^X(\eta_B) = H^X(G(f)\circ \eta_B)$ for each $X \in |\mathcal{C}|$ so that $H^{F(B)}(\eta_{B'}\circ F(f))(1_{F(B)}) = H^{F(B)}(G(f)\circ\eta_B)(1_{F(B)})$. Thus $\eta_{B'}\circ F(f) = \eta_{B'}\circ F(f)\circ 1_{F(B)} = G(f)\circ\eta_B\circ 1_{F(B)} = G(f)\circ\eta_B$.

<u>A.4 Remark.</u> For each $X \in |\mathcal{C}|$ the corresponding contravariant hom-functor $H_X : \mathcal{C} \longrightarrow$ <u>Set</u> assigning $H_X(Y) = [Y,X]_{\mathcal{C}}$ to each $Y \in |\mathcal{C}|$ can be considered as the covariant hom-functor $H^X : \mathcal{C}^* \longrightarrow$ <u>Set</u>. Thus every H_X converts colimits to limits.

<u>A.5 Definition.</u> A functor $F : \mathcal{C} \longrightarrow$ <u>Set</u> is called <u>representable</u> provided that F is naturally equivalent to a (covariant) hom-

functor H^X for a suitable $X \in |\mathcal{C}|$. A <u>representation</u> of $\mathcal{F}$ is a pair (X,η) where X is a $\mathcal{C}$-object and $\eta = (\eta_C) : H^X \longrightarrow \mathcal{F}$ is a natural equivalence.

<u>A.6 Yoneda Lemma.</u> Let $\mathcal{F} : \mathcal{C} \longrightarrow \underline{Set}$ be a functor, $X \in |\mathcal{C}|$ and let $[H^X, \mathcal{F}]$ be the conglomerate of all natural transformations from H^X to $\mathcal{F}$. Then the Yoneda map

$$Y : [H^X, \mathcal{F}] \longrightarrow \mathcal{F}(X)$$

defined by $Y(\eta) = \eta_X(1_X)$ for each $\eta = (\eta_C)_{C \in |\mathcal{C}|} \in [H^X, \mathcal{F}]$ is bijective.[a1)]

<u>Proof.</u> At first a map $Y' : \mathcal{F}(X) \longrightarrow [H^X, \mathcal{F}]$ is defined by $Y'(a) = \eta = (\eta_C)$ with $\eta_C(f) = \mathcal{F}(f)(a)$ for each $f \in H^X(C) = [X,C]_{\mathcal{C}}$; for η is a natural transformation: Namely, if $g : C \longrightarrow C'$ is a $\mathcal{C}$-morphism, then the diagram

$$\begin{array}{ccc} H^X(C) & \xrightarrow{\eta_C} & \mathcal{F}(C) \\ {\scriptstyle H^X(g)}\downarrow & & \downarrow{\scriptstyle \mathcal{F}(g)} \\ H^X(C') & \xrightarrow{\eta_{C'}} & \mathcal{F}(C') \end{array}$$

is commutative since for each $f \in H^X(C)$, $(\mathcal{F}(g) \circ \eta_C)(f) = \mathcal{F}(g)(\eta_C(f)) = \mathcal{F}(g)(\mathcal{F}(f)(a)) = \mathcal{F}(g \circ f)(a) = \eta_{C'}(g \circ f) = \eta_{C'}(H^X(g)(f)) = (\eta_{C'} \circ H^X(g))(f)$. Then the following are satisfied:

(1) $(Y \circ Y')(a) = a$ for each $a \in \mathcal{F}(X)$;

for $Y(Y'(a)) = \eta_X(1_X) = \mathcal{F}(1_X)(a) = 1_{\mathcal{F}(X)}(a) = a$.

(2) $(Y' \circ Y)(\eta) = \eta$ for each $\eta \in [H^X, \mathcal{F}]$;

for we have $Y'(Y(\eta)) = Y'(\eta_X(1_X)) = \eta' = (\eta'_C)$ with $\eta'_C(f) = \mathcal{F}(f)(\eta_X(1_X))$ and since η is a natural transformation, the diagram

$$\begin{array}{ccc} H^X(X) & \xrightarrow{\eta_X} & \mathcal{F}(X) \\ {\scriptstyle H^X(f)}\downarrow & & \downarrow{\scriptstyle \mathcal{F}(f)} \\ H^X(C) & \xrightarrow{\eta_C} & \mathcal{F}(C) \end{array}$$

---

a1) Then $[H^X, \mathcal{F}]$ can be considered as a set.

is commutative so that $F(f)(\eta_X(1_X)) = \eta_C(H^X(f)(1_X)) = \eta_C(f)$, i.e. $\eta'_C = \eta_C$ and thus $\eta = \eta'$.
From (1) and (2), it follows that Y is bijective (Y' is the inverse map of Y!).

A.7 Definition. Let $F : C \longrightarrow \underline{Set}$ be a functor. A pair (U,u) with $U \in |C|$ and $u \in F(U)$ is said to be a universal point of F provided that for any pair (A,a) with $A \in |C|$ and $a \in F(A)$ there is a unique C-morphism $f : U \longrightarrow A$ with $F(f)(u) = a$.

A.8 Theorem. Let $F : C \longrightarrow \underline{Set}$ be a functor, $X \in |C|$ and $\eta : H^X \longrightarrow F$ a natural transformation. Then the following are equivalent:

(1) (X,η) is a representation of F.

(2) $(X,Y(\eta))$ is a universal point of F.

Proof. (1) $\Rightarrow$ (2): If $A \in |C|$ and $a \in F(A)$, then $\eta_A : H^X(A) \longrightarrow F(A)$ is an isomorphism and $\eta_A^{-1}(a) = f \in H^X(A) = [X,A]_C$. Since η is a natural transformation, the diagram

$$(*) \qquad \begin{array}{ccc} H^X(X) & \xrightarrow{\eta_X} & F(X) \\ {\scriptstyle H^X(f)}\downarrow & & \downarrow{\scriptstyle F(f)} \\ H^X(A) & \xrightarrow{\eta_A} & F(A) \end{array}$$

is commutative so that $F(f)(Y(\eta)) = F(f)(\eta_X(1_X)) = \eta_A(H^X(f)(1_X)) = \eta_A(f) = \eta_A(\eta_A^{-1}(a)) = a$. The uniqueness of f is evident.

(2) $\Rightarrow$ (1): We have to show that $\eta_A : H^X(A) \longrightarrow F(A)$ is an isomorphism for each $A \in |C|$. If $a \in F(A)$, then there exists a unique C-morphism $f_a : X \longrightarrow A$ with $F(f_a)(Y(\eta)) = a$ since $(X,Y(\eta))$ is a universal point of F. Now a map $\eta'_A : F(A) \longrightarrow H^X(A)$ is defined by $\eta'_A(a) = f_a$ for each $a \in F(A)$. Then we have:

(1) For each $a \in F(A)$,
$(\eta_A \circ \eta'_A)(a) = \eta_A(f_a) = F(f_a)(Y(\eta)) = a$. Hence $\eta_A \circ \eta'_A = 1_{F(A)}$.

(2) For each C-morphism $f : X \longrightarrow A$,
$(\eta'_A \circ \eta_A)(f) = \eta'_A(\eta_A(f)) = f_{\eta_A(f)}$ where $f_{\eta_A(f)} = f$ by the uniqueness of $f_{\eta_A(f)}$ [note: $F(f)(Y(\eta)) = \eta_A(f)$ because of (*)].

Hence $\eta'_A \circ \eta_A = 1_{H^X(A)}$.

Thereby proving the theorem.

<u>A.9 Remarks.</u> ① The preceding theorem has shown that representable functors and universal points are connected by the Yoneda Lemma.

② If $F : C \longrightarrow \underline{Set}$ is a functor, then the universal points of F can be considered as the initial[a2)] objects of that category whose objects are pairs (X,x) with $X \in |C|$ and $x \in F(X)$ and whose morphisms $f : (X,x) \longrightarrow (Y,y)$ are those C-morphisms $f : X \longrightarrow Y$ for which $F(f)(x) = y$ holds.

<u>A.10 Proposition.</u> If $F : C \longrightarrow \underline{Set}$ is a functor, then the following are satisfied:

(1) If (U,u) and (V,v) are universal points of F, then there exists a unique isomorphism $f : U \longrightarrow V$ with $F(f)(u) = v$.

(2) If (U,η) and (V,μ) are representations of F, then there exists a unique isomorphism $f : U \longrightarrow V$ such that the diagram

$$\begin{array}{ccc} H_C(U) = H^U(C) & \xleftarrow{H_C(f)} & H^V(C) = H_C(V) \\ & {\scriptstyle \eta_C} \searrow \quad \swarrow {\scriptstyle \mu_C} & \\ & F(C) & \end{array}$$

commutes for each $C \in |C|$.

For the <u>proof</u> note A.9②, A.8 and the fact that initial objects are unique (up to isomorphisms).

<u>A.11 Theorem.</u> Let $F : A \longrightarrow B$ be a functor, $A \in |A|$, $B \in |B|$ and $r : B \longrightarrow F(A)$ a B-morphism. Then the following are equivalent:

(1) (r,A) is a universal map for B with respect to F.

(2) (A,r) is a universal point of $H^B \circ F$.

<u>Proof.</u> Obvious.

---

a2) Dual notion: terminal [cf. 33)]; i.e. an object X of a category C is called an <u>initial</u> object provided that for each $Y \in |C|$, $[X,Y]_C$ is a singleton.

<u>A.12 Corollary.</u> An object $B \in |\mathcal{B}|$ has a universal map with respect to $F : \mathcal{A} \longrightarrow \mathcal{B}$ if and only if $H^B \circ F$ is representable.

<u>Proof.</u> 1) If $H^B \circ F$ is representable, then there exists $A \in |\mathcal{A}|$ and a natural equivalence $\eta : H^A \longrightarrow H^B \circ F$, i.e. (A,η) is a representation of $H^B \circ F$. By A.8, $(A,Y(\eta))$ is a universal point of $H^B \circ F$ and thus by A.11, $(Y(\eta),A)$ is a universal map for B with respect to F.

2) Let (r,A) be a universal map for $B \in |\mathcal{B}|$ with respect to $F : \mathcal{A} \longrightarrow \mathcal{B}$. Then by A.11 (A,r) is a universal point of $H^B \circ F$. If Y' denotes the inverse of the Yoneda map $Y : [H^A, H^B \circ F] \longrightarrow H^B(F(A))$, then $(A,Y'(r))$ is a representation of $H^B \circ F$ by A.8 (i.e. $Y'(r)$ is a natural equivalence).

<u>A.13 Theorem.</u> If $F : \mathcal{A} \longrightarrow \mathcal{B}$ is a functor, then the following are equivalent:

(1) $F : \mathcal{A} \longrightarrow \mathcal{B}$ has a left adjoint $G : \mathcal{B} \longrightarrow \mathcal{A}$.

(2) Each $B \in |\mathcal{B}|$ has a universal map with respect to F.

(3) $H^B \circ F$ is representable for each $B \in |\mathcal{B}|$.

<u>Corollary.</u> If (G,F) is a pair of adjoint functors, then F preserves limits and G preserves colimits.

<u>Proof.</u> The equivalence of (1) and (2) has been shown by the theorems 2.1.7 (including the corollary) and 2.1.12. The equivalence of (2) and (3) has been proved by A.12.

If $(L,(l_{A'}))$ is a limit of $\mathcal{D} : \mathcal{A}' \longrightarrow \mathcal{A}$ ($\mathcal{A}'$ is a small category), then we have to show that $(F(L),(F(l_{A'})))$ is a limit of $F \circ \mathcal{D}$. By A.2 it suffices to prove that $(H^B(F(L)),(H^B(F(l_{A'}))))$ is a limit of $H^B \circ F \circ \mathcal{D}$ for each $B \in |\mathcal{B}|$. By assumption (note A.13(3)) for each $B \in |\mathcal{B}|$, there exists an $A_B \in |\mathcal{A}|$ such that $H^B \circ F \approx H^{A_B}$. Therefore, using A.2, F preserves limits. By means of the duality principle for adjoint functors the second assertion of the corollary is also proved (obviously $G^*: \mathcal{B}^* \longrightarrow \mathcal{A}^*$ preserves limits if and only if $G: \mathcal{B} \longrightarrow \mathcal{A}$ preserves colimits).

EXERCISES

CHAPTER 1

1. a) Show that the category <u>Haus</u> of topological Hausdorff spaces and continuous maps is not topological.

   b) Let $\mathcal{C}$ be a concrete category. A family of $\mathcal{C}$-morphisms $(f_i: (X,\xi) \to (X_i,\xi_i))_{i\in I}$ indexed by a class I is called a source in $\mathcal{C}$. A source $(f_i: (X,\xi) \to (X_i,\xi_i))_{i\in I}$ is called initial provided that ξ is the initial structure with respect to $(X,f_i,(X_i,\xi_i),I)$. Consider the initial sources in <u>Haus</u> and give a characterization of these sources.

2. Let the following be defined for a topological space $(X,\mathcal{X})$:

   $A \subset X$ is bounded iff $\begin{cases}\text{for each open cover } \mathcal{U} \text{ of } X \text{, there} \\ \text{exist } U_1,\dots,U_n \in \mathcal{U} \text{ with } A \subset \bigcup_{i=1}^{n} U_i .\end{cases}$

   $A \subset X$ is relatively compact iff $\bar{A}$ is compact.

   Further let $\mathcal{B}_1 = \{A \subset X: A \text{ is bounded}\}$

   $\mathcal{B}_2 = \{A \subset X: A \text{ is relatively compact}\}$.

   Determine whether or not $\mathcal{B}_1$ and $\mathcal{B}_2$ are bornologies on X and show that $\mathcal{B}_2$ is a proper subset of $\mathcal{B}_1$.

3. Let $\mathcal{C}$ be a topological category, let X be a set and let $\mathcal{C}_X$ be the set of all $\mathcal{C}$-structures on X . Show that $\mathcal{C}_X$ endowed with the order "$\leq$" defined in 1.1.4 is a complete lattice.

4. Show that any ordered set is already a complete lattice if each subset has an infimum. What are the consequences of this theorem for topological categories?

5. A non-empty subclass K of $|\underline{\text{Top}}|$ is called a *component class* provided that the following are satisfied:

   (1) If $X \in K$, $Y \in |\underline{\text{Top}}|$ and there exists a surjective continuous map $f: X \to Y$, then $Y \in K$.

   (2) For each $X \in |\underline{\text{Top}}| \setminus \{\emptyset\}$, $\mathcal{M} = \{M \subset X: M \in K\} \neq \{\emptyset\}$ and each linearly ordered subset of $\mathcal{M}$ has an upper bound.

(3) If $A \in K$ and $B \in K$ are subspaces of a topological space X and $A \cap B \neq \emptyset$, then $A \cup B \in K$.

A topological space X is called a *local K-space* provided that for each $x \in X$ the neighbourhoods of x belonging to K (considered as subspaces) form a neighbourhood base at x. Show that for each component class K, the local K-spaces together with the continuous maps form a topological category (Hint. Use theorem 1.2.1.1).

[Note that the class of connected topological spaces (resp. pathwise connected topological spaces) is a component class.]

6. Let C be a topological category. Show that if $(f_i: X \to X_i)$ is an initial source in C and all X_i are indiscrete, then X is indiscrete. Find the dual assertion.
7. Determine the discrete and indiscrete objects of the categories under 1.1.6.
8. Let C be a topological category. Show that the composition of extremal monomorphisms (resp. extremal epimorphisms) in C is again an extremal monomorphism (resp. extremal epimorphism).
9. Let C be a topological category and let f be a C-morphism. Then f is called a *regular epimorphism* (resp. *regular monomorphism*) provided that there are C-morphisms g and h such that f is the coequalizer (resp. equalizer) of g and h. Show that the following are equivalent for any epimorphism (resp. monomorphism) f in C:

   (1) f is extremal.

   (2) f is regular.
10. Let C be a topological category and let $(f_i: X \to X_i)_{i \in I}$ be a source in C (written $(X,f_i)_I$). Then $(X,f_i)_I$ is called a mono-source in C provided that for any pair α,β of C-morphisms such that $f_i \circ \alpha = f_i \circ \beta$ for each $i \in I$, it follows that $\alpha = \beta$; and $(X,f_i)_I$ is called extremal in C provided that for each source $(Y,g_i)_I$ in C and each C-epimorphism e such that $f_i = g_i \circ e$ for each $i \in I$, it follows that e is an isomorphism. Show that $(X,f_i)_I$ is

    (a) a mono-source in C if and only if it separates points, i.e. iff for any two distinct points x and y there exists some $i \in I$ with $f_i(x) \neq f_i(y)$.

(b) an extremal source in C if and only if it separates points and X is endowed with the initial structure with respect to $(f_i)_{i\in I}$.

11. Let C be a topological category, I a set and $(X,f_i)_I$ an extremal source in C . Show that X can be embedded in a suitable product.

12. Determine the connected objects (according to 1.3.8) in the category Simp .

13. Prove the following assertion by giving an example: In the category Top quasicomponents do not coincide with components in general.

14. Prove: Let (K,K) be a simpicial complex. Then the components of (K,K) coincide with the quasicomponents.

15. a) Give a characterization of connectedness in Unif and prove the following "surprising" theorem: If Y is a dense subspace of X (in Unif) , then the following are equivalent:

    (1) Y is connected.

    (2) X is connected.

    b) Show that $\mathbb{Q}$ (endowed with the uniformity induced by the usual uniformity on $\mathbb{R}$) is connected in Unif (in contrast to the situation in Top) by means of (a).

CHAPTER 2

16. Let $\mathcal{C}$ be a topological category. Show that the forgetful functor $F_u: \mathcal{C} \to \underline{\text{Set}}$ has a left adjoint as well as a right adjoint.
17. Does there exist another Near-structure on a two-point set besides the discrete and the indiscrete Near-structure?
18. Let $\mathcal{A}$ and $\mathcal{B}$ be categories and let $F: \mathcal{A} \to \mathcal{B}$ and $G: \mathcal{B} \to \mathcal{A}$ be functors. Consider $\hom_{\mathcal{A}}(G(\cdot),\cdot): \mathcal{B}^* \times \mathcal{A} \to \underline{\text{Set}}$ and $\hom_{\mathcal{B}}(\cdot,F(\cdot)): \mathcal{B}^* \times \mathcal{A} \to \underline{\text{Set}}$ (where $\mathcal{B}^*$ is the dual category of $\mathcal{B}$ and $\mathcal{B}^* \times \mathcal{A}$ is the product category [i.e. $|\mathcal{B}^* \times \mathcal{A}| := |\mathcal{B}^*| \times |\mathcal{A}|$, Mor $\mathcal{B}^* \times \mathcal{A}$:= Mor $\mathcal{B}^* \times$ Mor $\mathcal{A}$ and $(f',g')\circ(f,g) := (f'\circ f, g'\circ g)$] and e.g. $\hom_{\mathcal{A}}(G(\cdot),\cdot)$ assigns to each object $(B,A) \in |\mathcal{B}^* \times \mathcal{A}|$, the set $\hom_{\mathcal{A}}(G(B),A) = [G(B),A]_{\mathcal{A}}$ and to each morphism $(g,f): (B',A) \to (B,A')$, the morphism $\hom_{\mathcal{A}}(G(g),f): \hom_{\mathcal{A}}(G(B'),A) \to \hom_{\mathcal{A}}(G(B),A')$ assigning $\hom_{\mathcal{A}}(G(g),f)(a) = f\circ a\circ G(g)$ to each $a \in \hom_{\mathcal{A}}(G(B'),A))$. Show that:

    I. $\hom_{\mathcal{A}}(G(\cdot),\cdot): \mathcal{B}^* \times \mathcal{A} \to \underline{\text{Set}}$ and $\hom_{\mathcal{B}}(\cdot,F(\cdot)): \mathcal{B}^* \times \mathcal{A} \to \underline{\text{Set}}$ are functors.

    II. If $u = (u_B): I_{\mathcal{B}} \to F\circ G$ is a natural transformation, then $U = (U_{(B,A)}): \hom_{\mathcal{A}}(G(\cdot),\cdot) \to \hom_{\mathcal{B}}(\cdot,F(\cdot))$ is likewise a natural transformation where $U_{(B,A)}: \hom_{\mathcal{A}}(G(B),A) \to (B,F(A))$ assigns to each $b \in \mathrm{Hom}_{\mathcal{A}}(G(B),A)$, the morphism $U_{(B,A)}(b) = F(b)\circ u_B$.

    III. Conversely, given a natural transformation $U = (U_{(B,A)}): \hom_{\mathcal{A}}(G(\cdot),\cdot) \to \hom_{\mathcal{B}}(\cdot,F(\cdot))$. Then a natural transformation $u = (u_B): I_{\mathcal{B}} \to F\circ G$ is defined by $u_B: B \to F(G(B))$ with $u_B = U_{(B,G(B))}(1_{G(B)})$ for each $B \in |\mathcal{B}|$.

    IV. The assignments $U \mapsto u$ and $u \mapsto U$ given in II and III are bijective.

    V. Let $U = (U_{(B,A)})$ and $u = (u_B)$ be natural transformations corresponding to each other by the bijection described above. Further let $V = (V_{(B,A)}): \hom_{\mathcal{B}}(\cdot,F(\cdot)) \to \hom_{\mathcal{A}}(G(\cdot),\cdot)$ and $v = (v_A): G\circ F \to I_{\mathcal{A}}$ be natural transformations corresponding to each other similarly. Then

a) $U_{(B,A)} \circ V_{(B,A)} = 1_{[B,F(A)]_{\mathcal{B}}}$ for each $(B,A) \in |\mathcal{B}^* \times \mathcal{A}|$

if and only if

$F(v_A) \circ u_{F(A)} = 1_{F(A)}$ for each $A \in |\mathcal{A}|$,

b) $V_{(B,A)} \circ U_{(B,A)} = 1_{[G(B),A]_{\mathcal{A}}}$ for each $(B,A) \in |\mathcal{B}^* \times \mathcal{A}|$

if and only if

$v_{G(B)} \circ G(u_B) = 1_{G(B)}$ for each $B \in |\mathcal{B}|$.

VI. (G,F) is a pair of adjoint functors if and only if $\hom_{\mathcal{A}}(G(\cdot),\cdot) \approx \hom_{\mathcal{B}}(\cdot,F(\cdot))$.

19. Show that in the category Haus the extremal monomorphisms are precisely the closed embeddings.

20. Prove that there is no *proper* subcategory of Top which is simultaneously epireflective and monocoreflective.

21. Let $\mathcal{C}$ be a topological category and $\mathcal{A}$ a full and isomorphism-closed subcategory of $\mathcal{C}$. Then the following are equivalent:

(a) $\mathcal{A}$ is bireflective in $\mathcal{C}$.

(b) $\mathcal{A}$ is closed under formation of products (in $\mathcal{C}$) and initial subobjects (in $\mathcal{C}$) .

($Y \in |\mathcal{C}|$ is called an initial subobject of $X \in |\mathcal{C}|$ provided that there is a $\mathcal{C}$-morphism $f: Y \to X$ such that Y carries the initial structure with respect to f) .

22. Determine the coseparators for each topological category $\mathcal{C}$.

23. Characterize the objects in the bireflective hull of a full subcategory of a topological category $\mathcal{C}$.

24. An object X of a topological category $\mathcal{C}$ is called a T_0-object provided that each $\mathcal{C}$-morphism

$$f: I_2 \to X$$

is constant (where I_2 is the set $\{0,1\}$ endowed with the indiscrete $\mathcal{C}$-structure). Determine the T_0-objects of Top .

25. A topological category is called <u>universal</u> iff it is the bireflective hull of its T_0-objects. Show that <u>Top</u> is universal and give an example of a topological category which is not universal.

26. Let P be a class of Hausdorff spaces. A topological space $(X,\mathcal{X})$ is said to have a P-<u>compactification</u> iff there is a dense embedding of $(X,\mathcal{X})$ in a P-compact space $(Y,\mathcal{Y})$. Show that each P-regular space has a P-compactification via the theory of reflections (the converse is trivial!).
Well-known constructions are reobtained, e.g. the Stone-Čech compactification for $P = \{[0,1]\}$, the Hewitt realcompactification for $P = \{\mathbb{R}\}$ and the Banaschewski zero-dimensional compactification for $P = \{D_2\}$.

CHAPTER 3

27. Let $(\mathbb{R},\mu_R)$ be the nearness space induced by the Sorgenfrey line (cf. 3.1.1.4 ③) and let $(\mathbb{R}\times\mathbb{R},\mu_R\times\mu_R)$ be the product of $(\mathbb{R},\mu_R)$ with itself in <u>Near</u>. Show that

$$\{\mathbb{R}^2\setminus A,\mathbb{R}^2\setminus B\} \in (\mu_R\times\mu_R)_t \ ,$$

$$\{\mathbb{R}^2\setminus A,\mathbb{R}^2\setminus B\} \notin \mu_R\times\mu_R \quad \text{where}$$

$$A = \{(x,-x): x\in\mathbb{Q}\} \quad \text{and} \quad B = \{(x,-x): x\in\mathbb{R}\setminus\mathbb{Q}\} \ .$$

This completes the proof of 3.1.1.4 ③.

28. Let $\mathcal{U}$, $\mathcal{V}$, $\mathcal{W}$ be covers of a set X. Prove the following:

$$\mathcal{U}\ \Delta\ \mathcal{V} \ \text{and}\ \mathcal{V}\ \Delta\ \mathcal{W} \ \text{imply}\ \mathcal{U} \overset{*}{<} \mathcal{W} \quad \text{(cf. footnote 12))}$$

29. Show that the underlying topological space of a uniform nearness space is a completely regular space (Without using the known isomorphism between <u>Unif</u> and <u>U-Near</u>).

30. Prove that a topological nearness space (X,μ) is uniformizable if and only if μ is the coarsest of all <u>T-Near</u>-structures η on X with

$$(\eta)_u = (\mu)_u$$

31. For the nearness spaces $\mathbb{R}_f$ and $\mathbb{R}_t$ defined in 3.1.3.9 ④ and ② whose underlying topological spaces are the real numbers together with the usual topology the following are equivalent:

(1) $f: \mathbb{R}_f \to \mathbb{R}_t$ is uniformly continuous.

(2) $f: T(\mathbb{R}_f) \to T(\mathbb{R}_t)$ is continuous and bounded.

32. Let (X,d) be a compact metric space and let $\mathcal{X} = \mathcal{X}_d$ be the topology induced by the metric d. Show that the nearness structure μ_d induced by the metric d coincides with the nearness structure $\mu_{\mathcal{X}}$ induced by the topology $\mathcal{X}$. Show that the compactness of (X,d) is essential.

33. Consider the nearness spaces defined in 3.1.3.9 and show that the following are satisfied:

 a) $\mathbb{R}_t$ is topological and uniform but not pseudometrizable.

 b) $\mathbb{R}_u$ is pseudometrizable but not topological.

 c) $\mathbb{R}_p$ is neither topological nor pseudometrizable but proximal.

 d) $\mathbb{R}_f$ is neither topological nor pseudometrizable but proximal.

34. The nearness space $\mathbb{R}_u$ is not grill-determined.

35. The nearness spaces $\mathbb{R}_f$ and $\mathbb{R}_p$ are grill-determined.

36. Let $X = \mathbb{N} \times \{1,2\}$ and μ be the set of all covers $\mathcal{U}$ of X satisfying the following condition:

 There exists some $n \in \mathbb{N}$ such that for each $m \geq n$ there is some $U \in \mathcal{U}$ with $\{(m,1),(m,2)\} \subset U$.

 Then the following are satisfied:

 (1) (X,μ) is a metrizable nearness space.

 (2) (X,μ) is not grill-determined.

37. The product $\mathbb{R}_t \times \mathbb{R}_t$ in <u>Near</u> is not grill-determined.

38. Every contigual seminearness space (X,μ) is grill-determined.

39. Let (X,μ) be a nearness space. Show that each convergent collection of subsets of X is a Cauchy system.

40. Let (X,μ) and (X',μ') be prenearness spaces and let $f: X \to X'$ be a map. Then the following are equivalent:

 (1) For every Cauchy system $\mathcal{U}$ in (X,μ), $f\mathcal{U}$ is a Cauchy system in (X',μ').

 (2) $f: (X,\mu) \to (X',\mu')$ is uniformly continuous.

41. Show that in a topological nearness space every Cauchy system converges. Give an example of a nearness space in which there is some non-convergent Cauchy system.

42. Show that in each of the following topological categories every quotient map is a <u>hereditary quotient map</u> (i.e. if $\mathcal{C}$ denotes one of these categories and $f: X \to Y$ is a quotient map in $\mathcal{C}$ then for every $Z \subset Y$ the map

$(f|_{f^{-1}[Z]})'$: $f^{-1}[Z] \to Z$ is again a quotient map in $\mathcal{C}$):

a) P-Near

b) S-Near

c) Grill

43. A filter-merotopic space is a pair (X,γ) where X is a set and γ is a set of filters on X such that the following axioms hold:

(1) If $\mathcal{F} \in \gamma$, and a filter $\mathcal{G}$ is finer than $\mathcal{F}$, then $\mathcal{G} \in \gamma$.

(2) For every $x \in X$, $\{A \subset X: x \in A\} \in \gamma$.

The elements of γ are called convergent filters. A map $f: (X,\gamma) \to (X',\gamma')$ between filter-merotopic spaces is called continuous provided that for every $\mathcal{F} \in \gamma$ the filter $f(\mathcal{F}) = (\{f[F] | F \in \mathcal{F}\})$ belongs to γ'. Show that there is an isomorphism between the category Fil of filter-merotopic spaces (and continuous maps) and the category Grill.

CHAPTER 4

44. An object A of a category C is called exponential in C provided that the functor $A \times -: C \to C$ has a right adjoint. Show that for a topological space A the following are equivalent:
   (a) A is exponential in Top.
   (b) $A \times -$ preserves quotient maps.
   (c) A is quasi locally compact, i.e. for each $x \in A$ and each neighborhood U of x there exists a neighborhood V of x such that each open cover of U contains a finite cover of V .
45. Prove that for a uniform space A the following are equivalent:
   (a) A is exponential in Unif.
   (b) $A \times -$ preserves coproducts.
   (c) A has a finest uniform cover.
46. Show that the category of sequential spaces (and continuous maps) [i.e. the coreflective hull of all metrizable spaces in Top] is cartesian closed.
47. None of the following categories contains a non-trivial (i.e. there is a non-indiscrete space) cartesian closed epireflective full subcategory: Top, Unif, Prox, Near. Try to prove at least one of these statements.
48. Let X be a locally compact Hausdorff space, let Y denote an arbitrary topological space, and let Y^X denote the set of all continuous mappings of X into Y endowed with the compact-open topology. Then the evaluation map $e_{X,Y}: X \times Y^X \to Y$ is continuous.

CHAPTER 5

49. Let K be an (E,M)-category in the sense of 5.1.1 and let A be a (full and isomorphism-closed) subcategory of K. A K-morphism $f: X \to Y$ is called <u>A-concentrated</u> if and only if f belongs to E and is A-extendable (i.e. the source $(X, F(X,A))$ of all K-morphisms with domain X and codomain in A can be factorized through f). A source (X,F) in K is called <u>A-dispersed</u> if and only if $(X, F \cup F(X,A))$ belongs to M. Show that (CA, DA) is a factorization structure on K where CA (resp. DA) denotes the class of all A-concentrated K-morphisms (resp. the conglomerate of all A-dispersed sources in K).
50. Let K and A be as above. If B is the E-reflective hull of A then $CA = CB$ and $DA = DB$.
51. A factorization structure (C,D) on an (E,M)-category K is called <u>dispersed</u>, or more precisely (E,M)-dispersed, if and only if there exists a subcategory A of K such that $C = CA$ and $D = DA$. Show that there is a bijection between the class of E-reflective subcategories of C and the class of dispersed factorization structures on C.
52. If I denotes the inclusion functor of a (full and isomorphism-closed) subcategory A of an (E,M)-category C into C then the following are equivalent:
    (a) I is (E,M)-topological.
    (b) If $(X,(m_i: X \to A_i)_{i \in I})$ belongs to M and all A_i belong to A, then $(X,(m_i)_{i \in I})$ belongs to A.
    (c) A is an E-reflective subcategory of C.
53. Let (A,T) be an initially structured category. Prove the following:
    (a) If $(g_i \circ f)_{i \in I}$ is a T-initial mono-source in A, then f is an embedding.
    (b) $(g_i \circ f)_{i \in I}$ is a T-initial source whenever $(g_i)_{i \in I}$ is a T-initial source and f is T-initial (considered as one-element source).

54. Let (A,T) be an initially structured category and B a non-trivial (i.e. $P \in |B|$) [full and isomorphism-closed] subcategory of A . Then the following are equivalent:
    (a) B is coreflective in A .
    (b) B is bicoreflective in A .
    (c) B is closed under formation of colimits.
    (d) B is closed under formation of coproducts and quotient objects.
    (e) B is closed under formation of final epi-sinks.
55. Let B be non-trivial (cf. 54.) [full and isomorphism-closed] coreflective subcategory of an initially structured category (A,T) . Then $(B, T \circ I)$ is initially structured where $I: B \to A$ denotes the inclusion functor.
56. If (A,T) is a cartesian closed initially structured category and B is a non-trivial (full and isomorphism-closed) coreflective subcategory of A such that the inclusion functor $I: B \to A$ preserves finite products, then B is cartesian closed.
57. Let (A,T) be a cartesian closed initially structured category and B the coreflective (=bicoreflective) hull in A of a non-trivial full subcategoy K of A such that B contains all finite products (formed in A) of K-objects. Then B is cartesian closed.

CHAPTER 6

58. Let $\mathcal{A}$ be the category of finite topological spaces (and continuous maps).
    a) The MacNeille completion of $\mathcal{A}$ is the category of finitely generated topological spaces.
    b) The universal initial completion of $\mathcal{A}$ is that concrete category, whose objects are pairs $(X,\mathcal{X})$ consisting of a set X and a collection $\mathcal{X}$ of subsets of X closed under finite unions and finite intersections, and whose morphisms $f: (X,\mathcal{X}) \to (Y,\mathcal{Y})$ are functions $f: X \to Y$ such that $f^{-1}[O] \in \mathcal{X}$ for each $O \in \mathcal{Y}$.
59. The cartesian closed topological hull of a concrete category $(\mathcal{A},\mathcal{F})$ over Set is - when it exists - the smallest cartesian closed final completion.
60. The full embedding of the category Ind of indiscrete spaces into the category Top of topological spaces is neither initially dense nor finally dense.
61. Show that the category of complete lattices (and join preserving maps) is reflective in the category of ordered sets (and join preserving maps). Examine the relationship to the categorical constructions of chapter 6.
62. a) A nearness space (X,μ) is connected (as object in Near) if and only if its canonical completion (X^*,μ^*) is connected. (Hint: Prove 15. a) for the category Near).
    b) Look for other properties P of nearness spaces such that a nearness space (X,μ) fulfills P if and only if (X^*,μ^*) fulfills P.
63. Let (X,μ) be a contigual N_1-space. Then the following are equivalent:
    (a) (X,μ) is separated.
    (b) (X,μ) is regular.
    (c) (X,μ) is uniform.
64. A nearness space (X,μ) is contigual if and only if its canonical completion (X^*,μ^*) is compact.

65. A nearness space (X,μ) is compact if and only if it is complete and contigual.

66. Show that the category $\underline{CSepNear}_1$ of complete separated N_1-spaces (and uniformly continuous maps) is epireflective in the category $\underline{SepNear}_1$ of separated N_1-spaces (and uniformly continous maps).

CHAPTER 7

67. A nearness space (X,μ) is connected (i.e. each uniformly continuous map of (X,μ) into the two-point discrete nearness space is constant) if and only if $\check{H}^0(X) \cong G$, where $\check{H}^0(X)$ denotes the zerodimensional Čech cohomology group of (X,μ) with coefficients in the group G.

68. Let (X,μ) be a nearness space and $j_X\colon (X,\mu) \to (X^*,\mu^*)$ its canonical completion. Then, for each q,

$$\check{H}^q((X,\mu)) \cong \check{H}^q((X^*,\mu^*)) .$$

69. Let (X,μ) be a proximal nearness space (=proximity space). Then (X,μ) is zerodimensional (i.e. $\dim(X,\mu) = \mathrm{Dim}(X,\mu) = 0$) if and only if, for each subspace (Y,μ_Y) of (X,μ),

$$\check{H}^1((X,\mu),(Y,\mu_Y)) = 0 .$$

70. Let (X,μ) be a nearness space and $(\mathbb{R},\rho)$ the reals endowed with the usual uniform nearness structure. Further let $f,g\colon (X,\mu) \to (\mathbb{R},\rho)$ be uniformly continuous. Prove the following statements:
   (1) If f and g are bounded, then $f+g$ and $f\cdot g$ are bounded and uniformly continuous.
   (2) If $f(x) \geq t$, $t > 0$, for all $x \in X$, then $\frac{1}{f}$ is uniformly continuous.

71. If $f,g\colon (X,\mu) \to S^n$ are uniformly continuous maps of a nearness space (X,μ) into an n-sphere (endowed with the usual uniformity) such that for each $x \in X$ there is some $z \in S^n$ with $\|f(x)-z\| < \frac{1}{4}$ and $\|g(x)-z\| < \frac{1}{4}$, then f and g are uniformly homotopic.

72. Let X_1, X_2 be nearness spaces. Prove the inequality $\mathrm{Dim}(X_1 \times X_2) \leq \mathrm{Dim}(X_1) + \mathrm{Dim}(X_2) + (\mathrm{Dim}(X_1)\cdot\mathrm{Dim}(X_2))$.

73. If (X,μ) is a dense subspace of a regular nearness space (Y,ν), then $\mathrm{Dim}(X,\mu) = \mathrm{Dim}(Y,\nu)$.

74. Let A be the set of all ordinals α which do not exceed the first uncountable ordinal ω_1 and let $\mathcal{X}_A$ be the order topology on A induced by the natural order of A. Further let B be the set of all ordinals β which do not exceed the first non-finite ordinal ω_0 and let $\mathcal{X}_B$ be the order topology on B induced by the natural order of B. Let $Y = (A\times B) \setminus \{(\omega_1,\omega_0)\}$ and let $\mathcal{Y}$ be the relative topology on Y with respect to the topological product $(A,\mathcal{X}_A) \times (B,\mathcal{X}_B)$. If ν denotes the nearness structure on Y induced by $\mathcal{Y}$, then (Y,ν) is a regular N_1-space. If (X,μ) is the subspace of (Y,ν) such that $X = (A\setminus\{\omega_1\}) \times (B\setminus\{\omega_0\})$, then the following are satisfied:

(a) $\operatorname{Dim}(Y,\nu) = \infty$

(b) $\operatorname{Dim}(X,\mu) = \infty$

(c) $\dim(Y,\nu) = 1$

(d) $\dim(X,\mu) = 0$

75. A nearness space (X,μ) is normal iff $C((X,\mu))$ is uniform where C denotes the contigual bireflector.

76. Let (X,μ) be a normal nearness space and (Y,η) a proximal nearness space. Then

$$\dim((X,\mu) \times (X,\eta)) \leq \dim(X,\mu) + \dim(Y,\eta)$$

[Hint. Note that

$$\operatorname{Dim}((\overline{X},\overline{\mu}) \times (\overline{Y},\overline{\eta})) \leq \operatorname{Dim}(\overline{X},\overline{\mu}) + \operatorname{Dim}(\overline{Y},\overline{\eta})$$

provided $(\overline{X},\overline{\mu})$ and $(\overline{Y},\overline{\eta})$ are uniform spaces (cf.[50])].

DIAGRAM OF RELATIONS BETWEEN
SUB- AND SUPERCATEGORIES OF THE
CATEGORY NEAR

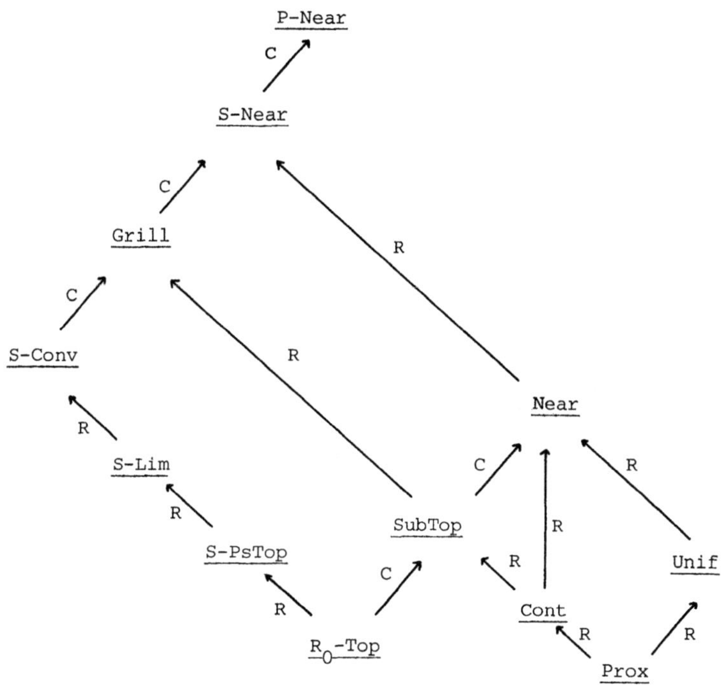

In the above diagram R (resp. C) stands for embedding as a bireflective (resp. bicoreflective) subcategory. S-Lim (resp. S-PsTop) denotes the category of symmetric limit spaces (resp. symmetric pseudotopological spaces).

BIBLIOGRAPHY

[1] Adámek, J., Herrlich, H., Strecker, G.E.: Least and largest initial completions. Comm. Math. Univ. Carolinae 20, 43 - 77 (1979).

[2] Adámek, J., Reiterman, J., Strecker, G.E.: Realizations of cartesian closed topological hulls. Preprint.

[3] Adámek, J., Strecker, G.E.: Construction of cartesian closed topological hulls. Comm. Math. Univ. Carolinae 22, 235 - 254 (1981).

[4] Arhangelskii, A.V., Wiegandt, R.: Connectednesses and disconnectednesses in topology. Gen. Top. Appl. 5, 9 - 33 (1975).

[5] Bahauddin, M., Thomas, J.: The homology of uniform spaces. Canad. J. Math. 25, 449 - 455 (1973).

[6] Banaschewski, B., Bruns, G.: Categorical characterization of the MacNeille completion. Arch. Math. 18, 369 - 377 (1967).

[7] Barón, S.: Reflectors as composition of epireflectors. Trans. Amer. Math. Soc. 136, 499 - 508 (1969).

[8] Bauer, H.: Wahrscheinlichkeitstheorie und Grundzüge der Maßtheorie. Berlin - New York: de Gruyter. 1968.

[9] Bentley, H.L.: Nearness spaces and extensions of topological spaces. In: Studies in Topology, pp. 47 - 66. New York: Academic Press. 1975.

[10] Bentley, H.L.: The role of nearness in topology. In: Categorical Topology, pp. 1 - 22. Lecture Notes Math. 540. Berlin - Heidelberg - New York: Springer. 1976.

[11] Bentley, H.L.: Homology and cohomology for merotopic and nearness spaces. Quaest. Math. 6, 27 - 47 (1983).

[12] Bentley, H.L., Herrlich, H.: Completion as reflection. Comm. Math. Univ. Carolinae 19, 541 - 568 (1978).

[13] Bentley, H.L., Herrlich, H.: The reals and the reals. Gen. Top. Appl. 9, 221 - 232 (1978).

[14] Bentley, H.L., Herrlich, H., Robertson, W.A.: Convenient categories for topologists. Comm. Math. Univ. Carolinae 17, 207 - 227 (1976).

[15] Binz, E.: Continuous convergence on C(X). Lect. Notes Math. 469. Berlin - Heidelberg - New York: Springer. 1975.

[16] Birkhoff, G.: Lattice theory. Am. Math. Soc. Coll. Publ. 25. Providence: Am. Math. Soc. 1940.

[17] Börger, R.: Kategorielle Beschreibungen von Zusammenhangsbegriffen. Thesis. Univ. Hagen. 1981.

[18] Bourbaki, N.: Théorie des ensembles, Ch. 4. Paris: Hermann. 1957.

[19] Brandenburg, H., Hušek, M.: A remark on cartesian closedness. In: Category Theory. Lect. Notes Math. 962, pp. 33 - 38. Berlin - Heidelberg - New York. 1982.

[20] Brümmer, G.C.L.: A categorical study of initiality. Thesis. Univ. Cape Town. 1971.

[21] Čzarcinski, D.: The Čech homology theory for nearness spaces. Thesis. University of Toledo (Ohio). 1975.

[22] Derdérian, J.-C.: A completion for partially ordered sets. J. London Math. Soc. 44, 360 (1969).

[23] Dowker, C.H.: Mapping theorems for non-compact spaces, Amer. J. Math. 69, 202 - 242 (1947).

[24] Dowker, C.H.: Homology groups of relations, Ann. of Math. 56, 84 - 95 (1952).

[25] Dugundji, J.: Topology. Boston: Allyn and Bacon. 1966.

[26] Eilenberg, S., Steenrod, N.: Foundations of Algebraic Topology. Princeton: University Press. 1952.

[27] Engelking, R., Mrowka, S.: On E-compact spaces. Bull. Acad. Pol. Sci. Ser. Sci. Math. Astr. Phys. 6, 429 - 436 (1958).

[28] Gähler, W.: Grundstrukturen der Analysis, Band 1, 2. Basel - Stuttgart: Birkhäuser. 1977/78.

[29] Guili , E. : Two families of cartesian closed topological categories. Quaest. Math. 6, 353 - 362 (1983).

[30] Herrlich, H.: Fortsetzbarkeit stetiger Abbildungen und Kompaktheitsgrad topologischer Räume. Math. Zeitschrift 96, 64 - 72 (1967).

[31] Herrlich, H.: Topologische Reflexionen und Coreflexionen. Lect. Notes Math. 78. Berlin - Heidelberg - New York: Springer. 1968.

[32] Herrlich, H.: Categorical topology, Gen. Top. Appl. 1, 1 - 15 (1971).

[33] Herrlich, H.: Topological functors. Gen. Top. Appl. 4, 125 - 142 (1974).

[34] Herrlich, H.: A concept of nearness. Gen. Top. Appl. 4, 191 - 212 (1974).

[35] Herrlich, H.: Cartesian closed topological categories. Math. Coll. Univ. Cape Town 9, 1 - 16 (1974).

[36] Herrlich, H.: Topological structures. In: Math. Centre Tracts 52, pp. 59 - 122. Amsterdam: Math. Centrum. 1974.

[37] Herrlich, H.: Initial completions. Math. Zeitschrift 150, 101 - 110 (1976).

[38] Herrlich, H.: Some topological theorems which fail to be true. In: Categorical Topology, pp. 265 - 285. Lect. Notes Math. 540. Berlin - Heidelberg - New York: Springer. 1976.

[39] Herrlich, H.: Topologische und uniforme Räume I, II. Skriptum. Fernuniv. Hagen. 1979/80.

[40] Herrlich, H.: Categorical topology 1971 - 1981. In: Gen. Topol. Rel. Modern Anal. and Algebra V, pp. 279 - 383. Berlin: Heldermann. 1982.

[41] Herrlich, H., Nel, L.D.: Cartesian closed topological hulls. Proc. Amer. Math. Soc. 62, 215 - 222 (1977).

[42] Herrlich, H., Salicrup, G., Vázquez, R.: Dispersed factorization structures. Can. J. Math. 31, 1059 - 1071 (1979).

[43] Herrlich, H., Salicrup, G., Vázquez, R.: Light factorization structures. Quaest. Math. 3, 189 - 213 (1979).

[44] Herrlich, H., Strecker, G.E.: Category Theory. Boston: Allyn and Bacon. 1973.

[45] Herrlich, H., Strecker, G.E.: Semi-universal maps and universal initial completions. Pac. J. Math. 82, 407 - 428 (1979).

[46] Hoffmann, R.-E.: Note on universal topological completion. Cahiers Top. Geom. Diff. 20, 199 - 216 (1979).

[47] Hogbe-Nlend, H.: Théories des bornologies et applications. Lecture Notes Math. 213. Berlin - Heidelberg - New York: Springer. 1971.

[48] Hurewicz, W., Wallman, H.: Dimension Theory. Princeton: University Press. 1941.

[49] Hušek, M.: Reflective and coreflective subcategories of Unif (and Top.). Sem. Unif. Sp. Prague 1973 - 1974, 113 - 126 (1975).

[50] Isbell, J.R.: Uniform spaces. Math. Surveys No. 12. Providence: Amer. Math. Soc. 1964.

[51] Ivanova, V.M., Ivanov, A.A.: Contiguity spaces and bicompact extensions, Izv. Akad. Nauk SSSR 23, 613 - 634 (1959).

[52] Katětov, M.: Allgemeine Stetigkeitsstrukturen. In: Proc. Intern. Congr. Math. Stockholm 1962, pp. 473 - 479. Djursholm: Inst. Mittag-Leffler. 1963.

[53] Katětov, M.: On continuity structures and spaces of mappings. Comm. Math. Univ. Carolinae 6, 257 - 278 (1965).

[54] Kelley, J.C.: Bitopological spaces. Proc. London Math. Soc. (3) 13, 71 - 89 (1963).

[55] Kennison, J.F.: Full reflective subcategories and generalized covering spaces. Illinois J. Math 12, 353 - 365 (1968).

[56] Kodama, Y.: Note on cohomological dimension for noncompact spaces II, J. Math. Soc. Japan 20, 490 - 497 (1968).

[57] Kuzminov, V., Švedov, J.: Cohomology and dimension of uniform spaces, Soviet Math. Dokl. 1, 1383 - 1386 (1960).

[58] Lee, R.S.: The category of uniform convergence spaces is cartesian closed. Bull. Austr. Math. Soc. 15, 461 - 465 (1976).

[59] Lee, R.S.: Function spaces in the category of uniform convergence spaces. Thesis. Carnegie-Mellon Univ. 1978.

[60] MacLane, S.: Categories for the Working Mathematician. Graduate Texts Math. 5. New York - Heidelberg - Berlin: Springer. 1971.

[61] Marny, T.: On epireflective subcategories of topological categories. Gen. Top. Appl. 10, 175 - 181 (1979).

[62] Mrowka, S., Pervin, W.: On uniform connectedness, Proc. Am. Math. Soc. 15, 446 - 449 (1964).

[63] Nagami, K.: Dimension Theory. New York: Academic Press. 1970.

[64] Nagata, J.: Modern Dimension Theory. New York: Wiley (Interscience). 1965.

[65] Nel, L.D.: Initially structured categories and cartesian closedness. Canad. J. Math. 27, 1361 - 1377 (1975).

[66] Niefield, S.B.: Cartesianness: Topological spaces, uniform spaces and affine schemes. J. Pure Appl. Algebra 23, 147 - 167 (1982).

[67] Porst, H.-E.: Characterizations of MacNeille completions and topological functors. Bull. Austr. Math. Soc. 18, 201 - 210 (1978).

[68] Preuß, G.: Allgemeine Topologie. Berlin - Heidelberg - New York: Springer. 1972.

[69] Preuß, G.: Relative connectednesses and disconnectednesses in topological categories. Quaest. Math. 2, 297 - 306 (1977).

[70] Preuß, G.: Some categorical aspects of simplicial complexes. Quaest. Math. 6, 313 - 322 (1983).

[71] Preuß, G.: Some point set properties of merotopic spaces and their cohomology groups. Mh. Math. 96, 41 - 56 (1983).

[72] Preuß, G.: A cohomological characterization of dimension for normal nearness spaces. In: Categorical Topology, pp. 441 - 452. Sigma Ser. Pure Math. 5. Berlin: Heldermann. 1984.

[73] Pust, H.: Normalität und Dimension für Nearness-Räume. Thesis. Freie Universität Berlin. 1977.

[74] Ringleb, P.: Untersuchungen über die Kategorie der geordneten Mengen. Thesis. Freie Universität Berlin. 1969.

[75] Salicrup, G., Vâzquez, R.: Connection and disconnection. In: Categorical Topology, pp. 326 - 344. Lect. Notes Math. 719. Berlin - Heidelberg - New York: Springer. 1979.

[76] Schröder, J.: The category of Urysohn spaces is not cowellpowered, Top. Appl. 16, 237-241 (1983).

[77] Schwarz, F.: Funktionenräume und exponentiale Objekte in punktetrennend initialen Kategorien. Thesis. Univ. Bremen. 1983

[78] Schwarz, F.: Powers and exponential objects in initially structured categories and applications to categories of limit spaces. Quaest. Math. 6, 227 - 254 (1983).

[79] Smirnov, Yu.M.: On the dimension of proximity spaces. Math. Sb. 38, 283 - 302 (1956) = Amer. Math. Soc. Transl. Ser. 2, 21, 1 - 20 (1962).

[80] Tukey, J.W.: Convergence and Uniformity in Topology. Princeton: University Press. 1940.

[81] Weil, A.: Sur les espaces à structure uniforme et sur la topologie générale. Paris: Hermann. 1938.

[82] Willard, S.: General Topology. Reading - Menlo Park - London - Don Mills: Addison-Wesley. 1970.

[83] Wyler, O.: Top categories and categorical topology. Gen. Top. Appl. 1, 17 - 28 (1971).

[84] Wyler, O.: Convenient categories for topology. Gen. Top. Appl. 3, 225 - 242 (1973).

[85] Wyler, O.: Function spaces in topological categories. In: Categorical Topology, pp. 411 - 420. Lect. Notes Math 719. Berlin - Heidelberg - New York: Springer. 1979.

INDEX

(The numbers behind the key-words refer to the corresponding pages of the text. The abbrevations A., E. and F. stand for appendix, exercise and footnote, respectively).

GPSR Compliance

The European Union's (EU) General Product Safety Regulation (GPSR) is a set of rules that requires consumer products to be safe and our obligations to ensure this.

If you have any concerns about our products, you can contact us on ProductSafety@springernature.com

In case Publisher is established outside the EU, the EU authorized representative is:

Springer Nature Customer Service Center GmbH
Europaplatz 3
69115 Heidelberg, Germany

Batch number: 10607320

Printed by Printforce, the Netherlands